Just Sustainabilities

Urban and Industrial Environments

Series editor: Robert Gottlieb, Henry R. Luce Professor of Urban and Environmental Policy, Occidental College

Maureen Smith, *The U.S. Paper Industry and Sustainable Production: An Argument for Restructuring*

Keith Pezzoli, *Human Settlements and Planning for Ecological Sustainability: The Case of Mexico City*

Sarah Hammond Creighton, *Greening the Ivory Tower: Improving the Environmental Track Record of Universities, Colleges, and Other Institutions*

Jan Mazurek, *Making Microchips: Policy, Globalization, and Economic Restructuring in the Semiconductor Industry*

William A. Shutkin, *The Land That Could Be: Environmentalism and Democracy in the Twenty-First Century*

Richard Hofrichter, ed., *Reclaiming the Environmental Debate: The Politics of Health in a Toxic Culture*

Robert Gottlieb, *Environmentalism Unbound: Exploring New Pathways for Change*

Kenneth Geiser, *Materials Matter: Toward a Sustainable Materials Policy*

Thomas D. Beamish, *Silent Spill: The Organization of an Industrial Crisis*

Matthew Gandy, *Concrete and Clay: Reworking Nature in New York City*

David Naguib Pellow, *Garbage Wars: The Struggle for Environmental Justice in Chicago*

Julian Agyeman, Robert D. Bullard, and Bob Evans, eds, *Just Sustainabilities: Development in an Unequal World*

Just Sustainabilities

Development in an Unequal World

Edited by
Julian Agyeman, Robert D. Bullard and Bob Evans

The MIT Press
Cambridge, Massachusetts

First MIT Press edition, 2003

This book was set in Garamond by MapSet Ltd, Gateshead, UK, and was printed and bound in the United States of America.

Library of Congress Cataloging-in-Publication Data

Just sustainabilities: development in an unequal world / edited by Julian Agyeman, Robert D. Bullard, and Bob Evans.
    p. cm. (Urban and industrial environments)
  Includes bibliographical references and index.
  ISBN-13: 978-0-262-01199-0 (alk. paper)—978-0-262-51131-5 (pbk. : alk. paper)
  1. Sustainable development. 2. Environmental justice. 3. Social justice.
  I. Agyeman, Julian. II. Bullard, Robert D. (Robert Doyle), 1946– III. Evans,
  Bob, 1947– IV. Series.

HC79.E5 J87 2003
338.9'27—dc21

2002032166

10 9 8 7 6 5 4

Contents

PART 1 – SOME THEORIES AND CONCEPTS

PART 2 – CHALLENGES

PART 3 – CITIES, COMMUNITIES AND SOCIAL AND ENVIRONMENTAL JUSTICE

PART 4 – SELECTED REGIONAL PERSPECTIVES ON SUSTAINABILITY AND ENVIRONMENTAL JUSTICE

List of Tables, Figures and Boxes

TABLES

FIGURES

BOXES

List of Contributors

Tunde Agbola is a Professor of Urban and Regional Planning in the University of Ibadan, Nigeria. An Economics graduate from Ahmadu Bello University, he obtained his Masters and Doctoral degrees in City and Regional Planning from the University of Pennsylvania, Philadelphia, US. An accomplished academic, he has to his credit five books, 20 chapters in different books and over 25 scholarly journal articles in national and international journals. He is a consultant to many international organizations including the United Nations Human Settlements Programme (UN-Habitat) and the World Health Organization (WHO). He is the Executive Director of the Institute for Human Settlement and Environment, an international NGO in Ibadan and an elected Chairman of the Network of NGOs in Human Settlement and Environment in Nigeria.

Julian Agyeman is Assistant Professor of Environmental Policy and Planning at Tufts University, Boston-Medford, US. His interests span environmental justice, community involvement in local environmental and sustainability policy, social marketing and sustainability and the development of sustainable communities. A long time activist, he was founder in 1987 of the UK Black Environment Network. He is founder and co-editor with Bob Evans of the international journal *Local Environment*. He has written widely on issues of sustainability and justice, including *Local Environmental Policies and Strategies* (Longman, 1994) which he co-edited and has published in the *Journal of Environmental Policy and Planning, Environmental Politics, Space and Polity*, the *Canadian Journal of Environmental Education* and *Environmental Education Research*. He is a Member of the Massachusetts Environmental Justice Advisory Committee, a Fellow of the UK Royal Society of the Arts and a Board Member of the North American Association for Environmental Education.

Moruf Alabi is a doctoral student in the Department of Geography, University of Ibadan, Nigeria. He holds a BSc in Geography and a Master of Urban and Regional Planning degree from the same university. He is currently a Project Officer with the Institute for Human Settlements and Environment, an international NGO in Ibadan. In conjunction with some senior academics, Mr Alabi has written papers on housing, poverty and the environment, and he has participated in some deep-seated foreign-agencies' sponsored national researches on biomedical waste and environmental health.

Andrew Blowers is Professor of Social Sciences (Planning) at the Open University, UK. He has published widely in the fields of environmental policy, politics and planning, specializing in radioactive wastes. Among his books are *The Limits of Power* (Pergamon Press, 1980), *Something in the Air* (HarperCollins, 1984), *The International Politics of Nuclear Waste* (co-author) (Palgrave Macmillan, 1991) and *Planning for a Sustainable Environment* (editor) (Town and Country Planning Association, 1993). Since 1991 he has been a member of the UK government's Radioactive Waste Management Advisory Committee and is also on the Board of UK Nirex. He served for nearly three decades on Bedfordshire County Council and is Chairman of the Bedford Hospital NHS Trust. In 2000 Professor Blowers was awarded the OBE for services to environmental protection.

Robert D Bullard is the Ware Professor of Sociology and Director of the Environmental Justice Resource Center at Clark Atlanta University, US. He is the author of ten books that address environmental justice, environmental racism, urban land use, facility permitting, community reinvestment, housing, transportation, suburban sprawl and smart growth. His book, *Dumping in Dixie: Race, Class and Environmental Quality* (Westview Press, 2000), is a standard text in the environmental justice field. He co-edited, with Glenn S Johnson and Angel O Torres, *Sprawl City: Race, Politics and Planning in Atlanta* (Island Press, 2000).

Alberto Costi researches and teaches in the fields of public international law and comparative law, including international and European environmental law and policy. He studied law at the Université de Montréal, the College of Europe in Bruges, Belgium and the University of Cambridge. He clerked at the Supreme Court of Canada before being called to the Bar of Québec. Before taking up his appointment at Victoria University of Wellington in July 2000, he taught at Keele University, UK and at the Central European University in Budapest, where he also acted as a consultant with the Hungarian government on the harmonization of Hungary's environmental law with European standards. He serves on the Board of Editors of the *Journal of International Wildlife Law and Policy* and on the Editorial Board of *European Environment*.

Andrew Dobson is Professor of Politics at Keele University, UK. Among his publications are *Justice and the Environment* (OUP, 1998), an edited collection, *Fairness and Futurity* (OUP, 1999), and the third edition of his *Green Political Thought* (Routledge).

Kevin Dunion OBE is Chief Executive of Friends of the Earth Scotland and Honorary Senior Research Fellow in the Department of Environmental Planning, Strathclyde University. He is a Member of the Scottish Executive's Ministerial Advisory Group on Sustainable Development. His publications include *Living in the Real World – The International Role for Scotland's Parliament* (Friends of the Earth Publications) and *Sustainable Development in a Small Country: The Global and European Agenda* (Friends of the Earth Publications). He was Chair of Friends of the Earth International for four years to 2000.

Veronica Eady was until recently Director of the Environmental Justice and Brownfields Programs for the Massachusetts Executive Office of Environmental Affairs. She is an environmental lawyer and Lecturer in Urban and Environmental Policy and Planning at Tufts University, US. She serves on EPA's National Environmental Justice Advisory Council as Vice-Chair of the Waste and Facility Siting Subcommittee as well as on the Boards of Directors for Earth Island Institute and the Massachusetts Black Women Attorneys.

Bob Evans is Professor and Director of the Sustainable Cities Research Institute at the University of Northumbria at Newcastle, UK. He is author of *Experts and Environmental Planning* (Avebury, 1995) and co-author of *Local Environmental Policies and Strategies* (Longman, 1994); *Environmental Planning and Sustainability* (Wiley, 1996) and *Town Planning into the 21st Century* (Routledge, 1997) in addition to many book chapters and articles on land use planning and environmental policy. He is co-founder and co-editor (with Julian Agyeman) of the international journal *Local Environment* and he has worked as a town planner in the public, private and community sectors.

Daniel R Faber is Director of the Philanthropy and Environmental Justice Research Project at Northeastern University, US. A long time environmental justice advocate and scholar, in 1984 he co-founded the Environmental Project on Central America (EPOCA). Based with the Earth Island Institute in San Francisco and the Environmental Policy Institute in Washington, DC, EPOCA worked to broaden and deepen the movement for peace, social justice and sustainable development by addressing the human and ecological impacts of US policy in the region, and involved US environmentalists in efforts to change that policy. His latest book is the edited collection, *The Struggle for Ecological Democracy: Environmental Justice Movements in the United States* (Guilford Press, 1998). He is currently an Associate Professor in the Department of Sociology and Anthropology and a faculty member of the Latino, Latin American and Caribbean Studies (LLACS) and Environmental Studies Programs at Northeastern University.

Joan Martinez-Alier is Professor of Economics and Economic History at the Universitat Autonoma de Barcelona, Spain. He is a founding member of the International Society for Ecological Economics, editor of the journal *Ecologia Politica*, author of *Ecological Economics: Energy, Environment and Society* (Blackwell, Oxford, 1987) and (with Ramachandra Guha) *Varieties of Environmentalism: Essays North and South* (Earthscan, London, OUP, Delhi, 1997).

Deborah McCarthy is an Assistant Professor in the Department of Sociology and Anthropology at the College of Charleston, US. She has been researching issues of environmental injustice for a number of years. From 1999 to 2001 she served as a Research Associate with the Philanthropy and Environmental Justice Research Project at Northeastern University, US and is the co-author of the report 'Green of Another Color: Building Effective Relationships Between Foundations and the Environmental Justice Movement' (with Dr Daniel Faber). In 2000 she was awarded the Thoreau Teaching Fellowship at the University of

Maine, US. She is currently working on a book entitled *The Color of Green: Lessons Learned from the Developing Relationship Between Foundations and the Environmental Justice Movement.*

Duncan McLaren is Head of Policy and Research at Friends of the Earth Trust Ltd, the charitable arm of Friends of the Earth England, Wales and Northern Ireland. His role is developing policy, and managing research and campaigns on national and international sustainable development issues. He has written widely on environmental policy issues, most notably *Tomorrow's World: Britain's Share in a Sustainable Future* (Earthscan, 1998), which sets sustainability targets for the UK and explains how we could achieve them.

Devon G Peña teaches anthropology and ethnic studies at the University of Washington, US, where he serves as coordinator of the PhD Program in Environmental Anthropology. Professor Peña is completing work on his next two books: *Tierra Vida: Mexican Americans and the Environment*, an introductory level college textbook for Chicana/o studies, and *Gaia en Aztlan: Endangered Landscapes and Disappearing People in the Politics of Place*, an environmental history and ethnographic study of ecological politics in acequia farming communities of the Upper Rio Grande. Professor Peña is also co-editing a forthcoming anthology of essays by acequia farmers and research scholars: *Voces de la Tierra: Five Hundred Years of Acequia Farming in the Rio Arriba, 1598–1998* (all books are forthcoming from the University of Arizona Press). As a grass-roots activist, Professor Peña is currently collaborating with the Colorado Acequia Association on the development of a land and water trust and a centre for acequia agroecosystems studies. He also serves on the national planning committee of the Second National People of Color EnvironmentalJustice Summit.

William Rees received his PhD in population ecology from the University of Toronto, Canada. He has taught at the University of British Columbia's School of Community and Regional Planning since 1970 and served as Director of the school from 1994 to 1999. As an applied human ecologist Professor Rees researches the policy implications of global environmental trends and the necessary ecological conditions for sustainable socio-economic development. Much of this work falls in the domain of ecological economics where he is best known for inventing the 'ecological footprint' concept. He is a founding member and past-President of the Canadian Society for Ecological Economics. A dynamic speaker, Professor Rees has been invited to lecture on his work across Canada and the US and in 20 other countries around the world. In 1997 William Rees was awarded a UBC Killam Senior Research Prize and in 2000 he was recognized by *The Vancouver Sun* newspaper as one of British Columbia's top 'public intellectuals'.

Stefanie S Rixecker is a Senior Lecturer in the Environmental Management and Design Division, Lincoln University, Aotearoa New Zealand. She was born in Germany, educated in Canada and the US and has resided in Aotearoa New Zealand since 1994. She is tauiwi (an immigrant or non-Maori), and her two sons are of Ngati Raukawa descent. Stefanie specializes in environmental policy and has contributed to academic, governmental and global policy initiatives in the areas of biodiversity, biotechnology, ecological gender analysis, cultural risk assessment and ecological justice. Her current research focuses on the sociocultural implications of genetic technologies, especially in relation to disenfranchized populations such as indigenous peoples, women and children and the queer community. In addition to presenting numerous international and national conference papers, she has contributed to various edited books and published in journals such as *Policy Sciences*, the *Canadian Journal of Environmental Education* and *World Archaeology*. She was recently invited to be the co-editor for the *Women's Studies Journal* (New Zealand) from 2002 to 2005, and she will be the guest editor for a Special Issue of the WSJ, 'Feminists Consider Biotechnology', to be published in late 2002. She is currently working on a book focused upon biopolitics and the queer community.

Debra Roberts is a biologist by training, with 19 years' experience in the field of urban environmental planning and management. She is currently employed as the Environmental Manager of the eThekweni Council in Durban, South Africa. In this position she and her team have been responsible for the initiation and development of Durban's Local Agenda 21 programme. The aim of this programme is to develop an environmental management system for the city that will lead Durban towards a more sustainable development path. Outside of the workplace – and her passion for Africa and its cities – Debra, and her life-partner Rosanna, enjoy travelling the globe and are aiming to notch up all the great urban skylines of the world before retirement!

Eurig Scandrett is Head of Community Action, Friends of the Earth Scotland, and Associate Lecturer at the Open University in Scotland. He has worked in ecological research and has over ten years experience in adult and community education. His publications include *Cultivating Knowledge: Education, the Environment and Conflict in Popular Education* (National Institute of Adult and Continuing Education).

Bevan Tipene Matua has extensive research and professional experience in environmental issues relating to the indigenous Maori of Aotearoa New Zealand and particularly the impacts of biotechnology, intellectual property rights and globalization on indigenous populations globally. He was the inaugural Maori Research Fellow at Crop & Food Research, has a Masters Thesis on the impacts of Biotechnology on Maori, has been a Senior Policy Analyst for the Ministry of Maori Development and the Environmental Risk Management Authority and a consultant to the Royal Commission on Genetic Engineering in Aotearoa New Zealand. Bevan is currently a lecturer at Canterbury University and is working towards a PhD on the impacts of genetic engineering on the Maori people of Aotearoa New Zealand.

Laura Westra received her PhD from the University of Toronto, Canada in 1983. She has taught at several Canadian and US universities, specializing in the philosophy of ecology, environmental ethics and applied ethics, including bioethics and business ethics. Dr Westra has published three sole-authored and nine co-edited books in the areas of her expertise. She is currently working on two co-edited volumes and a single-authored manuscript on eco-violence. In her most recent fulltime academic appointment, Dr Westra held the Chair of Environmental Studies at Sarah Lawrence College, US. Since 2000 she has been pursuing her D Jur Degree at Osgoode Hall Law School, York University, Toronto, Canada, and teaching philosophy at the university.

Anoja Wickramasinghe is Professor and Head of geography at the University of Peradeniya, Sri Lanka, and Coordinator of the Collaborative Regional Research Network in South Asia (CORRENSA). She has written widely on the topics of forestry, resource use, biodiversity conservation, land degradation and the role of women in environmental management and development. She is the author of several books in these areas including *Deforestation, Women and Forestry* (International Books, 1994), *People and the Forest: Management of the Adam's Peak Wilderness* (Sri Lanka Forest Department, 1995), *Gender Aspects of Woodfuel Flows in Sri Lanka: A Case Study in Kandy District* (RWEDP, FAO, Bangkok, 1999), and the editor of *Land and Forestry: Women's Local Resource-based Occupations for Sustainable Survival in South Asia* (CORRENSA, 1997) and *Development Issues Across Regions: Women, Land and Forestry* (CORRENSA, 1997).

Beverly Wright is a Professor of sociology and the founding Director of the Deep South Center for Environmental Justice (DSCEJ) at Xavier University of Louisiana, New Orleans, US. For more than a decade she has been a leading scholar, advocate and activist in the environmental justice arena. The DSCEJ is one of the few community/university partnerships that addresses environmental and health inequities in the Lower Mississippi River Industrial Corridor – the area commonly referred to as 'Cancer Alley'. She is a Member of the National Advisory Committee for the First National People-of-Color Environmental Leadership Summit, and the Planning/Protocol Committee for the National Institute for Environmental Health Sciences' Health and Research Needs to Ensure Environmental Justice National Symposium. In 1994 she was named to the EPA's National Environmental Justice Advisory Council (NEJAC). She was recently appointed to the US Commission of Civil Rights for the state of Louisiana and to the city of New Orleans' Select Committee for the Sewerage and Water Board.

List of Acronyms and Abbreviations

ACE	Alternatives for Community and Environment
ADC	Asociación por Derechos Civiles
AEA	American Economics Association
AFL-CIO	American Federation of Labor and Congress of Industrial Organizations
AMOS	Association of Minority Oil States
ANWR	Alaska National Wildlife Reserve
APEN	Asian Pacific Environmental Network
BEN	Black Environment Network
CAA	Colorado Acequia Association
CAMPO	Campesinos a la Mesa Politica
CBD	Convention on Biological Diversity
CCCD	Costilla County Conservancy District
CCHW	Citizens Clearinghouse on Hazardous Waste
CCSCLA	Concerned Citizens of South Central Los Angeles
CEE	Central and Eastern Europe
CERCLA	Comprehensive Environmental Response Compensation and Liability Act
CFC	Chlorofluoro carbon
CHEJ	Center for Health, Environment and Justice
CNT	Center for Neighborhood Technologies
CORRENSA	Collaborative Regional Research Network in South Asia
CSIR	Council for Scientific and Industrial Research
DARE	Direct Action for Rights and Equality
DETR	Department of the Environment, Transport and the Regions
DLC	Democratic Leadership Council
DMA	Durban Metropolitan Area
D'MOSS	Durban Metropolitan Open Space System
DSCEJ	Deep South Center for Environmental Justice
EBRD	European Bank for Reconstruction and Development
EC	European Commission
ECO	Environmental Careers Organization
EHC	Environmental Health Coalition
EIA	Environmental Impact Assessment
EIR	Environmental Impact Report
ELAN	Conservation de la Naturaleza Amazonica (Peru)
EM	Ecological Modernization
EOEA	Executive Office of Environmental Affairs

EOI	Export-oriented Industrialization
EPA	Environmental Protection Agency
EPOCA	Environmental Project on Central America
ERI	EarthRights International
ERMA	Environmental Risk Management Authority
EU	European Union
ESA	Endangered Species Act
FAO	Food and Agricultural Organization (United Nations)
FEPA	Federal Environmental Protection Agency
FOE	Friends of the Earth
FOEI	Friends of the Earth International
FTAA	Free Trade Area of the Americas
FWNEEJ	Farmworker Network for Economic and Environmental Justice
GAIA	Global Anti-Incinerator Alliance
GAO	Government Accounting Office
GATT	General Agreement on Tariffs and Trade
GDI	Gender Development Index
GDP	Gross Domestic Product
GE	Genetic engineering
GEF	Global Environmental Facility
GEFCO	George E Failing Company
GEO	Global Environment OutlookGGP Gross Geographic Product
GIS	Geographic Information Systems
GMO	Genetically Modified Organisms
GNP	Gross National Product
HSNO	Hazardous Substances and New Organisms Act
IA	Interfaith Action
ICREA	International Commodity Related Environmental Agreements
IEN	Indigenous Environmental Network
IFI	International Financial Institutions
IMF	International Monetary Fund
IPR	Intellectual Property Rights
IPTEP	Industrial Property Tax Exemption Program
IRC	International Red Cross
IUCN	International Union on Conservation and Nature (World Conservation Union)
LABI	Louisiana Association of Business and Industry
LDC	Less Developed Country
LDED	Louisiana Department of Economic Development
LDEQ	Louisiana Department of Environmental Quality
LGC	Land Grant Commission
LRC	Land Rights Council
LSF	La Sierra Foundation
LSRSA	La Sierra Regional Service Authority
MEJAC	Massachusetts Environmental Justice Advisory Committee
MEPA	Massachusetts Environmental Policy Act

MNC	Multinational corporation
MOSIEN	Movement for the Survival of Ijaw Ethnic Nationality
MOSOP	Movement for the Survival of Ogoni People
NAFTA	North American Free Trade Agreement
NDDS	Niger Delta Development Commission
NDES	Niger Delta Environment Survey
NEJAC	National Environmental Justice Advisory Committee
NEJN	Northeast Environment Justice Network
NEMA	National Environment Management Authority/Act
NEPA	National Environmental Policy Act
NEST	Nigerian Environmental Study/Action Team
NGO	Non-governmental Organization
NHS	National Health Service
NIC	Newly Industrializing County
NIABY	Not in Anyone's Back Yard
NIMBY	Not in My Back Yard
NJDEP	New Jersey Department of Environmental Protection
NNPC	Nigeria National Petroleum Development Corporation
NPL	National Priorities List
NRC	Nuclear Regulatory Commission
NRTEE	National Round Table on the Environment and the Economy
NTC	National Toxics Campaign
OECD	Organisation for Economic Co-operation and Development
OEJ	Office of Environmental Justice
OMPADEC	Oil Minerals Producing Areas Development Commission
OPEC	Organization of the Petroleum Exporting Countries
OPM	Organisasi Papua Merdeka
PCB	Polychlorinated Biphenyls
PODER	People Organized in Defense of Earth and Her Resources
PPP	Polluter Pays Principle
RCGM	Royal Commission on Genetic Modification
REC	Regional Environmental Center
RMA	Resource Management Act
RSA	Regional Service Authority
RWEDP	Regional Wood Energy Development Program
SCP	Sustainable Cities Programme
SDCEA	South Durban Community Environmental Alliance
SEA	Strategic Environmental Assessment
SEO	Sociedad Española Ornitología
SIP	Sustainable Ibadan Project
SNEEJ	Southwest Network for Economic and Environmental Justice
SNP	Scottish National Party
SOC	Southern Organizing Committee for Economic and Social Justice
SPDC	Shell Petroleum Development Corporation
SPEAC	St Paul Ecumenical Alliance of Congregations
TEK	Traditional Environmental Knowledge

TRI	Toxics Release Inventory
TURI	Toxics Use Reduction Institute
TVA	Tennessee Valley Authority
UHP	Urban Habitat Program
UNDP	United Nations Development Programme
UNECE	United Nations Economic Commission for Europe
UNEP	United Nations Environment Programme
UNESCO	United Nations Educational, Scientific and Cultural Organization
UN-Habitat	United Nations Human Settlements Programme
UFW	United Farm Workers
WDM	World Development Movement
WEACT	West Harlem Environmental Action
WECD	World Commission on Environment and Development
WHO	World Health Organization
WRI	World Resources Institute
WTO	World Trade Organization

Acknowledgements

Thanks to Tufts University students Briony Angus, Adrienne Ralph, Victoria Gellis and Megan Amundson for their help with this book.

Introduction

Joined-up Thinking: Bringing Together Sustainability, Environmental Justice and Equity[1]

Julian Agyeman, Robert D Bullard and Bob Evans

INTRODUCTION

In recent years it has become increasingly apparent that the issue of environmental quality is inextricably linked to that of human equality. Wherever in the world environmental despoilation and degradation is happening, it is almost always linked to questions of social justice, equity, rights and people's quality of life in its widest sense. There are three related dimensions to this.

First, it has been shown by Torras and Boyce (1998) that globally, countries with a more equal income distribution, greater civil liberties and political rights and higher literacy levels tend to have higher environmental quality (measured in lower concentrations of air and water pollutants, access to clean water and sanitation) than those with less equal income distributions, fewer rights and civil liberties and lower levels of literacy. Similarly, in a survey of the 50 US states, Boyce et al (1999) found that states with greater inequalities in power distribution (measured by voter participation, tax fairness, Medicaid access and educational attainment levels) had less stringent environmental policies, greater levels of environmental stress and higher rates of infant mortality and premature deaths. At an even more local level, a study by Morello-Frosch (1997) of counties in California showed that highly segregated counties, in terms of income, class and race, had higher levels of hazardous air pollutants. From global to local, human inequality is bad for environmental quality.

The second and related dimension is that environmental problems bear down disproportionately upon the poor. While the rich can ensure that their children breathe cleaner air, that they are warm and well housed and that they do not suffer from polluted water supplies, those at the bottom of the socio-economic ladder are less able to avoid the consequences of motor vehicle exhausts, polluting industry and power generation or the poor distribution of essential facilities. This unequal distribution of environmental 'bads' is, of course, compounded by the fact that globally and nationally the poor are not the major polluters. Most environmental pollution and degradation is caused by the actions of those in the rich high-consumption nations, especially by the more affluent groups within those societies. Even recent optimism about the Kyoto Protocol post-Marrakech hides a stark reality: affluent countries in the North are avoiding or delaying any real reduction in their greenhouse gas emissions through the so-called 'flexible mechanisms': emissions trading, the Clean Development Mechanism and Joint Implementation (The Corner House, 2001). The emergence of the environmental justice movement in the US over the last two decades was in large part a response to these distributional inequities, as are the growing international calls for environmental justice (Adeola, 2000).

The third dimension is that of sustainable development. The 'new policy agenda' of sustainability emerged after the publication of the World Commission on Environment and Development's report in 1987, but more fully after the 1992 United Nations Conference on Environment and Development (UNCED) in Rio de Janeiro and its successor, the 2002 World Summit on Sustainable Development in Johannesburg. Sustainability is clearly a contested concept, but our interpretation of it places great emphasis upon precaution: on the need to *ensure a better quality of life for all, now, and into the future, in a just and equitable manner, while living within the limits of supporting ecosystems*. In addition, we fully endorse Middleton and O'Keefe's (2001, p16) point that 'unless analyses of development begin not with the symptoms, environmental or economic instability, but with the cause, social injustice, then no development can be sustainable'. Sustainability, we argue, cannot be simply an 'environmental' concern, important though 'environmental' sustainability is. A truly sustainable society is one where wider questions of social needs and welfare, and economic opportunity, are integrally connected to environmental concerns.

This emphasis upon greater equity as a desirable and just social goal, is intimately linked to a recognition that, unless society strives for a greater level of social and economic equity, both within and between nations, the long term objective of a more sustainable world is unlikely to be secured. The basis for this view is that sustainability implies a more careful use of scarce resources and, in all probability, a change to the high-consumption lifestyles experienced by the affluent and aspired to by others. It will not be easy to achieve these changes in behaviour, not least because this demands acting against short term self-interest in favour of unborn generations and 'unseen others' who may live on the other side of the globe. The altruism demanded here will be difficult to secure and will probably be impossible if there is not some measure of perceived equality in terms of sharing common futures and fates.

The chapters that follow in *Just Sustainabilities: Development in an Unequal World* seek to address different aspects of these three dimensions of the multiscalar links between environmental quality and human equality (Torras and Boyce, 1998, Boyce et al, 1999 and Morello-Frosch, 1997) and those between sustainability and environmental justice more generally. Each chapter extends and develops issues that address these core themes including: *anthropocentrism; biotechnology; bioprospecting; (bio)cultural assimilation; Deep and Radical ecology; ecological debt; ecological democracy; ecological footprints; ecological modernization; environmental space; feminism and gender; globalization; participatory research; place, identity and legal rights; precaution; risk society; selective victimization; valuation.*

In so doing, the chapters contribute to an important and emerging realization that a sustainable society must also be a just society, locally, nationally and internationally, both within and between generations and species. We are fully aware of the extensive and contested debates surrounding the concept and conceptions of justice (Rawls, 1971, Low and Gleeson, 1998, Middleton and O'Keefe, 2001). It is not our intention, however, to contribute to this debate. Rather, we are concerned to examine the linkages that can be made between the political and policy processes surrounding environmental justice and sustainability. Similarly, we make no claims to being geographically nor theoretically comprehensive in our treatment of this vast, interdisciplinary area, nor conclusive in our thoughts. We asked our chapter authors, who were selected on the basis of their research and interests in this area, to map what *they* considered to be some of the key conceptual and practical challenges confronting both the ideas of *sustainability and environmental justice* to understand if and how we might see greater linkages between these ideas and their practical actions in the future. In effect, we asked them to explore the nexus between the ideas of sustainability and environmental justice.

What became apparent on receiving submissions was that the linkages between the ideas of sustainability and environmental justice are clearer with respect to the *problems* identified by our authors than the *solutions* currently possible within the dominant social paradigm (Milbrath, 1989), despite the increasing challenges to its hegemony. But this does not mean that, as Dobson (see Chapter 4 in this book) argues, 'ne'er the twain shall meet'. We attribute this to there being a far greater ease in identifying and (co-)organizing around problems (ie being reactive), than around solutions (ie being *proactive*). Proactivity in this case requires the sharing of values, visions and vocabularies among groups of people who are for the most part from different social locations: it is difficult to do, but not impossible. The structures required to build bridges, which represent a direct challenge to the dominant social paradigm (Milbrath, 1989), are only now being identified by organizations and individuals around the world and put into place. This is the single greatest challenge to developing greater theoretical and practical co-activism between the two areas.

However, despite the very real challenges facing greater linkages, and the arguments of those such as Dobson (in Chapter 4) who believe the two ideas to be politically incompatible (he is, though, talking only about *environmental* sustainability and *social* justice), it is felt by most of our authors that *justice* and *sustainability* are intimately linked and mutually interdependent, certainly at the *problem* level and increasingly at the *solution* level through issues like toxics use reduction; waste reduction, re-use and recycling; guaranteed public access to information and public involvement; and sustainable, equitable and just consumption patterns. This is a theme we shall revisit in our Conclusion. Indeed, there is some evidence that this linkage of *problems* and practical *solutions* is already happening at the local level in the US in organizations such as Alternatives for Community and Environment (ACE) in Boston with their work on sustainable transportation and transit inequity; the Center for Neighborhood Technologies (CNT) in Chicago, who are undertaking an Inter-religious Sustainability Project and developing Partnerships for Regional Livability; and the Urban Habitat Program (UHP) in San Francisco, who are operating within an environmental justice framework (Bullard, 1994a) but who are also exploring the emerging terrain of sustainable development and the development of sustainable communities (their mission is 'building multicultural, urban environmental leadership for socially just and sustainable communities in the San Francisco Bay Area' (ECO/UHP, 1998, 21)).

While the environmental justice movement started in the US, and the organizations above may be considered to be in the vanguard, we hope to showcase both its expression and interaction with sustainability in different countries. But the varied discussions in this book, from local issues to global ones, from theoretical debates to policy implementation, come at a critical time when the first tangible successes of sustainability policies, and movements for social and environmental justice, are being eroded by both the processes and products of globalization. Due to the increasingly competitive nature of the global economy, multinational corporations (MNCs) are maintaining profits by relying on unsustainable forms of production. The enormous financial gains that are being made by those fortunate enough to benefit from neo-liberal economic policies come with large social and ecological costs in terms of higher pollution levels, greater resource exploitation, less protection for workers and massive social and cultural dislocation.

However, not all peoples bear these costs equally. As has traditionally been the case, companies usually locate their dirtiest businesses in areas that offer the path of least political resistance. Thus, in the US, Europe and around the world it is the least politically powerful and most marginalized sectors of the population who are being selectively victimized to the greatest extent by environmental crises. The causes and effects of environmental injustice and unsustainable production are becoming increasingly related in places as far apart as the Mississippi Chemical Corridor in Louisiana, Papua New Guinea, West Papua in Indonesia, the Niger Delta in Nigeria, the Brazilian Amazon and Durban's South Basin in South Africa, spurring coalitions between advocates of sustainable development on one hand and environmental justice and human rights on the other.

In the remainder of this introductory chapter we want to lay the foundations for the rest of the book by briefly tracing the emergence and development of ideas about sustainability and environmental justice, and the nexus, or common ground, in between them. Both these ideas have moved to the fore in public policy circles in many countries in recent years. However, while the sustainability agenda has been advanced in the more formal policy-making arenas of government at all levels from municipal and local authorities through state and regional to national governments, as well as a growing number of businesses, environmental justice agendas have achieved prominence as a result of grass-roots organizing, advocacy and action. An examination of the development of the two agendas will prove useful in determining the potential for, benefits of and the obstacles to more broadly based ideological and practical linkages, which many of our authors discuss.

SUSTAINABILITY

The surge in material in recent years dealing with the concepts of sustainability and its action-oriented variant sustainable development, has led to competing and conflicting views over what the terms actually mean and what is the most desirable means of achieving the goal. According to Redclift (1987), sustainability as an idea can be traced back to the 'limits to growth' debates of the 1970s and the 1972 UN Stockholm Conference. The single most frequently quoted definition of sustainable development comes from the World Commission on Environment and Development (WCED) (1987) who argued that 'sustainable development is development that meets the needs of the present without compromising the ability of future generations to meet their own needs' (WCED, 1987, p43). This definition implied an important shift away from the traditional, conservation-based usage of the concept as developed by the 1980 World Conservation Strategy (IUCN, 1980) to a framework that emphasized the social, economic and political context of 'development'. By 1991 the IUCN had modified its definition. Along with that of the WCED it is the most used definition: 'to improve the quality of life while living within the carrying capacity of ecosystems' (IUCN, 1991). However, unlike our working 'definition' mentioned earlier (*the need to ensure a better quality of life for all, now, and into the future, in a just and equitable manner, whilst living within the limits of supporting ecosystems*), neither the WCED or IUCN definitions specifically mentions justice and equity, which we hold to be of fundamental importance.

McNaghten and Urry (1998, p215) argue that 'since Rio, working definitions of sustainability have been broadly accepted by governments, NGOs and business. These tend to be cast in terms of living within the finite limits of the planet, of meeting needs without compromising the ability of future generations to meet their needs and of integrating environment and development'. More recent thinking, according to Jacobs (1999) and McNaghten and Urry (1998, p215), is the 'growing impetus within the policy-making community to move away from questions of principle and definition. Rather they have developed tools and approaches which can translate the goals of sustainability into specific actions, and assess whether real progress is in fact being made towards achieving them'. Prominent among these tools, they argue, are sustainability indicators. Within the sustainability discourse itself there has also emerged two divergent trends – that of strong/hard sustainability versus weak/soft sustainability (Jacobs, 1992). Hard or strong sustainability implies that renewable resources must not be drawn down faster than they can be replenished in that natural capital must not be spent – we must live off the income produced by that capital. Soft or weak sustainability accepts that certain resources can be depleted as long as they can be substituted by others over time. Natural capital can be used up as long as it is converted into manufactured capital of equal value. The problem with weak sustainability is that it can be very difficult to assign a monetary value to natural materials and services, and it does not take into account the fact that some of these cannot be replaced by manufactured goods and services. Strong sustainability thus maintains that there are certain functions or ecosystem services that the environment provides that cannot be replaced by techno-fixes.

Sustainability is at its very heart a political rather than a technical construct. It represents a *belief* in the need for societies to adopt more sustainable patterns of living, and it is both a focus for political mobilization by individuals and organized interests, and a policy goal for governments. As with other over-arching societal values such as democracy and freedom, there are many interpretations of what sustainability might be and how societies might make progress towards it (Agyeman and Evans, 1995).

ENVIRONMENTAL JUSTICE

The roots of the US environmental justice movement can be traced to citizen revolts against the siting of toxic waste or hazardous and polluting industries in areas inhabited by predominantly minority populations. A 1983 Government Accounting Office report (GAO, 1983) indicating that African-Americans comprised the majority population in three of the four communities of the south-eastern US where hazardous waste landfills were located, and the landmark report 'Toxic Wastes and Race in the United States' (United Church of Christ Commission for Racial Justice, 1987), contributed significantly to the development of a public awareness of 'environmental racism' (Bullard, 1993, 1994a). Thus arose the traditional definition of environmental injustice – that certain minority populations are forced, through their lack of access to decision-making and policy-making processes, to live with a disproportionate share of environmental 'bads' – and suffer the related public health problems and quality of life burdens. Environmental justice activists claim that the 'path-of-least-resistance' nature of locational choices within economies functions to the detriment of minorities and, moreover, that this disproportionate burden is an intentional result (Portney, 1994).

In addition, studies have shown that not only are minority populations more likely to live in environmentally degraded and dangerous places, but the amount of environmental and public health protection afforded these groups by the US Environmental Protection Agency (EPA) is substantially less than that generated for whites and more wealthy people (Lavelle and Coyle, 1992). Furthermore, claims of environmental racism point to the limited participation of non-whites in environmental affairs and the lack of public advocates who represent minority and low income communities (Pulido, 1996, Camacho, 1998). Advocates of environmental justice argue that the victims of environmental inequities will only be afforded the same protection as others when they have access to the decision-making and policy-making processes that govern the siting of hazardous materials and polluting industries (Faber, 1998). Environmental justice advocates go beyond so-called 'fair share' principles, which maintain that every municipality should have an equal share of environmental 'goods' and 'bads', regardless of the race or class of its population (distributional justice). Instead, they argue that environmental bads should be eliminated at the source (procedural or process justice) (Faber, 1998).

The environmental justice movement in the US has benefited from several successes in recent years. Though traditionally a movement built upon the grass-roots organizing and activism characteristic of other social action movements, environmental justice advocates have of late been able to secure official government responses to their demands. The establishment of an EPA Office of Environmental Justice, the National Environmental Justice Advisory Council (NEJAC) and the issuance of Executive Order 12898 by President Clinton's administration, directing federal agencies to ensure that minority communities are not disproportionately affected by environmental burdens, all signal the recognition of the movement (Sandweiss, 1998). The development of environmental justice policy federally, and at the state level (see Eady, Chapter 8), is reflective of a growing coalition between mainstream environmental movements, traditionally concerned about wilderness preservation and conservation, and minority groups with concerns for basic civil rights such as employment security and public health. The inherent organizational and ideological conflicts between the elitist, preservationist goals of mainstream environmental movements and the greater social equity concerns of the environmental justice movement have meant however that this linkage has been slow to develop (Camacho, 1998). While the Bush administration has not tampered with the institutional successes of the Clinton years, it remains to be seen how the new political climate will affect the future development of environmental justice in policy-making in the US.

One explanation for the success of the environmental justice movement can be seen in the mutual benefits of a coalition between environmental and social concerns. As the mainstream conservationist ideologies of wilderness preservationists received criticism for being elitist, the benefits of adopting more of a social justice perspective on environmental issues, and the broadened base of support that this perspective would enable, became recognized (Ringquist, 1998). Similarly, advocates of a social justice perspective on environmental issues realized the increased credibility, resources and support that would result from a coalition with the mainstream organizations. Perhaps the key reason for the success of the environmental justice movement however, can be attributed to its ability to tap into the discourse and rhetoric of the civil rights movement (Camacho, 1998, Sandweiss, 1998, Taylor, 2000). Sandweiss shows how the collective action frame of the civil rights movement – which emphasized such values as individual rights, equal opportunities, social justice, human dignity and self-determination – provided a master frame through which victims of disproportionate exposure could articulate their concerns (Sandweiss, 1998). The success of the environmental justice movement in linking environment, labour and social justice into a master frame through which to communicate claims and clarify goals and grievances to others, and to create a powerful 'environmental justice paradigm', has been extensively analyzed by Taylor (2000). This, she argues 'is the first paradigm to link environment and race, class, gender and social justice concerns in an explicit framework' (Taylor, 2000, p42).

Emerging voices for justice

The environmental justice movement developed in the US in response to concerns about the uneven distribution of environmental risks among certain groups of people, especially those of colour. However, because of the increasingly broad usage of the term environmental justice outside the US, it has come to be used to include poor and disadvantaged groups as well. As Cutter (1995, p113) notes, 'environmental justice ... moves beyond racism to include others (regardless of race or ethnicity) who are deprived of their environmental rights, such as women, children and the poor'. In many other European countries, such as Scotland (see Dunion and Scandrett, Chapter 15), class issues and issues of exclusion, not race, are to the fore in environmental justice debates.

However, the race dimension of environmental justice is of crucial importance, as several of our authors indicate. The environmental degradation and exploitation of the world's resources by rich Northern nations is often underpinned by an implicit racism which legitimizes differential life chances for people of colour compared with whites. The Bhopal disaster, the exploitation of West African fishing grounds, the dumping of toxic wastes and the issues of bio-piracy and bio-cultural assimilation (see Rixecker and Tipene-Matua, Chapter 12) are some examples of this environmental racism.

The international dominance of the US experience within the environmental justice discourse is unsurprising given the history and experience of its grass-roots communities and its linkages with the civil rights movement. Nevertheless, it is clear from the cases we illustrate in this book that other countries have recognized the links between environmental exclusion and social exclusion, and between environmental degradation and economic exploitation (see also ESRC Global Environmental Change Programme (2001) for an in-depth discussion of the differences between environmental justice in the US and elsewhere). In some cases, such as in the UK, there have been initiatives that have sought to explicitly link race and racism to environmental issues. The Black Environment Network (BEN), for example, created in the late 1980s, was an early UK initiative bringing together black and ethnic minority organizations to address environmental concerns. Its major campaign at the time was about black access to the British countryside, which was, and still is, seen as an exclusive, ecological and predominantly white space (Agyeman and Spooner, 1997) which restricts black recreational access because of fears of racism. Some would argue that this is an access issue, not strictly an 'environmental justice issue', but according to one definition of environmental justice, it includes 'the equal distribution of environmental benefits' which includes 'the provision of access to open space' (Commonwealth of Massachusetts, 2000, p2). In addition, a case could be made for access being an 'environmental right' (Cutter, 1995). Increasingly therefore, environmental justice is being seen not only as being about stopping 'bads', but about promoting 'goods' and being able to experience quality environments and environmental quality.

Although there are examples of explicit environmental justice actions in other parts of the world outside the US – the UK, the Indian sub-continent,

Nigeria, South Africa and Australia are clear examples – few if any places have developed an 'environmental justice paradigm' (Taylor, 2000) including the environmental justice vocabulary and the range of organizations which exist in the US. This does not mean, of course, that there is no experience of, or understanding of, US-style environmental justice issues, or that political activity is somehow non-existent. Far from it. There are political struggles around the world which are clearly evident in terms of both traditional siting issues, and also in areas such as housing, work or opposition to new developments. In this sense, the discourse of environmental justice may be seen as a unifying process, bringing together diverse situations and sharing understandings and experiences. In some countries an environmental justice discourse is emerging. In the UK context, Agyeman (2000, p8) argues that:

> *What is clearly happening under New Labour, as it was in the US when Clinton took office, is that environmental and sustainability policy discourses and claims are beginning to be re-framed and this is being driven by NGO activists and policy entrepreneurs. Instead of being framed within a 'green', or predominantly 'environmental' agenda, these discourses are being refocused around quality of life, using the notions of justice, rights and equity.*

SUSTAINABILITY AND ENVIRONMENTAL JUSTICE

As we have said, it is the intent of this book to map the nexus, the common ground, between both the concepts of sustainability and environmental justice and their practical actions. This means, as Haughton (1999, p64) has argued, 'acknowledging the interdependency of social justice, economic well-being and environmental stewardship. The social dimension is critical since the unjust society is unlikely to be sustainable in environmental or economic terms in the long run'. Despite this criticality, both contemporary and historical equity and justice aspects of sustainability are only now beginning to be better understood (Boardman et al, 1999, Agyeman, 2000) in terms of 'environmental space' (McLaren et al, 1998, and Chapter 1 in this book, Carley and Spapens, 1997), 'ecological debt' (McLaren et al, 1998, and Chapter 1 in this book), 'ecological footprinting' (Wackernagel and Rees, 1996, Rees and Westra, Chapter 5 in this book) and 'asset-based' approaches (Boyce and Pastor, 2001).

However, lest we should become too optimistic that change is happening quickly, Warner (2002, p37) 'found that more than 40 per cent of the largest cities (33 of 77) in the US had sustainability projects on the web, but only five of these dealt with environmental justice on their web pages'. He continued that 'few communities were building environmental justice into local definitions of sustainability. Only five local sustainability projects made these connections: Albuquerque, New Mexico; Austin, Texas; Cleveland, Ohio; San Francisco, California; and Seattle, Washington' (p38). Warner's (2002, p38) conclusion was that 'while environmental justice seemed to be having an impact on mainstream environmental organizations and on government agencies, this did not apparently extend to groups working on sustainability projects'.

Rights

International calls for justice considerations to be incorporated into sustainability policies usually focus on inter-generational equity, and *intra*-generational equity between what have been traditionally referred to as the *core*, industrialized nations in the North, and the *peripheral*, developing nations in the South. The notion of justice can be seen in reactions against the social costs of unsustainable natural resource extraction such as the case of the destruction of mangrove swamps by shrimp exporters (Guha and Martinez-Alier, 1999), the environmentally degrading practices of oil companies in West Africa (Adeola, 2000, Agbola and Alabi, Chapter 13 in this book) and the bio-prospecting of agricultural biotechnology companies (Smith, 1999, Rixecker and Tipene-Matua, Chapter 12 in this book). Activists claim that the injustices caused by MNCs represent a major human rights violation against the local people who are caught in the path of globalization and describe these violations as 'ecological imperialism' (Adeola, 2000). They argue that the right to a clean and safe environment is an important and essential human right that should not be denied on the basis of race, class, ethnicity or position in the global economic system (Adeola, 2000, Sachs, 1995, Hartley, 1995, Johnston, 1995). This argument is reflective of greater concerns for social justice issues in the environmental movement, which is often blamed for paying too much attention to issues of environmental quality, such as nature and biodiversity (important though these are) while other basic human rights, equity issues and needs remain unprotected (Agyeman, 2001).

Following the Universal Declaration of Human Rights proclaimed by the General Assembly of the United Nations on 10 December 1948, the Council of Europe passed the Convention for the Protection of Human Rights and Fundamental Freedoms in Rome on 4 November 1950. Of especial relevance in environmental justice and sustainability terms is Article 14 on 'The Prohibition of Discrimination' which states that 'the enjoyment of the rights and freedoms set forth in this Convention shall be secured without discrimination on any ground such as sex, race, colour, language, religion, political or other opinion, national or social origin, association with a national minority, property, birth or other status' (Council of Europe, 1950).

Liepietz (1996, p223) has argued that we need 'new rights and obligations to be incorporated within social norms'. This, he continued, would involve 'the recognition, at first moral, of new rights, new bearers of rights and new objects of rights'. Adeola (2000, p687) talks of the 'need to frame environmental rights as a significant component of human rights'. The recognition of 'human environmental rights' has led to an overlap of international environmental and human rights law, as can be seen especially in the 1999 Aarhus Convention on 'Access to Information, Public Participation in Environmental Decision-Making and Access to Justice in Environmental Matters', which recalls Principle 1 of the 1972 Stockholm Declaration on the Human Environment and Principle 10 of the 1992 Rio Declaration on Environment and Development. The Convention, which came into effect on 30 October 2001, is therefore unique in being the first to ensure citizens' rights in the field of the environment. It implies *substantive rights* (the right to a cleaner environment) and guarantees *procedural rights* (the right to participate) to European citizens. It states, as the objective of

Article 1, that: 'in order to contribute to the protection of the right of every person of present and future generations to live in an environment adequate to his or her health and well-being, each party shall guarantee the rights of access to information, public participation in decision-making and access to justice in environmental matters in accordance with the provisions of this Convention' (UNECE, 1999).

Environmentalists are also increasingly questioning the justice and equity implications of other international agreements, especially those related to trade or economic development. There is great (and under-researched) potential for the notions of environmental justice, human rights and sustainability to permeate environmental regimes and international policy and agreements. Indeed, it is being increasingly recognized that one of the best ways to protect environmental rights is to uphold the basic civil, human and political rights of the individual (Sachs, 1995, Anderson, 1996). Furthermore, once these ideas become enshrined in policy, they have the capability to enable legal challenges to existing practices upon which they could not make a previous impact. An example of this confluence of activism, policy and law can be seen in recent developments surrounding the use of Title VI of the 1964 Civil Rights Act to prove discriminatory intent in environmental injustice cases in the US.

The ideas of sustainability and justice are also being used to influence policy change at the global level. The recent Earth Charter (2000) represents an initiative to form a global partnership that hopes to recognize the common destiny of all cultures and life forms on Earth and to foster a sense of universal responsibility for the present and future well-being of the living world. The Earth Charter Initiative was launched in 1994 by the Earth Council and Green Cross International, and is now overseen by the Earth Charter Commission in Costa Rica. The current draft of the charter was produced at the Earth Charter meeting at UNESCO headquarters in Paris in March 2000, after revisions by over 40 national Earth Charter committees. The charter stresses the need for a shared vision of basic values to provide an ethical foundation for the emerging world community (Earth Charter, 2000).

The set of principles that are outlined in the document reflect the necessary and inherent linkages between the ideas of sustainability and justice that will enable the development of this shared vision. The four principles that constitute the basis of the document include: respect and care for the community of life; ecological integrity; social and economic justice and democracy; non-violence and peace. It is hoped by the members of the Commission that the Earth Charter principles find expression in individual lifestyles, professional and organizational work ethics, educational curricula, religious teachings, public policy and government practices and that the principles will be endorsed by the United Nations by the year 2002 (Earth Charter, 2000).

Security

One final consideration is the relationship of the environmental security agenda to sustainability and justice concepts. In recent years, a huge literature on the subject of conflicts arising from environmentally degraded communities or scarcity of natural resources has permitted the environmental security paradigm to enter into post-Cold War mainstream policy discussions. In the absence of a Cold War enemy, military strategists have identified 'the environment' as a flashpoint and a potential source of conflict. Authors writing on the subject have documented the conflicts that can emerge as a result of ecological degradation and scarcity, particularly in developing countries where there is great competition for land and resources (Tuchman-Mathews, 1989, Homer-Dixon, 1994, Kaplan, 1994). Deudney and Matthew (1999) however argue that much of the empirical evidence for scarcity-induced conflict upon which the environmental security literature is based rests on 'contested grounds' while Hartmann (1998) draws attention to the lack of attention paid in the environmental security literature to the role of MNC interests in environmental degradation.

CONCLUSION

Despite Warner's (2002) reality check regarding the (lack of) integration of environmental justice into sustainability programmes in US cities, mapping the terrain at the nexus of environmental justice and sustainability provides a rich environment for policy research around co-activism. At the more local level, Schlosberg (1999, p194) argues that there are a growing number of 'examples of cooperative endeavours between environmental justice groups and the major organizations'. This emerging co-activism on sustainability and environmental justice issues, which we think represents the next stage for environmental justice movements worldwide, can be found, for example, in local fights for just transportation (for instance, Heart and Burrington, 1998, Bullard and Johnson 1997, Patel and Sharma, 1998), community food security (Gottlieb and Fisher, 1996, Perfecto, 1995), sustainable communities and cities (Roseland, 1998, Rees, 1995, Haughton, 1999, Satterthwaite, 1999, Patel and Sharma, 1998) and in Friends of the Earth Scotland's environmental justice campaign whose slogan – 'no less than our right to a decent environment; no more than our fair share of the Earth's resources' – both localizes, and globalizes the issues (Friends of the Earth Scotland, 2000). Suffice it to say here that we agree with Goldman's (1993, p27) visionary assessment that goes so far as to suggest that 'sustainable development may well be seen as the next phase of the environmental justice movement'. Essentially therefore, our book contributes to this process by setting up the arguments as to why this is important, and why it needs to happen now.

THE CHAPTERS

The book is organized into four sections: *Some Theories and Concepts* in which our authors lay some of the theoretical and conceptual ground upon which this book is built; *Challenges* in which Dobson outlines his contention that the two ideas of social justice and environmental sustainability are politically incompatible; *Cities, Communities and Social and Environmental Justice* in which communities from the village to the (US) state level are considered; and *Selected Regional Perspectives on Sustainability and Environmental Justice* in which our authors reflect on practical issues in a wide range of regional, national and sub-national contexts.

NOTES

1   The discussions in this chapter are based upon an article by the editors which appeared in *Space and Polity*, vol 6 (1), pp77–90.

REFERENCES

Adeola, F O (2000) 'Cross-national environmental justice and human rights issues – A review of evidence in the developing world', *American Behavioral Scientist*, vol 43 (4), pp686–706

Agyeman, J and Evans, B (1995) 'Sustainability and democracy: Community Participation in Local Agenda 21', *Local Government Policy Making*, vol 22 (2), pp35–40

Agyeman, J and Spooner, R (1997) 'Ethnicity and the rural environment' in P Cloke and J Little (eds) *Contested Countryside Cultures*, Routledge, London

Agyeman, J (2000) *Environmental Justice: From the Margins to the Mainstream?*, Town and Country Planning Association, London

Agyeman, J (2001) 'Ethnic minorities in Britain: Short change, systematic indifference and sustainable development', *Journal of Environmental Policy and Planning*, vol 3 (1), pp15–30

Anderson, M (1996) 'Human rights approaches to environmental protection: An overview' in A Boyle and M Anderson, *Human Rights Approaches to Environmental Protection*, Clarendon Press, Oxford

Boardman, B, Bullock, S and McLaren, D (1999) *Equity and the Environment. Guidelines for Socially Just Government*, Catalyst/Friends of the Earth, London

Boyce, J K, Klemer, A R, Templet, P H and Willis, C E (1999) 'Power distribution, the environment, and public health: a state level analysis', *Ecological Economics*, vol 29, pp127–40

Boyce, J K and Pastor, M (2001) *Building Natural Assets*, University of Massachusetts Amherst Political Economy Research Institute/Ford Foundation, New York

Bullard, R D (1993) 'Anatomy of environmental racism', in R Hofrichter (ed) *Toxic Struggles: The Theory and Practice of Environmental Justice*, New Society Publishers, Gabriola Island, British Columbia

Bullard, R D (ed) (1994a) *Unequal Protection: Environmental Justice and Communities of Colour*, Sierra Club Books, San Francisco

Bullard, R D (1994b) 'Environmental racism and the environmental justice movement' in C Merchant (ed) *Ecology: Key Concepts in Critical Theory*, Humanities Press, New Jersey

Bullard, R D & Johnson, G S (1997) *Just Transportation: Dismantling Race and Class Barriers to Mobility*, New Society Publishers, Gabriola Island, British Columbia

Camacho, D E (ed) (1998) *Environmental Injustices, Political Struggles: Race, Class and the Environment*, Duke University Press, Durham

Carley, M and Spapens, P (1997) *Sharing the World*, Earthscan, London

The Corner House (2001) *Democracy or Carbocracy? Intellectual Corruption and the Future of the Climate Debate*, Briefing 24, The Corner House, Sturminster Newton, Devon

Commonwealth of Massachusetts (2000) *Draft Environmental Justice Policy of the Executive Office of Environmental Affairs*, EOEA, Boston, (12/7/00 version)

Council of Europe (1950) *Convention for the Protection of Human Rights and Fundamental Freedoms,* Council of Europe, Strasbourg.

Cutter, S (1995) 'Race, Class and Environmental Justice', *Progress in Geography*, vol 19 (1), pp111–22

Deudney, D H and Matthew, R A (eds) (1999) *Contested Grounds: Security and Conflict in the New Environmental Politics,* State University of New York Press, Albany

Dobson, A (1999) *Justice and the Environment: Conceptions of Environmental Sustainability and Dimensions of Social Justice,* Oxford University Press, Oxford

Earth Charter (2000) *The Earth Charter,* The Earth Council, Costa Rica.

Environmental Careers Organization and Urban Habitat Program (1998) *Working for Sustainable Communities,* Environmental Careers Organization and Urban Habitat Program, San Francisco

ESRC Global Environmental Change Programme (2001) *Environmental Justice: Rights and Means to a Healthy Environment for All,* Special Briefing Number 7, University of Sussex

Faber, D (ed) (1998) *The Struggle for Ecological Democracy: Environmental Justice Movements in the United States,* The Guilford Press, New York

Friends of the Earth Scotland (2000) *The Campaign for Environmental Justice,* Friends of the Earth Scotland, Edinburgh

General Accounting Office (1983) *Siting of Hazardous Waste Landfills and their Correlation with Racial and Economic Status of Surrounding Communities,* Government Printing Office, Washington, DC

Goldman, B (1993) *Not Just Prosperity, Achieving Sustainability with Environmental Justice,* National Wildlife Federation, Washington, DC

Gottlieb, R and Fisher, A (1996) '"First feed the face": Environmental justice and community food security', *Antipode,* vol 28 (2), pp193–203

Guha, R and Martinez-Alier, J (1999) 'Political ecology, the environmentalism of the poor and the global movement for environmental justice', *Kurswechsel*, Heft 3 pp27–40

Hartley W (1995) 'Environmental justice: An environmental civil rights value acceptable to all world views', *Environmental Ethics*, vol l7 (13), pp277–89

Hartmann, B (1998) 'Population, environment and security: a new trinity', *Environment and Urbanization*, vol 10 (2), pp113–27

Haughton, G (1999) 'Environmental justice and the sustainable city', *Journal of Planning Education and Research*, vol 18 (3), pp233–43

Heart, B and Burrington, S (1998) *City Routes, City Rights: Building Livable Neighborhoods and Environmental Justice by Fixing Transportation,* Conservation Law Foundation, Boston, MA

Homer-Dixon, T (1994) 'Environmental scarcities and violent conflict: Evidence from cases', *International Security*, vol 19 (1), pp5–40

IUCN (1980) *World Conservation Strategy* , IUCN, Gland

IUCN (1991) *Caring for the Earth*, IUCN, Gland

Jacobs, M (1992) *The Green Economy: Environment, Sustainable Development and the Politics of the Future,* Pluto Press, London

Jacobs, M (1999) 'Sustainable development: A contested concept' in A Dobson (ed) *Fairness and Futurity: Essays on Environmental Sustainability and Environmental Justice,* Oxford University Press, Oxford, pp211–45

Johnston, B (1995) 'Human rights and the environment', *Human Ecology*, vol 23 (2), pp111–23

Kaplan, R (1994) 'The Coming Anarchy', *The Atlantic Monthly*, vol 273 (2), pp44–76

Lavelle, M and Coyle, M (1992) (eds) 'The racial divide in environmental law: Unequal protection' (Special Supplement), *National Law Journal*, 21 September

Lipietz, A (1996) 'Geography, ecology, democracy', *Antipode,* vol 28 (3), pp219–28

Low, N and Gleeson, B (1998) *Justice, Society and Nature*, Routledge, London

McLaren, D, Bullock, S and Yousuf, N (1998) *Tomorrow's World: Britain's Share in a Sustainable Future*, Earthscan, London

McNaghten, P and Urry, J (1998) *Contested Natures*, Sage, London

Middleton, N and O'Keefe, P (2001) *Redefining Sustainable Development*, Pluto Press, London

Milbrath, L (1989) *Envisioning a Sustainable Society*, SUNY Press, Albany

Morello-Frosch, R (1997) 'Environmental justice and California's "riskscape": the distribution of air toxics and associated cancer and non cancer risks among diverse communities', unpublished dissertation, Department of Health Sciences, University of California, Berkeley

O'Connor, J (1988) 'Introduction', *Capitalism, Nature, Socialism*, vol 1 (1)

Patel, S and Sharma, K (1998) 'One David and three Goliaths: avoiding anti-poor solutions to Mumbai's transport problems, *Environment and Urbanization*, vol 10 (2), pp149–60

Perfecto, I (1995) 'Sustainable agriculture embedded in a global sustainable future: Agriculture in the United States and Cuba' in B Bryant (ed) *Environmental Justice: Issues, Policies and Solutions*, Island Press, Washington, DC

Portney, K E (1994) 'Environmental justice and sustainability: Is there a critical nexus in the case of waste disposal or treatment facility siting?', *Fordham Urban Journal*, Spring, pp827–39

Pulido, L (1996) 'A critical review of the methodology of environmental racism research', *Antipode*, vol 28 (2), pp142–59

Rawls, J (1971) *A Theory of Justice*, Harvard University Press, Cambridge, MA

Redclift, M (1987) *Sustainable Development*, Routledge, London

Rees, W E (1995) 'Achieving Sustainability: Reform or Transformation?', *Journal of Planning Literature*, vol 9 (4), pp343–61

Ringquist, E J (1998) 'Environmental Justice: Normative concerns and empirical evidence' in N J Vig and M E Kraft (eds) *Environmental Policy in the 1990s: Toward a New Agenda*, Congressional Quarterly Press, Washington, DC

Roseland, M (1998) *Toward Sustainable Communities: Resources for Citizens and Their Governments*, New Society Publishers, Gabriola Island, BC

Sachs, A (1995) *Eco-Justice: Linking Human Rights and the Environment*, Worldwatch Paper 127, Washington, DC

Sandweiss, S (1998) 'The social construction of environmental justice' in D E Camacho (ed) *Environmental Injustices, Political Struggles: Race, Class and the Environment*, Duke University Press, Durham

Satterthwaite, D (1999) *The Earthscan Reader in Sustainable Cities*, Earthscan, London

Schlosberg, D (1999) *Environmental Justice and the New Pluralism: The Challenge of Difference for Environmentalism*, Oxford University Press, Oxford

Smith, L T (1999) *Decolonizing Methodologies: Research and Indigenous Peoples*, University of Otago Press, Dunedin; Zed Books, London

Taylor, D E (2000) 'The rise of the environmental justice paradigm: Injustice framing and the social construction of environmental discourses', *American Behavioral Scientist*, vol 43 (4), pp508–80

Torras, M and Boyce, J K (1998) 'Income, inequality and pollution: a reassessment of the environmental Kuznets Curve', *Ecological Economics*, vol 25, pp147–60

Tuchman-Mathews, J (1989) 'Redefining security', *Foreign Affairs*, vol 68 (2), pp162–77

United Church of Christ Commission for Racial Justice (1987) *Toxic Wastes and Race in the United States*, United Church of Christ Commission for Racial Justice, New York

United Nations Economic Commission for Europe (1999) *Convention on Access to Information, Public Participation in Decision Making and Access to Justice in Environmental Matters*, UNECE, Geneva

Wackernagel, M and Rees, W (1996) *Our Ecological Footprint*, New Society Publishers, Gabriola Island, BC

Warner, K (2002) 'Linking local sustainability initiatives with environmental justice', *Local Environment*, vol 7 (1), pp35–47

World Commission on Environment and Development (1987) *Our Common Future*, Oxford University Press, Oxford

Part 1

Some Theories and Concepts

Both sustainability and environmental justice consist of eclectic theoretical and knowledge bases. In addition, as theory and practice evolve in each, new ideas, approaches and techniques are emerging. This is especially so in sustainability where there is an industry around the development, use and evaluation of tools and techniques for use at different scales and in different contexts. Part 1 presents just some of the key theoretical and conceptual issues surrounding sustainability and social and environmental justice.

McLaren examines how equity considerations at all scales can be incorporated in global conceptions of sustainable development. He focuses on the concepts of 'environmental space' and 'ecological debt'. Chapter 1 then explores the international and intra-national equity consequences of sustainability in such terms. Environmental space intrinsically embraces equity, but it is future facing, and thus needs the addition of a historical perspective to incorporate justice. This is provided by the concept of ecological debt, which leads inevitably to an analysis of the global actors and institutions that have created unsustainability and injustice. This examination of power relations lays down a 'just sustainability' challenge to globalization, with which the chapter concludes.

In Chapter 2, Faber and McCarthy examine how the linkages between sustainability and environmental justice can further the goals of 'ecological democracy'. This is based on the premise that the only way nature and labour will be protected is if environmental justice activists and workers are afforded the same access to political decision-making and policy-making processes as are those who control capital.

Chapter 3 attempts to demonstrate the limitations of contemporary discourses of ecological modernization and 'risk society', which, Blowers argues, notably neglect inequality and community and their implications for changing power relations. He goes on to identify how these deficiencies are addressed by alternative approaches, namely through the environmental justice movement which, both in principle and in terms of practical political action, emphasizes the distributional implications of environmental change. He then gives an empirical focus by drawing on evidence from so-called 'peripheral communities' to demonstrate how power relations may shift over time to mitigate disadvantage. This leads into a discussion of the principles that must inform society's efforts to shift towards sustainability and the implications for policy-making.

* * *

Chapter 1

Environmental Space, Equity and the Ecological Debt

Duncan McLaren

'No less than a decent quality of life for all, no more than a fair share of the Earth's resource'

INTRODUCTION

This chapter sketches out the relationships between sustainability and equity. It offers the environmental space approach as a helpful framework. This conceptualizes sustainability in terms of access for all to fair shares in the resources on which healthy quality of life depends.

SUSTAINABLE DEVELOPMENT

In Rio in 1992 and in New York in 1997 world leaders affirmed their intent to pursue sustainable development: delivering improved quality of life whilst living within the carrying capacity of supporting ecosystems (UNEP et al, 1990). Today environmentalists remain alarmed at the continued rapid erosion of that carrying capacity through pollution and resource exploitation, fuelled by escalating consumption. As the UN Environment Programme (UNEP) put it, humanity's inefficient use of resources and patterns of wasteful consumption are *'driving us towards an environmental precipice'* (UNEP, 1997).

Equity considerations are embedded in all conceptualizations of sustainable development but are rarely unpacked. All interpretations of sustainable development including that of the widely cited Brundtland Commission (World Commission on Environment and Development, 1987) agree that it involves some form of redistribution from current to future generations. This is because our current rates of consumption are depriving future generations of consumption opportunities, or generating impacts which they will bear (see Box

---

**BOX 1.1 INTER-GENERATIONAL ISSUES IN SUSTAINABILITY –
SOME EXAMPLES**

Biodiversity loss: From a human perspective, every species that is made extinct eliminates knowledge and reduces future options for society. Currently unknown or unvalued species may offer great future potential as sources of drugs, chemicals, fibre and food. The contemporary rate of species extinction – largely caused by humans through destruction of habitat – is estimated to be between 50 and 100 times the average 'background rate', the highest since the dinosaur extinction (UNEP, 1995).

Climate change: Climate change is expected to be one of the driving forces of habitat loss and extinction in this millennium. It is also expected to have devastating effects on human livelihoods through flooding, storms, changing rainfall patterns, changing disease habits and habitats, and sea level rise. Global reductions of greenhouse gas emissions of at least 50 per cent in the next 50 years are needed to stabilize the climate (Intergovernmental Panel on Climate Change, 1996).

PCBs and other endocrine disrupting chemicals: Evidence is growing that the effects of such persistent and bioaccumulative industrial chemicals – which now lace our environment – can reach through not just to the next generation but beyond that to affect our children's children. In faster reproducing species the wide-ranging health effects of endocrine disrupters – the so-called 'gender benders' – are already being seen (Colborn et al, 1996). Emissions of all such chemicals need to be eliminated.

Ozone depletion: The worst emissions of ozone depleting substances – chemicals used as cheap propellants and refrigerants for consumer goods – have been tackled. But even if the Montreal Protocol – which seeks to eliminate emissions of ozone-depleting substances – is fully implemented, the ozone layer will not return to its pre-depletion state before 2100. Hundreds of thousands of additional cases of skin cancer can be expected every year – along with impacts on other species (UNEP, 1994).

Nuclear wastes are not only an environmental health hazard now, they will persist in dangerous states for centuries. Even the financial costs of decommissioning and waste management will fall predominantly on future generations. The OECD estimates that the costs of waste management for the UK alone will total £45 billion by 2100 (OECD, 1996), whilst the direct costs of the Chernobyl disaster to Belarus alone are estimated at US$235 billion (Nuclear Engineering International, 1996).

1.1). Over-use of environmental resources is at the heart of the challenge of sustainable development. Humankind is consuming (and wasting) resources at a rate faster than the ecological systems of the planet can tolerate. Substantial cuts in resource use and emissions are required to deliver inter-generational equity. From this, as will be elaborated below, flow dramatic implications for intra-generational equity. Where there are limited resources, distributional questions cannot be avoided. But these are less commonly considered in debates on sustainability.

But the question is more complex even than this suggests. Current economic inequality and the marginalization of the poor by the wealthy also drive resource exploitation. For example, the immediate needs of the poor for fuelwood tend to outweigh their future need for forests to protect the watershed and prevent erosion. Similarly, poor countries' need to earn foreign exchange to repay financial debts leads to regimes for management of forests and minerals which encourage over-exploitation by multinational investors – generating impacts which are normally much larger than those arising from the activities of poor people meeting their immediate needs. Thus inequality is a driving force behind unsustainability.

The benefits of current rates of resource use go mainly to a rich minority, and the costs – where they impinge on human beings, in terms of poor health and loss of homes and livelihoods – are borne disproportionately by poorer people. In many countries women especially suffer from such impacts (see Wickramasinghe, Chapter 11). This is environmental injustice. This injustice is multiplied when we develop so-called solutions to our environmental problems whose costs are also borne by the poor. Environmental justice demands that the poor do not bear the costs of overconsumption by the rich (or of cleaning up after them)! Indeed environmental justice demands more generally that these costs are not offloaded onto other groups. But prescriptions for substantial cuts in resource use – essential at the aggregate level for environmental sustainability – not only seem unrealistic for the majority of the world's population, they appear to threaten just such an unfair distribution of costs.

If dramatic cuts were achieved in the current system, who would bear the costs? Whose personal consumption would be cut? Would it be the rich and powerful, or the poor and weak? The social dimension of sustainability is inescapable. The implication might appear to be that we must trade off our environmental objectives for social ones. If poorer groups must increase consumption, surely we must accept a higher level of environmental damage.

But this conclusion is fallacious – it is based on at least two erroneous assumptions. First is the belief that sustainable development is about extending Northern levels of wealth, consumption and well-being to Southern countries (albeit without triggering economic, social or environmental disruption). This belief assumes that sustainable development is simply a more efficient, better managed process of conventional economic development – with further extension of the Anglo-Saxon cultural model of business, markets, indicators and aspirations. This would leave power relations – a key issue in current resource distribution to which I return at some length below – almost untouched.

The second false assumption is that higher wealth and consumption directly translate to higher well-being and quality of life. Empirical evidence demonstrates that in reality there is not a direct correlation between income and quality of life at all levels (Veenhoven, 1987, Seabrook, 1994). Above a certain level of income any correlation breaks down, and other factors, notably health, come to play a critical role in quality of life (Oswald, 1996, Wilkinson, 1996). This creates scope for policies which can reduce inequalities in well-being without increasing aggregate material consumption.

A parallel concern is that to make dramatic cuts in environmental resource use would require undemocratic measures. This is a legitimate concern, as some environmentalist advocates of cuts in consumption have paid little attention to this aspect. However, most proponents of sustainability recognize that without democratic support it is neither possible to deliver such dramatic changes, nor to sustain them. Thus there has been growing interest in rights-based approaches to sustainability, through, for example, the Aarhus Declaration.

This chapter explores how a more rigorous approach to sustainable development can demonstrate that justice and environmental sustainability are not only compatible, but interdependent, and that sustainable development poses a fundamental challenge to existing economic systems and institutions at both national and global scales.

ENVIRONMENTAL SPACE

Stimulated by global debate over sustainable development (World Commission on Environment and Development, 1987, UN Conference on Environment and Development, 1992), the international environmental network Friends of the Earth has led the development of global measures of environmental capacity or 'environmental space'. These combine our growing understanding of the nature of environmental limits with an equity-based approach to environmental justice (Spangenberg, 1995, Carley and Spapens, 1998). The basic idea behind environmental space is simple: for each individual it is possible to calculate a maximum rate of consumption of environmental resources – a fair share of the maximum available within global limits; while recognizing the existence of a minimum determined by need and human dignity. Within this space, living and lifestyles can be truly sustainable. The methodology for determining practical targets based on this approach is outlined in Box 1.2. Below I outline key elements of this methodology with the aim of demonstrating how it can underpin the emerging concept of 'just sustainability'.

Focused on outputs

This programme of work has revealed the difficult judgements and severe limitations of data involved in establishing such measures – but has also generated meaningful quantitative estimates in a small number of key areas. This has been achieved by focusing on the flows of resources used to support human societies rather than on the waste and pollution that has driven much environmental policy. The impacts of such outputs are taken into account as a potential limiting factor on rates of resource use. For example, the rate of use of fossil energy resources is limited not by how quickly they are being depleted, but by the climate change associated with the resulting carbon dioxide emissions.

The most significant resources are all included

The validity of the environmental space methodology relies, at least in part, on the resources it covers. Fossil fuels, water, timber, steel, aluminium, cement, foodstuffs (land) and chlorine (the set most widely assessed) together account for over 95 per cent of resource consumption by weight in the UK and similar developed economies (McLaren et al, 1998). They include the fundamental

> **BOX 1.2 SETTING ENVIRONMENTAL SPACE TARGETS –**
> **A SUMMARY**
>
> For each resource:
>
> • Determine the 'sustainable global use' of that resource by identifying the limiting constraint and – in a precautionary fashion – estimating the annual rate of use which avoids breaching that constraint. Limiting constraints may be renewable harvests; the rate at which non-renewable resources can be replaced by renewable substitutes; or the ecological or health impacts of exploitation, use and disposal.
> • Calculate a per person 'fair share' in that resource use, based on a forecast global population of 10 billion in 2050 and multiply by national forecast population for 2050 to obtain a national 'fair share'.
> • Estimate current national consumption, taking account of trade in the resource (and, insofar as possible, in goods embodying the resource).
> • Compare the current consumption with the national 'fair share' to estimate the reduction needed, or increase desirable, in percentage terms.
>
> *Source:* Based on McLaren et al, 1998

resources required to provide for a healthy quality of life: building materials for shelter, land for food production, water for drinking and cleansing, power and materials for transport, warmth, light, education and leisure. Nor does the choice ignore the qualitative health or ecological impacts of pollutants such as toxic chemicals (which account for only a tiny fraction of resource use by mass). Chlorinated organic compounds are assessed as an example of the need to eliminate such toxic, persistent and bioaccumulative materials outside of use in closed systems.

A precautionary approach

More broadly, the impacts of toxic materials are taken into account in the precautionary assessment of limits, which applies to most resources. For energy resources, the work of the Intergovernmental Panel on Climate Change (1500 scientists working for several years) means that we can be quite confident in our estimates that use of fossil fuels must be cut, *globally*, by at least 50 per cent by 2050. However, in the absence of such intensive research effort, estimates for other resources are less certain (see Table 1.1) and reflect a precautionary approach.

Limits and distribution

To apply a principle of equity, such precautionary assessments are combined with a distributional analysis of current resource consumption. Optimists on both right and left of the political spectrum (Ross, 1994, North, 1995) tend to reject or avoid such implied or actual 'limits to growth'. But the physical and environmental resources available to human societies are limited in practical terms – as indeed world leaders have already agreed with respect to stratospheric

Table 1.1 *Beyond the limits: Required global reductions in environmental resource consumption*

Resource	Limiting factor	Sustainable level – reduction needed
Fossil Energy	CO_2 emissions and climate impacts	50 to 75%
Agricultural Land	Impacts on soils, forests, water and biodiversity	up to 15%[1]

Note: 1 An increase of up to 35 per cent may be possible through the development and widespread adoption of efficient sustainable harvesting techniques.
Source: Based on McLaren et al (1998)

ozone, climate stability and fish stocks in a series of global and regional treaties. Such a conclusion inevitably raises the question of distribution – is current distribution fair and just? If not, what would be fair and just? At present, consumption of environmental space resources (on a per capita basis) is broadly five times greater in Northern 'developed' nations, than in Southern 'developing' nations.[1] For some resources – such as fossil fuels and chemicals – the ratio is closer to ten times or more. Table 1.2 shows that these disparities in basic resource use are reflected in even broader disparities in end consumption.

Valid assumptions

The environmental space approach establishes a safe global level of resource use – and then divides this equally between the world's people. It would be inappropriate to simply use the current population of the world for the simple reason that population trends are radically different in different parts of the world – and cannot be changed overnight. Various authors working with the methodology have applied different time frames, but the present author has previously used the projected world population for 2050 as a benchmark for three reasons: first, this time frame provides an opportunity for rapidly growing populations to be brought under control without coercion; second (and linked), there is reason to hope that the planet's overall population may have stabilized by then, as the UN suggest (UN Population Division 1994); and third, such a time period allows for a realistic process of achieving reductions in the resource use of wealthier societies (McLaren et al, 1998).

Like women's movements in the global South, the environmental space approach rejects the crude assumption that population growth is at the root of the problem, and that a coercive approach to population control can be justified to solve environmental problems. On the other hand it assumes that increasing justice for women, including access to education, livelihoods, resources and fertility control, will help reduce future population growth, as well as reducing inequity.

Practical outcomes

Having determined a fair share of a sustainable level of resource use per person, it is then relatively simple for any country (or region, or even community) to compare its average consumption of these key environmental resources with

Table 1.2 *Global consumption disparities*

Product	Consumption of richest 20% in comparison to poorest 20%
Meat	11 times
Energy	17 times
Fish	7 times
Telephone lines	49 times
Paper	77 times
Cars	145 times

Source: UNDP (1998)

that level and derive targets for either necessary reductions (the case in all Northern countries) or for acceptable or even desirable increases (the case in many Southern countries). Targets for the UK are shown in Table 1.3.

A distinctive approach

Thus the environmental space approach is distinctive from most other interpretations of sustainable development in several respects, including its focus on resource impacts, its emphasis on consumption, rather than population and technology, and its integral incorporation of the equity principle. One further distinction is its relationship to the concepts of 'sufficiency' and demand management. Analyses that focus predominantly on outputs (waste and pollution) tend to lead to prescriptions based on end of pipe and efficiency measures to reduce these outputs. Environmental space's input orientation tends to deprioritize end of pipe measures. It also emphasizes sufficiency measures, which seek ways to directly improve quality of life by consuming less.

In summary, environmental space is a rights-based approach that conceptualizes sustainable development in terms of access for all to a fair share in the limited environmental resources on which healthy quality of life depends. It implies eliminating at least international inequalities in aggregate resource consumption as well as inter-generational ones.

Table 1.3 *Environmental space targets for the UK*

Resource	Reduction required (per cent) by 2050
Aluminium	88
Carbon dioxide (emissions)	88
Cement	72
Chlorine	100
Construction aggregates	50
Land (UK average quality)	27
Steel	83
Water	15
Wood (Wood Raw Material Equivalent)	73

Source: McLaren et al (1998)

ENVIRONMENTAL SPACE AND
INTER-GENERATIONAL EQUITY

In this section I rehearse some of the arguments raised to resist reductions in inequalities, and I challenge conventional thinking and practices in this context.

In practice, even whether (not merely how) we should actively seek to reduce international inequalities in resource consumption is still a topic of political debate. The evidence of global environmental limits such as climate stability adds a new urgency to this debate. But many economists are unconcerned by inequalities in income and resource consumption and some even see them as necessary to the effective functioning of the economy. Inequalities increase overall savings and thus investment, or provide incentives, they argue (see for example Galor and Tsiddon, 1994 or Birdsall and Graham, 1999). But such arguments translate economic dogmas from the national scale – where inequalities can at least theoretically be argued to encourage economic effort and risk taking – to the international, where there is no evidence for the relevance of such arguments.

Others raise practical concerns. Even were there moral reasons to reduce inequalities, they say, developing countries have little scope to consume more resources so it would not matter if the North continues to take the lion's share. Such apologists for environmental colonialism are generally more concerned about population growth as a driving force of environmental damage, rather than consumption levels. And although content to raise the spectre that resource limits might lead to real falls in per capita consumption for the world's poorest people, few actively support reducing consumption in richer countries to allow redistribution of access to resources. As a result, their arguments echo those of business critics who cite the potential economic (and, with remarkable hypocrisy, social) impacts of cutting resource use in the North as reasons to delay, or entirely avoid action. This ignores the historical evidence that increased resource consumption rates in developing countries are associated with declining population growth rates – as part of a complex process involving increased economic security, greater equality between the sexes and improved education (see for example Mazur, 1994).

In practice, the policies of Northern countries and Northern dominated global institutions continue to widen inequalities – increasing Northern resource consumption and maintaining flows of resources and money (in 'debt' repayments) from South to North. The arguments aired above lead to advocacy of the distribution technique known as grandfathering (where, for example, rights to burn fossil fuels would be allocated in proportion to current rates of use). Most proposals for implementation mechanisms in the Kyoto Protocol to the Framework Convention on Climate Change (UN, 1997), such as 'joint implementation' and the 'clean development mechanism', implicitly assume a distribution of emissions rights on this basis. The equitable alternative would be to allocate consumption or pollution rights according to population, or in accordance with a planned transition to equal consumption. An example of this for fossil fuel use is the 'contraction and convergence' scenario promoted by the

Global Commons Institute, and supported by the GLOBE group of parliamentarians (Global Commons Institute, undated, Meyer, 2000). One implication of the analysis presented below is that 'compensation' should be added to 'contraction and convergence'.

There is a very pragmatic imperative for pursuing such strategies – which is that the participation of developing countries is essential in global programmes to deal with problems such as climate change – and politically such participation will not be obtained without a commitment by the North to increase equity. Moreover, the needs of developing countries to increase economic consumption to escape poverty cannot be met simply by increasing efficiency – they will need to increase real levels of material resource use. In the absence of voluntary reductions in the North, some of these limited resources – such as oil and water – may well become, even more than at present, a reason for conflict and war (Myers, 1996).

Equity-based strategies for cutting overall global consumption levels would have wider implications, by directly challenging the legitimacy of export-led development and globalization policies which increase resource exploitation. At present the global economy is dominated by policies and practices which drive in this direction. For example, structural adjustment progammes – promoted by the International Monetary Fund (IMF) and the World Bank as a condition for debt rescheduling – stress measures to earn foreign currency and improve the balance of payments. This basically means increasing commodity exports, whilst cutting social expenditure (Watkins, 1995). In parallel, Northern countries support and even subsidize exports of the equipment and infrastructure needed to exploit forests, fisheries, minerals and petroleum through export credit agencies (Hildyard, 1999).

So it should be no surprise that equity-based strategies challenge vested interests in the system – particularly multinational companies and their shareholders, who benefit most from current policies. Yet in the current context reducing Northern consumption actually threatens Southern well-being further, because of the dependence of Southern economies on exports of commodities. Thus, environmental space targets need to be accompanied by measures to reduce such dependence or support commodity producers. For example, to ensure that lower rates of commodity consumption are accompanied by higher prices for producers, or to rebuild local markets based in local control of resources.

ENVIRONMENTAL SPACE AND INTRA-NATIONAL SOCIAL EXCLUSION

Such targets also raise serious equity considerations at national scales. In this section I explore what the implications are and how they can be managed.

In the South, where inequality within nations is often severe, the implication of environmental space targets in many cases is that overall consumption can rise, especially in the poorest countries where large majorities are below the poverty line. In the North, while no one is proposing that *every* individual in the

UK or similar countries should make 80 per cent cuts in their *personal* consumption, some measures designed to cut the overall level of resource use (eg water charging, domestic energy taxes, closing coal-mines) could be regressive in their impacts. At the same time, the environmental space analysis largely eliminates the trickle down option for tacking poverty and emphasizes redistributive mechanisms.

Redistribution is also supported by accumulating evidence that suggests that economic inequalities (at least at contemporary levels) are economically damaging to national economies (OECD, 1996a, b, Deininger and Squire, 1996) and that quality of life and health too are directly damaged by inequalities within societies (Wilkinson, 1996, Kaplan et al, 1996, Chomsky, 2000, Jacobs, 1996). Reducing such inequalities could therefore contribute directly to the delivery of sustainability, while reducing some of the aspirational driving forces behind consumerism and conspicuous consumption. Nonetheless, at present there is a broad correlation between resource consumption and well-being, so policies and measures to reduce overall consumption of environmental resources will need to be carefully designed if they are not to maintain or even exacerbate existing inequalities (Boardman et al, 1999).

A key part of the solution is to design policies so that the correlation between resource consumption and well-being is weakened. If we recognize that economic well-being is a poor surrogate for quality of life or overall human well-being, then we can more effectively direct scarce resources (such as environmental resources) to increase the latter. International and inter-state comparison in the US suggest that even the correlation between rates of resource consumption and *economic well-being* can be broken (Barker, 1993, Templet, 1996). At the individual level this is more questionable – but it is increasingly clear that any relationship between resource consumption and *quality of life* is very tenuous above certain levels of resource use, due especially to effects on health (Wilkinson, 1996).

Designing policies to reduce environmental resource use to globally equitable levels (far beyond the minor shifts currently envisaged by Northern governments) while delivering social justice within those countries will be a major challenge of this century. But the first steps in this direction are easier than most governments believe. This is because the environmental improvements delivered by systemic, radical environmental policies would, broadly, benefit the poor most (Boardman et al, 1999). For example, research by Friends of the Earth on the location of polluting factories in the UK suggests (unsurprisingly) that they are predominantly located in poorer communities (see Box 1.3) mirroring similar evidence from the US (Bullard, 1999). More generally, poor environments – in terms of air pollution, traffic congestion, lack of green space and presence of contaminated land – are a significant factor in social exclusion.

Reconceiving environmental improvement as a social right rather than a middle-class luxury is a vital step for policy-makers. This cannot, however, justify unthinking imposition of regressive taxes or repressive regulatory regimes to deliver such improvements. Packages of redistributive policy reforms will be needed which include progressive taxes (ie gas guzzler taxes), socially just pricing

---

BOX 1.3 POLLUTION INJUSTICE

All across England and Wales the poorest families (reporting average household incomes below £5000) are twice as likely to have one of the country's 700 or so most polluting factories (one whose emissions are licensed by the Environment Agency under the UK's Integrated Pollution Control regulations) close by than those with average household incomes over £60,000.

Over 90 per cent of London's most polluting factories are located in communities of below average income. A similar but less extreme pattern is found throughout England and Wales. Overall, almost two-thirds of the most polluting industrial facilities are to be found in areas (defined in terms of postcode 'sectors' of around 2500 households) of below average income. The location of such factories also closely reflects the official Index of Multiple Deprivation at a ward scale. Of carcinogenic emissions 66 per cent come from factories in the most deprived 10 per cent of wards. Only 8 per cent of such emissions are in the least deprived 50 per cent of wards.

This is a clear cut issue of 'environmental injustice' in which poorer people are subjected to greater risks and impacts of pollution, and have less control over their environment while the benefits of the industrial activity largely accrue elsewhere. Measures to reduce industrial pollution from these factories would be clearly socially progressive.

Sources: McLaren et al, 1999, Bullock, 2001

---

structures (ie free allowances and escalating tariffs for energy), capital investment in efficiency (ie in domestic energy use) and measures to increase employment and target the provision of new livelihoods (ie cuts in employment taxes targeted at low-paid jobs) (Boardman et al, 1996). By combining such measures with increases in resource efficiency and investments in renewable technologies, resource use in Northern countries could be practically reduced towards environmental space targets (McLaren et al, 1998).

Similar measures may be needed in Southern countries, where small elites often consume well above fair share levels of resources. But large majorities tend to underconsume. Access to land is a typical and pressing issue. For example, in Brazil, the lack of access to land has spawned an activist movement, Sem Terra, dedicated to squatting land on large estates allocated for cash-cropping plantations and using it for subsistence livelihoods. Despite land reform progress in many countries, very uneven land distribution is not an unusual situation. The impacts of this are experienced most severely by women, as their ability to provide subsistence crops is lost, while it tends to be men who gain the limited cash incomes from labouring on plantations. This highlights the importance of economic rights and meeting basic needs as a priority for a country's share in environmental space.

The specific challenges vary from country to country, but the lessons are similar – reducing inequalities will help deliver social and environmental goals together, and progressive redistribution should be a policy option considered.

ECOLOGICAL DEBT

While the environmental space framework benchmarks what must be done to tackle existing environmental injustice between countries, there remains an issue of the historic injustice imposed on the poor and the poor countries of the planet. There is a vast historic ecological debt, which Southern campaigners argue that the rich morally owe to the poor (Accion Ecologica, 1999). Does redistribution – or perhaps compensation or reparation – provide a guide for us here also?

Centuries of political and environmental colonialism mean that the economies of rich nations have been built (in many ways literally) on resources such as timber, minerals and oil extracted from poorer ones (see Box 1.4). As Joan Martinez-Alier notes, 'the occupation of an environmental space larger than one's own territory, gives rise to ecological debt' (cited by Agarwal and Narain, 1998). The impacts of such resource exploitation are also distributed unevenly within Southern countries – women often bearing the brunt, especially of the effects of logging and timber plantations.

India's Centre for Science and Environment has long challenged the assumption that rights to global sinks for carbon dioxide, such as forests, should be 'grandfathered' along with emissions. Agarwal and Narain (1998) argue: 'the South needs ecological space to grow, which has already been colonized by the North. The poor are not even using a small share of their legitimate share of the global commons like the atmosphere, thus, permitting the North to pollute over the last century at little cost and build up its economy and industrial base extremely cheaply and rapidly.'

Theoretically it may be possible to put a money value on the ecological debt – by calculating the value of the environmental and social externalities associated with historic resource extraction and adding an estimated value for the share of global pollution problems borne by poor countries as the result of higher consumption levels in rich ones. In practice such an approach would suffer from the same shortcomings as efforts to value the external costs associated with climate change (Meyer and Cooper, 1995) and most campaigners sensibly resist it. However, even a preliminary assessment of the 'carbon debt' from the imbalance of carbon emissions between rich and poor countries since 1850 suggests that it heavily outweighs the current financial third world debt.[2]

Such an analysis raises hard political and ethical questions. Should the poorer countries therefore get a bigger share of resource consumption in the future to compensate? Similarly, should poor communities not have the same chance to consume that richer ones have had? Black communities in Chicago, using such a rationale, have demanded the right to 'consume for a hundred years' (Thorley, 1987). On the other hand, is it just to ask the current generation in rich countries to 'pay for the sins of their fathers' (using the masculine gender advisedly)?

Although these questions are not simple, two elements of a response can be suggested. First, the rich and developed countries of the North have the infrastructure, the capital and the educational and social investments necessary to allow them to bear a much greater share of the transitional costs. Therefore it is not practically unreasonable to ask for more than fair shares in resource use

BOX 1.4 ADDING TO THE ECOLOGICAL DEBT – EXAMPLES OF INJUSTICE BETWEEN CONSUMER AND SUBSISTENCE CLASSES

Prawn farming in Thailand – increasingly dominated by large commercial trading firms – has led to destruction of mangrove swamps (critical breeding grounds for fish), thus undermining the livelihoods of fishing communities, while the benefits (cheap prawns) arrive in Japanese sushi bars and on the shelves of European supermarkets after being air freighted halfway round the planet.

Palm oil plantations in Indonesia are being established to gain hard currency (to repay debt) by exporting to Australia and Japan. Rainforests are cleared by forest fires to make way for these plantations – eliminating sustainable livelihoods for the sake of repaying already wealthy bankers.

Oil drilling in Nigeria has led to vast environmental pollution and degradation, while those who try to protest have been subjected to gross abuses of human rights. Around the world many more communities and natural environments are threatened for the sake of cheap oil for the gas-guzzling automobiles of the US and Europe, and profits for the world's biggest companies.

Mechanized soya plantations in Brazil have expanded to cover 26 million hectares (an area larger than Great Britain) from virtually nothing over the past 30 years, displacing smallholders (who have been forced to carve out new farms from the rainforest), and triggering massive soil erosion. The soya has been exported as cattle feed (5 kilos of high protein soya are needed to produce 1kg of beef) to Europe and the US.

Sea levels are rising around the world, but the impacts will be worst in countries like Bangladesh and Egypt, where most agricultural land is low lying, and the Small Island States such as Kiribatu, where the islands of Tebua Tarawa and Abanuea are already being overwhelmed. Sea level rise is just one of the negative consequences of climate change, driven by Northern consumers' insatiable demands for energy and the multinationals, such as Shell and Exxon, that profit from oil extraction, car manufacture and energy generation.

where needed for the poorer countries to help them through a transition to sustainable development. Second, a just society must be a democratic one – giving political as well as economic and social equality. In such a framework, there will be a limit to how much cost we can ask richer countries to bear (and how rapidly). Having said this, as we saw earlier, the scope for sustainability strategies that bring parallel benefits in terms of quality of life and more inclusive societies (and thus better health) is such that with political leadership even British or American voters could be persuaded to take on a much greater share of the economic costs than they bear now. So rapid repayment of ecological debt may not be feasible, but immediate action to cancel financial debts could be linked to a more gradual repayment of the ecological debt.

However, the example targets set out in Table 1.3 do not account for any repayment of ecological debt – merely a reduction to zero in its rate of accumulation over the next five decades. In practice, whether or not ecological debts are taken into account, the direction and magnitude of the changes needed are similar. To repay ecological debt the reductions simply need to exceed the targets for a period of time until the debt has been repaid – the smaller the surplus, the longer the period. But the prescriptions for how to deliver them can differ significantly when we take ecological debt into account. The ecological debt concept sharpens our understanding of sustainable development, not just by adding a historical dimension but by bringing power and justice to centre stage. In particular, control over resources – fundamental to the levels and patterns of consumption – is an issue of power relations. Choices over whether to exploit resources, and who consumes the products, are being made by Northern dominated institutions and multinational corporations, not by Southern communities. The deliberate links drawn by campaigners between ecological and financial debt highlight this severe structural imbalance.

This focus is particularly valuable because it directs our attention beyond the economic and technological possibilities stressed in most current writing on sustainable development from an environmental perspective (Hawken et al, 1999, Fussler and James, 1997), to the institutions and actors in the global economy whose rapid growth in the last half century has been paralleled by a doubling of overall resource use from broadly sustainable levels to today's clearly unsustainable ones (Jackson, 1996, Hinterberger et al, 1994). The rate at which resources have been extracted from Southern countries has often grown much faster. For example, just 'between 1980 and 1995, the volume of exports from Latin America increased by 245 per cent. Between 1985 and 1996, 2706 million tons of basic resources, most of them non-renewable, were extracted and exported. The amount of resources that were transformed, destroyed or moved in order to produce these exports has not been calculated, nor has the number of people affected or displaced. Meanwhile, between 1982 and 1996, Latin America has repaid US$740 billion in debt … Yet the debt has not diminished, but has rather increased to US$607 billion' (Accion Ecologica, 1999).

GLOBABLIZATION, GLOBAL INSTITUTIONS AND 'JUST SUSTAINABILITY'

The driving forces behind increasing resource exploitation, consumption and growing inequality lie both at national levels, where the excesses of the Reagan and Thatcher years have yet to be fully reversed, and in global economic policy. Policies of economic liberalization and deregulation, reinforced by structural adjustment programmes under one name or another, are exacerbating inequalities not only between, but also within countries. Indeed, as we saw earlier, the only development model presently open for 'developing' countries involves exploiting more resources, more rapidly – in direct opposition to the imperatives for reduced global resource use suggested by the environmental space analysis – and export the majority of them to 'developed' countries.

This dependence on the export of commodities has arisen from a combination of debt – often incurred through Northern export credit agencies as the result of project failures (Hildyard, 1999); structural adjustment prescriptions; and tariff escalation under General Agreement on Tariffs and

Trade (GATT) and now World Trade Organization (WTO) rules which hamper the ability of poorer countries to develop processing and manufacturing exports (House of Commons Environment Committee, 1996, Environmental Audit Committee, 1999). The prescription today is more liberalization, based in the increasingly discredited theory of comparative advantage that assumes capital immobility, when in fact capital moves increasingly freely. The shortcomings of this prescription are also practical, as commodity markets are not freely competitive, but dominated by oligopolistic multinational corporations.

In such circumstances greater liberalization risks leaving commodity exporters – even those with honest democratic governments – structurally disadvantaged on an uneven playing field. As modern economic theory suggests, such asymmetries in power, like those in information, cannot be corrected by more liberalization and deregulation (for a simple overview see Sweeting, 1998). Such an approach also opens the door to further increases in corruption (Hawley, 2000). Alternative prescriptions – such as the establishment of international commodity related environmental agreements (ICREAs) to support commodity prices in return for environmental standards in production, or developed country taxation regimes with linked transfers to poor commodity exporting countries – have so far remained on the fringes of the debate (Rahim, 1994).

Globalization in the real world has extended economic inequalities at both ends of the scale. Global financial and currency markets have permitted the emergence of a wealthy, rootless global speculator class and corrupt elites. Globalization has extended inequalities based on 'winner takes all' markets, where the rewards to the 'best' far exceed those to the rest (Frank and Cook, 1995). The threat of capital mobility has demolished the economic security of many working-class people. Budget discipline and structural adjustment programmes have slashed the resources available in the poorest countries for even rudimentary welfare programmes. Even in many richer countries there are concerns over the growth or emergence of a poor, uneducated and unhealthy underclass in which life-chances are severely curtailed.

Globalization is exacerbating inequalities between countries and regions too. The failed experiments in rapid structural adjustment in the former Soviet Union highlight such problems and offer a clear indictment of the institutions promoting the process. The gains from trade liberalization under the Uruguay round of GATT went mainly to the US, EU and Japan. Africa was a net loser (Goldin et al, 1993). Such differences are effectively threatening the creation of a global underclass of countries unable to establish national sustainable development strategies because they lack access to the necessary resources.

But the alternative to liberalization and deregulation at the global scale is not a retreat to protectionism – at least not as long as social justice is a concern. Protectionism risks institutionalizing existing inequalities between nations and groups through fossilizing access to resources, technology and capital. Worse still it can culturally exacerbate xenophobia and legitimate the worst excesses of racism, in the hands (and rhetoric) of populist politicians. This path would be equally bad for the underclasses – both people and countries – leaving billions out in the cold, or perhaps more literally 'out in the heat' of global warming.

Fortunately, uncritical rapid liberalization is not the only alternative to protectionism, whatever popular politics might imply. There are a host of emerging proposals for global and regional structures and agreements on investment and trade which support environmental and social goals – including scope for capital and exchange controls to regulate and control 'hot money' and destabilizing speculation, investment conditions and corporate regulation to manage and direct foreign direct investment to support sustainability, and commodity agreements which increase prices, reduce volumes and promote fair trade (see for example WDM, 1999, FOE-US et al, 1999, Hines, 2000, FOEI, 2000).

CONCLUSIONS

Environmental space (sustainability with equity) and ecological debt (environmental space with history and justice) offer valuable tools, not only to the campaigner and activist, but also to the academic and policy-maker. Fundamentally they offer a joined-up framework for understanding and promoting both sustainable development and environmental justice.

At a basic level, they show us that neither pollution outputs nor resource inputs should be considered independent of an analysis of the power relations they reflect. Both are distributed not only spatially, but also socially. In this light, the concentration of toxic contaminated sites in American black communities is an expression of the same unsustainability as the expropriation of indigenous subsistence resources by logging companies in Indonesian forests. Both are facilitated by the processes of liberalization and globalization that have shaped economies – national and global – over the past decades. Both reflect the relative increase in the power of corporations relative to citizens. For both explanatory and prescriptive purposes such perspectives are essential.

For the academic these concepts also suggest a setting in which a legitimate interdisciplinary field of research can be established. They offer useful and informed metrics of sustainability. They help redefine questions of sustainability as questions of justice and vice versa. They identify the need for new subjects and approaches to research in economics, politics and other social sciences.

For the activist they offer a foundation for global campaigns which can unite poor and excluded communities in North and South, rather than dividing them, and similarly unite environmental and social campaigners. They help identify the common roots in the problems shared by such groups, common obstacles and targets and can even help suggest common solutions.

For the policy-maker, they suggest how sustainable development policies can explicitly incorporate equity concerns, as some, such as the UK's sustainable development strategy, have begun to do (DETR, 1999). Environmental space and ecological debt reinforce the arguments for many economic strategies to help achieve sustainable development, such as eco-efficiency, ecological tax reform, industrial ecology and so forth, but they also make it clear that these are not sufficient to deliver sustainable development. They highlight the burning moral and environmental need for democratic national and international policy initiatives to establish:

- Stringent targets for global resource consumption (starting with stricter climate targets), supported by innovative mechanisms to deliver them that allow poorer countries to profit from their previous under-consumption.
- Redistributive policies which share jobs, goods and incomes more equally to 'improve the quality of life for everyone by simultaneously improving the social fabric and slowing the pace of environmental damage' (Wilkinson, 1996).
- Environmental rights for all to a clean, safe and healthy environment, backed by legal protections to permit citizens to defend their rights.
- Accountability, liability and transparency measures to create a framework of civic accountability for multinational and other corporations, and international institutions.

NOTES

1   The terms 'South' and 'North' are, of course, just a crude shorthand to represent the main geographical division between the vast population in what Alan Durning (1992) calls the subsistence class and the smaller consumer class into which most Europeans and North Americans fall.
2   An estimate of US$1500 billion is arrived at, based on an increased atmospheric carbon stock of 160 gigatonnes, 80 per cent of which is from developed countries, which on average over this period account for around one third of global population, with a value of US$20 per tonne of emissions.

REFERENCES

Accion Ecologica (1999) *Initial Proposal,* http://www.cosmovisiones.com/Deuda Ecological/c_propuin.html

Agarwal, A and Narain, S (1998) 'Globalization: an agenda to tame the tiger', http://www.ms-dan.dk/uk/Politics_press/Articles/cse_globe.htm

Barker, T (1993) 'Is green growth possible?', *New Economy,* vol 1 (1), pp20–25

Birdsall, N and Graham, C (eds) (1999) *New Markets, New Opportunities: Social Mobility in a Changing World,* Carnegie Foundation for International Peace and Brookings Institution, Washington, DC

Boardman, B with Bullock, S and McLaren, D (1999) *Equity and the Environment: Guidelines for Green and Socially Just Government,* Catalyst, London

Bullard, R (1999) 'Dismantling environmental racism in the USA', *Local Environment,* vol 4 (1), pp5–20

Bullock, S (2001) *Pollution and Poverty: Breaking the Link,* Friends of the Earth, London

Carley, M and Spapens, P (1998) *Sharing the World,* Earthscan, London

Chomsky, A (2000) '"The threat of a good example": Health and revolution in Cuba' in J Kim, J Millen, A Irwin and J Gershman (eds) *Dying for Growth: Global Inequality and the Health of the Poor,* Common Courage Press, Monroe, ME

Colborn, T, Dumanoski, D and Myers, J (1996), *Our Stolen Future: Are We Threatening our Fertility, Intelligence and Survival?,* Abacus, London

Deininger K and Squire, L (1996) 'Measuring income inequality a new data-base', *World Bank Economic Review,* September

Department of the Environment, Transport and the Regions (DETR) (1999) *A Better Quality of Life,* The Stationery Office, London

Durning A (1992) *How Much Is Enough? The Consumer Society and the Future of the Earth,* WorldWatch Institute, Washington, DC

Environmental Audit Committee (1999) *Second Report: World Trade and Sustainable Development: An Agenda for the Seattle Summit,* HC45, House of Commons, London

Friends of the Earth International (FOEI) (undated) http://www.foei.org/campaigns/EcologicalDebt/what_is_ecodebt.htm

FOEI (Friends of the Earth International) (2000) *Towards Sustaianble Economies: Alternatives to Neo-liberal Economic Globalization,* FOEI, Amsterdam

Friends of the Earth-US (FOE-US) International Forum on Globalization and Third World Network, (1999) *Toward a Progressive International Economy,* Conference Report, FOE-US, Washington, DC

Frank, R and Cook, P (1995) *The Winner Take All Society,* Free Press, New York

Fussler, C and James, P (1997) *Driving Eco-Innovation: A Breakthrough Discipline for Innovation and Sustainability,* Pitman, London

Galor, O and Tsiddon, D (1994) 'Human capital distribution, technological progress, and economic growth', Center For Economic Policy Research, *Discussion Paper Series,* 971 p1–52

Global Commons Institute (undated) *Equity for Survival,* http://www.gci.org.uk

Goldin, I, Knudsen, O and van der Mensbrugghe, D (1993) *Trade Liberalization: Global Economic Implications,* OECD, Paris and World Bank, Washington, DC

Hawken, P, Lovins, A and Lovins, L (1999) *Natural Capitalism: Creating the Next Industrial Revolution,* Earthscan, London

Hawley, S (2000) 'Exporting corruption: Privatization, multinationals and bribery', *Cornerhouse Briefing,* 19

Hildyard, N (1999) 'Snouts in the trough: Export credit agencies, corporate welfare and policy incoherence', *Cornerhouse Briefing,* 14

Hinterberger, F, Kranendonk, S, Wlefens, M and Schmidt-Bleek, F (1994) 'Increasing resource productivity through eco-efficient services', *Wuppertal Paper,* 13, Wuppertal Institute

Hines, C (2000) *Localization – A Global Manifesto,* Earthscan, London

House of Commons Environment Committee (1996) *World Trade and the Environment,* HMSO, London

Intergovernmental Panel on Climate Change (1996) *Second Assessment Report,* Cambridge University Press, Cambridge

Jackson, T (1996) *Material Concerns: Pollution, Profit and Quality of Life,* Routledge, London

Jacobs, M (1996) *The Politics of the Real World,* Earthscan, London

Kaplan, G, Pamuk, E, Lynch, J, Cohen, R and Balfour, J (1996) 'Inequality in income and mortality in the United States', *British Medical Journal,* 312, pp999–1003

Mazur, L (ed) (1994) *Beyond the Numbers: A Reader on Population, Consumption and the Environment,* Island Press

McLaren, D, Bullock, S and Yousuf, N (1998) *Tomorrow's World: Britain's Share in a Sustainable Future,* Earthscan, London

McLaren, D, Cottray, O, Taylor, M, Pipes, S and Bullock, S (1999) *Pollution Injustice: The Geographic Relation Between Household Income and Polluting Factories,* Friends of the Earth, London

Meyer, A (2000) *Contraction and Convergence: The Global Solution to Climate Change,* Schumacher Briefing, No 5

Meyer, A and Cooper, T (1995) *GCI Critique of 'IPCC WG3 Social Costs of Climate Change': A Recalculation of the Social Costs of Climate Change,* Global Commons Institute, Cambridge

Myers, N (1996) *Ultimate Security: The Environmental Basis of Political Stability,* Island Press, Washington, DC

North, R (1995) *Life on a Modern Planet,* Manchester University Press, Manchester

Nuclear Engineering International (1996) *Chernobyl's Legacy,* July

OECD (1996a) *Future Financial Liabilities of Nuclear Activities,* OECD, Paris

OECD (1996b) *Employment Outlook,* July, OECD, Paris

Oswald, A (1996) 'GDP can't make you happy', *New Economy,* vol 3 (1), pp15–19

Rahim, K (1994) *Unilateral Environmental Regulations and the Implications for International Commodity-Related Environmental Agreements,* Paper to Nautilus Institute Workshop on Trade and Environment in Asia-Pacific: Prospects for Regional Cooperation, East-West Center, Honolulu

Ross, A (1994) *The Chicago Gangster Theory of Life: Nature's Debt to Society,* Verso, London

Seabrook, J (1994) 'Consumerism and happiness', *Ethical Consumer,* 27, January, pp12–23

Spangenberg, J (ed) (1995) *Towards Sustainable Europe,* FOE Europe, Brussels

Sweeting, A (1998) 'Discuss the reasons why asymmetric information can be a source of market failure', *Economic Research and Analysis* http://www.tommy.iinet.net.au/essays/asinfo.html

Templet, P (1996) 'The energy transition in international economic systems: an empirical analysis of change in development', *International Journal of Sustainable Development and World Ecology,* 3, pp13–30

Thorley, B (1987) Speech at the launch of the Black Environment Network, London

UN (1997) *Kyoto Protocol to the United Nations Framework Convention on Climate Change,* FCCC/CP/1997/7/Add 1

UN Conference on Environment and Development (1992) *Agenda 21,* http://www.un.org/esa/sustdev.agenda21.htm

UN Development Programme (UNDP) (1998) *Human Development Report 1998,* Oxford University Press, Oxford

UN Environment Programme (UNEP), International Union for the Conservation of Nature and World Wildlife Fund (1990) *Caring for the Earth: A Strategy for Sustainable Living,* UNEP, IUCN and WWF, Gland

UNEP (1994) *Environmental Effects of Ozone Depletion: 1994 Assessment,* UNEP, Nairobi

UNEP (1995) *Global Biodiversity Assessment,* UNEP, Nairobi

UNEP (1997) *Global Environmental Outlook 1,* UNEP, Nairobi

UN Population Division (1994) *World Population 1994,* UN Department for Economic and Social Information and Policy Analysis, Geneva

Veenhoven, R (1987) 'National wealth and individual happiness' in K Grunert and F Olander (eds) *Understanding Economic Behaviour,* Kluwer Academic, London

Watkins, K (1995) *The Oxfam Poverty Report,* Oxfam Publications, Oxford

World Commission on Environment and Development (1987) *Our Common Future,* Oxford University Press, Oxford

World Development Movement (WDM) (1999) *Making Investment Work for People,* Consultation paper, WDM, London

Wilkinson, R (1996) *Unhealthy Societies: The Afflictions of Inequality,* Routledge, London

Chapter 2

Neo-liberalism, Globalization and the Struggle for Ecological Democracy: Linking Sustainability and Environmental Justice

Daniel R Faber and Deborah McCarthy

INTRODUCTION: LINKING SUSTAINABILITY AND ENVIRONMENTAL JUSTICE

To sustain economic growth and higher profits in the new global economy, American companies are increasingly adopting ecologically unsustainable systems of production. Motivated by the growing costs of doing business and threat of increased international competition in the era of globalization, corporate America initiated a political movement beginning in the early 1980s for 'regulatory reform', ie the rollback of environmental laws, worker health and safety, consumer protection, and other state regulatory protections seen as impinging upon the 'free' market and the profits of capital. Termed 'neo-liberalism', the recent effect has been a general increase in the rate of exploitation of both working people (human nature) and the environment (mother nature), as witnessed by the assaults upon labour, the ecology movement and the welfare state. Coupled with increased trade advantages brought about by corporate-led globalization and significant innovations in high technology and service related industries in the 'new economy', the US experienced a record-breaking economic boom under the Clinton administration during the 1990s. However, this economic 'prosperity' was to a large degree predicated upon the increased *privatized-maximization* of profits via the increased *socialized-minimization* of the costs of production, ie the increased displacement of potential business expenses onto the American public in the form of pollution, intensified natural resource exploitation and other environmental problems. Though progress was made on a number of critical issues, the ecological crisis continued to deepen during the 1990s.

Since the 2000 election of George W Bush to the presidency and the slowdown in the US economy, the war against the environment has greatly intensified. Heavily supported by the most polluting sectors of American business – including campaign contributions of more than US$1.86 million from the oil and gas industry and US$1.25 million from the automotive industry (contender Al Gore received only US$131,764 and US$115,790 respectively) – the Bush administration is implementing sweeping measures aimed at delaying and/or dismantling programmes and policies designed to protect public health and the environment. President Bush has already backed down on a promise to curb US emissions of greenhouse gases, blocked efforts to protect a third of national forests from roads and logging, rescinded a key ergonomics workplace safety rule that was years in the making, and repealed tough scientific-based standards for removing poisonous arsenic in drinking water, among other assaults on environmental protection. Furthermore, following the terrorist attacks of 11 September 2001 on New York and Washington, the administration has also led reinvigorated attempts to open the Arctic National Wildlife Refuge in Alaska to energy development by invoking the cause of 'national security'.

Not all citizens, however, equally bear the 'externalized' social and ecological costs of these assaults by American business. In order to bolster profits and competitiveness, companies typically adopt strategies for the exploitation of nature that are not only economically 'efficient' but politically 'expedient' (that offer the path of least social resistance). The less political power a community of people commands, the fewer resources a community possesses to defend itself; the lower the level of community awareness and mobilization against potential ecological threats, the more likely they are to experience arduous environmental and human health problems at the hands of capital and the state. In the US (as elsewhere in the world), it is the most politically oppressed segments of the population, or the *subaltern* – dispossessed peoples of colour, industrial labourers, the underemployed and the working poor (especially women), rural farmers and farm workers, and undocumented immigrants – whom are being *selectively victimized* to the greatest extent by corporate practices (Johnston, 1994, p11; see also Agbola and Alabi, Chapter 13). The disenfranchised of America are serving as the dumping ground for American business, a fact that is often blatantly advertised. A 1984 report by Cerrell Associates for the California Waste Management Board, for instance, openly recommended that polluting industries and the state locate hazardous waste facilities in 'lower socio-economic neighbourhoods' because those communities had a much lower likelihood of offering political opposition (Roque, 1993, p25–28). In this respect, the prosperity of the American business community is predicated on specific forms of unsustainable production that *disproportionately impact oppressed peoples of colour and the working poor.*

It is now clear that the economic crisis tendencies of the 1970s–1980s have become increasingly displaced to the realm of nature in the 1990s–2000s, assuming the form of ecological crisis tendencies; while the short term economic health of the salariat and corporate owners is being increasingly secured through the long term sacrifice of the environmental health of the subaltern – peoples of colour and the poor (including developing world peoples). In this respect, the process of global economic restructuring, which neo-liberalism has helped facilitate, is thus responsible for the deterioration in ecological and working/living conditions of the poor and people of colour. The increased hardships of both the subaltern and their environment are thus two sides of the same political-economic coin and are now so dialectically related (if not essential) to each other as to become part of the same historical process. As a result, the issues of sustainable development and social/environmental justice have surfaced together as in no other period in world history. This chapter will explore the challenges confronting the environmental justice movement as it tries to forge a truly participatory ecological democracy capable of building a more just and sustainable society.

NEO-LIBERALISM, GLOBALIZATION AND THE RESTRUCTURING OF AMERICAN CAPITALISM

The new millennium has witnessed the triumph of a distinctly hard-nosed brand of American capitalism in the world economy. Spurred by a booming stock market, low interest rates (some 30 to 50 per cent lower than during the 1980s), 30-year lows in inflation and unemployment, record governmental budget surpluses, higher corporate earnings relative to Japan and Western Europe and other apparent signs of financial health, the US economy soared during the 1990s. The country's decade-long economic expansion became the longest in the nation's history. In the three-year period 1997–1999, economic growth averaged over 4 per cent, well above the 2.6 per cent growth rate experienced in the first half of the decade. Aggravated by the growing costs of energy, declining consumer confidence and spending, falling profits and corporate earnings and wild fluctuations and devaluations in the stock market, only recently has the US economy demonstrated significant drops in the rate of growth, falling into a recession immediately following the events of 11 September 2001.

Perhaps the most significant forces transforming the nature of American capitalism reside in the profound changes taking place in the global economy. Fuelled by innovations in communications, transportation and production technologies, huge investments in infrastructure, as well as major improvements in the educational, skill and productivity levels of labour power, multinational corporations and domestic industries located in the newly industrializing countries (NICs) have rapidly expanded in recent years to capture a growing share of the world market. This process of globalization, which is being facilitated in great part by a host of 'free-trade' agreements brokered by the Clinton and both Bush administrations, spurred many sectors of the US economy, particularly industries exporting high-tech and other capital goods and services of all kinds to both developed and newly industrializing countries overseas. As a result, semi- and highly skilled workers associated with these industries in the 'new economy' have witnessed a tremendous growth in demand for their services, with substantially higher salaries, lucrative stock options and rich opportunities for advancement.

On the other hand, industries that have traditionally served as the backbone of the US economy, as well as the trade union movement, have seen their competitive position for mass-produced consumer goods and processed raw materials (such as steel) steadily eroded by overseas producers (Ross and Trachte, 1990, Dicken, 1992). With an increased ability to relocate to low-wage havens and utilize 'job blackmail' strategies against unskilled or semi-skilled blue/pink-collar workers, the labour movement has been significantly weakened. Where union membership once comprised 36 per cent of all private-sector employees in 1953, today the figure has plunged to just above 9 per cent. As a result, union and non-union workers alike are under increased pressure to accept reduced wages, benefits and other programmes. Fearing that increased costs to business will undermine its ability to compete in the world market, US capital has become unwilling to abide by the traditional accords brokered by the liberal wing of the Democratic Party on its behalf with the labour, civil rights, women's and other progressive social movements.

Instead, the rise of neo-liberals committed to less governmental control of industry, as embodied in the Democratic Leadership Council (Clinton, Gore, and Lieberman all served in key DLC leadership positions), as well as the Republican Party and George W Bush, have become hegemonic. As a result, the defining characteristics of liberal capitalism that have traditionally enlisted the mass loyalty of working people with high wages, good benefits, job security and advancement, affirmative action, universal entitlements, civil rights and liberties and welfare protections are being eroded – a process further accelerating under the new bi-partisan consensus to increase resources for national security as part of America's War against Terrorism. For not only has the triumph of the 'Third Way' neo-liberalism model of globalization undermined traditional 'New Deal' liberalism and welfare state capitalism in North America, but it also dealt a death blow to bureaucratic state socialism in the East, nationalist-based models of dependent development in the South and severely weakened Keynesian social democratic regimes in the West (seen especially in the rise of Tony Blair's New Labour Party in the UK).

Without an adequate rate of profit in the global marketplace, and hence rate of capital accumulation, corporate America would lapse into economic stagnation. With the globalization of capital and the increased competition brought about by the adoption of Export-oriented Industrialization (EOI) economic policies in almost every corner of the planet, transnational corporations are less able to boost profits by passing along their increased costs to consumers in the form of higher prices that, along with a restrictive monetary policy implemented by the Federal Reserve at the behest of Wall Street, has maintained relatively low inflation rates in the 1990s–2000s. American consumers have kept the world market afloat and facilitated globalization by serving as the supermarket for European, Japanese and much of Latin American and Asian businesses. In the 1990s, cheap and easy credit allowed working Americans to spend far more than they earned and eventually run up unparalleled personal debts. This was clearly economically unsustainable and would require a more systemic solution than offered by the Bush administration 2000–2001 tax cut (most of the big savings have gone to the rich, while the relatively small tax rebates were utilized by working families to pay down their debts). Since the 11 September attacks, the bubble has burst, and domestic consumer spending has contracted by more than US$300 million annually. As a result, the world economy is now sinking, as businesses throughout the world are struggling to survive in a contracting market. As a result, the first imperative of capital in the new global economy is not to increase prices or even production but rather to lower production costs. Because domestic and world export markets are becoming more cut-throat, *cost minimization* strategies now lie at the heart of business strategies for *profit maximization* for all nations. Greater efficiency (greater output per unit of input) becomes more important precisely because it leads to more profits. Increases in sales matched by increases in cost of production are no longer viable for global capital given the gross contraction of consumer spending since the terrorist attacks.

Greater cost containment by American capital is thus being achieved through a process of capital restructuring. The aim of this restructuring is to re-establish the necessary economic, social, political and cultural conditions for renewed profitability, including new institutional arrangements congruent with the development of new technologies, production processes, work relations and changing patterns of commodity demand. So, for example, by closing higher-cost facilities and moving to lower-cost production facilities offshore more rapidly than competing nations, particularly West German and Japanese market-share maximizers (who were left with so few profits in the 1990s that they found it difficult to finance expansions even when more profitable opportunities presented themselves), American business has been able to recapture some of the markets in the 1990s and 2000s they had lost in the 1970s–1980s.

The most important goal of capital restructuring for American business in the current period is to re-establish corporate 'discipline' over trade unions and other social movements that are cutting into profits. Along with labour costs (which include health insurance and other benefits), environmental protection measures are considered by many industries to be some of the most expensive and burdensome. Companies are therefore seeking to protect profits not only by 'downsizing' the labour force but also by cutting investments in pollution control, environmental conservation, and worker health and safety. Simply put, the key to cost containment lies in processes of capital restructuring that have enabled American businesses to *extract more value from labour power and nature in less time and at less cost*. And in the 1990s–2000s, capital restructuring and deep cuts in labour and environmentally related costs are *boosting the earnings of American business at a much faster rate than revenue growth or increased sales*.

Thus, the primary force behind the profitability of American corporations has been the increased economic exploitation of working people (labour power) and nature. Generally speaking, increased rates of labour exploitation are being achieved by extracting more work (surplus-value) out of the American working class in shorter periods of time and at less cost. American business is achieving this result through a general assault on the past gains of the labour movement and other social justice movements, which is taking numerous forms: the business offensive against unions; increased layoffs of permanent workers and the increased use of temporary or contingent workers at less pay; greater job insecurity, stagnant or falling wages, benefits, and living standards for broad sectors of the workforce; longer hours, mandatory overtime, and a speedup of the production process; attacks on the minimal protections offered by the welfare state; deteriorating worker health and safety conditions; and a general assault on those private and public programmes and policies that serve the interest of lower and middle income working families (Gordon, 1996). The success of these assaults can be seen in the year 2000 labour productivity levels, which surged ahead at over 5 per cent – the fastest pace in 17 years. At the same time, labour costs declined for the first time since 1984. In fact, after more than two decades of lacklustre gains in productivity from 1973 to 1995, which averaged only 1.4 per cent a year, increases since 1996 have been over double that rate.

On the other hand, increased rates of environmental exploitation are being achieved by such measures as: extracting greater quantities of natural resources of greater quality more quickly and at less cost; cutting production costs by spending less on pollution prevention and control, as well as environmental restoration; adopting new production processes (such as biotechnology in agriculture) that increase productivity but are also more polluting or destructive of the environment; and so forth. American business is producing these results through a general assault on the past gains of the ecology movement and a general offensive upon the policies and programmes that make up the environmental protection state. The result is increased dumping of ever more toxic pollution into the environment, particularly in poor working-class neighbourhoods and communities of colour; more destructive extraction of raw materials from this country's most unique and treasured landscapes, especially Native lands and natural resources belonging to other subaltern groups; a deterioration in consumer product safety (and attempts to limit corporate liability for defective or damaging products); the disappearance of ever more natural species and habitats; suburban sprawl; and a general assault on those programmes and policies designed to protect the environment. In short, to sustain the process of capital accumulation and higher profits in the new global economy, American capital is increasingly relying on ecologically unsustainable forms of production which disproportionately impact communities of colour and lower income members of the working class – sectors, which are underrepresented in the traditional environmental movement.

For instance, under the devolution policies of 'new federalism' and the rhetoric of 'states-rights', governmental responsibilities are being shifted from the federal government to the states. The neo-liberal hope is that many states will neglect their responsibilities to engage in bidding wars with other states to attract capital to their home regions by offering more favourable investment conditions, including less worker and environmental regulation and enforcement (ie to aid in efforts at cost minimization). One reason that economic problems in the northern 'rust-belt' are deeper than in most of the rest of the country has been the disproportionate relocation of capital to the 'sun-belt' in search of cheaper labour, lower taxes and real estate costs and less stringent environmental regulations. Increased capital mobility is thus a primary mechanism by which

American business is restructuring itself to minimize costs. Hence, the political-economic power base since the 1980s has shifted to the south (through such figures as Carter, Perot, Bush, Clinton and Gore) and west (Reagan, Cheney and McCain).

Lax enforcement of environmental and worker health and safety laws, along with cheaper, non-union labour statutes, are key factors in the rise of cowboy capitalism in the sun belt. Fifteen southern states alone account for 33 of the 50 most polluting plant sites in the nation. Under the tutelage of former Governor George W Bush, the state of Texas possesses five of the ten most polluted zip code areas in the country and leads the nation in total air, water and land releases of carcinogenic pollution. A 1995 report by the Environmental Defense Fund showed that refineries in Texas, Mississippi, West Virginia and Kansas are the nation's most environmentally inefficient (in terms of pollution releases and waste produced per barrel of oil refined per day). Refineries in northern states such as New Jersey, which have some of the country's toughest pollution laws, are among the best. Furthermore, an emissions-to-jobs ratio report by environmental science professor Paul Templet of Louisiana State University in Baton Rouge showed that Louisiana's chemical plants, especially those located in poor African-American parishes in the corridor between New Orleans and Baton Rouge known as 'Cancer Alley', released nearly ten times as much pollution per worker as such plants in New Jersey and California, where law enforcement and industry spending for pollution control and abatement are greater (Selcraig, 1997, p38–43). 'Dumping in Dixie' is therefore part of a general pattern in which toxic waste dumps, polluting industries, incinerators and other ecologically hazardous facilities are becoming increasingly concentrated in communities of colour in the sun-belt (Bullard, 1990).

The Clean Air Act of 1990 is another such example. Supported by the Tennessee Senator Albert Gore, a key aspect of that legislation involves the commodification of pollution (which can be bought and sold on the stock market), which has allowed enterprises such as the Tennessee Valley Authority (TVA) to buy millions of dollars in 'pollution credits' from Wisconsin Power and Light. These pollution credits allow the TVA to exceed federal limitation on sulphur dioxide and other toxic emissions in older facilities which would otherwise be costly to upgrade, and are located mostly in poor working-class communities of colour in the south and west. The Act is therefore a powerful reminder of the manner in which neo-liberal, free-market environmentalism is exacerbating, rather than resolving, the profound social and environmental injustices fostered by traditional regulatory approaches over the past 30 years (Tokar, 1996, p24–29). These discrepancies are now beginning to be addressed through EPA's Office of Environmental Justice and the National Environmental Justice Advisory Council (NEJAC).

So, if increased profits are the economic engine pulling the train of American business in the world economy across what former President Bill Clinton termed 'the bridge to the 21st century', then unsustainable increases in the rate at which nature (both human and non-human) is being exploited is providing the energy powering the locomotive. Neo-liberal politicians stand at the controls, having engineered a loss of political power by the more progressive sectors of organized labour, environmentalists, and other social movements. The process of capital restructuring, which neo-liberalism has helped facilitate, is thus responsible for the deterioration in ecological and working/living conditions. The hardships of both the American working class, oppressed peoples of colour and their environments are thus different sides of the same political-economic coin and are now so dialectically related (if not essential) to each other as to become part of the same historical process of the restructuring and globalization of American capitalism. As a result, the issues of sustainable development and environmental justice have surfaced together as in no other period in American history.

THE EVOLUTION OF THE ENVIRONMENTAL JUSTICE MOVEMENT

In reaction to the growing economic and ecological disparities accentuated by the rise of neo-liberalism and corporate-led globalization, as well as the neglect of the mainstream environmental movement, a new wave of grassroots environmentalism has been building in the US. In Latino and Asian-Pacific neighbourhoods in the inner cities, small African-American townships, depressed Native American reservations, Chicano farming communities and white working-class districts all across the country, peoples traditionally relegated to the periphery of the ecology movement are now challenging the wholesale depredation of their land, water, air and community health by corporate polluters and indifferent governmental agencies and non-governmental organizations. At the forefront of this new wave of grass-roots activism are hundreds of community-based environmental justice organizations working to reverse the ecological and economic burdens borne by people of colour and poor working-class families (Schwab, 1994). Since the 1991 First National People of Colour Environmental Leadership Summit, the single most important event in the movement's history, these local and sometimes isolated community-based groups have become increasingly integrated into a number of strategic, regionally based networks, as well as national constituency-based and issue-based networks for environmental justice.

The diversity of people participating in these local and regional movements is matched by the diversity of political paths and approaches taken to achieving environmental justice. For the most part, environmental justice activists have primarily emerged out of six other popularly based political movements to embrace the mantra of environmental protection and sustainability. These independent movements have been present for decades, and are:

1   the civil rights movement as led by African-Americans and other disenfranchised people of colour;
2   the occupational health and safety movement, particularly that wing devoted to protecting non-union immigrants and undocumented workers;
3   the indigenous land rights movement, particularly that wing devoted to the cultural survival and sovereignty of Native peoples;

4   the public health and safety movement, particularly that wing devoted to tackling issues of lead poisoning and toxics;
5   the solidarity movement for promoting human rights and the self-determination of developing world peoples; and
6   the social/economic justice movement involved in multi-issue grass-roots organizing in oppressed communities of colour and poor working-class neighbourhoods all across the country.

The community-based organizations and regional/national networks for environmental justice established by these activists often bear the distinctive political imprints of the original movements from which they emerged, so it may appear to the casual observer that there is no united national movement at all. Although most organizations or *movements* for environmental justice are distinct from one another in a number of rather profound ways (the constituency served, unique cultural legacy and experiences of activists, core issues of emphasis, political strategies, set of challenges, etc), it should be emphasized that *all are united in the larger struggle for ecological democracy* (Faber, 1998). For the organizations within these various wings all share a passion for linking grass-roots activism and participatory democracy to problem-solving the issues of environmental abuse, unsustainable economic development, racial oppression, social inequality and community disempowerment (Bastian and Alston, 1993, p1–4). In this respect, there is occurring a steady and undeniable sublation of these various political heritages into a larger environmental justice body politic, whereby these differing elements are achieving a deeper appreciation and understanding of the other wings and merging it with their own political consciousness and movement-building strategies.

As witnessed by the creation of a number of new organizational entities, including: the Environmental Justice Fund; regionally based environmental justice networks such as the Southern Organizing Committee (SOC), the Southwest Network for Economic and Environmental Justice (SNEEJ), and the Northeast Environmental Justice Network (NEJN); national constituency-based networks such as the Asian Pacific Environmental Network (APEN), the Indigenous Environmental Network (IEN), and the Farmworker Network for Economic and Environmental Justice (FWNEEJ); the National Environmental Justice Advisory Council (NEJAC); the National People of Colour Environmental Leadership Summits in 1991 and 2002; and so forth; there is thus emerging a national, multi-racial environmental justice movement which is greater than the sum of its parts (Lee, 1992, Alston, 1992, pp30–31). The Fund and strategic networks are particularly important in serving to create a new infrastructure for building inter-group collaboration and coordinated programmatic initiatives that are taking the movement beyond the local level to have a broader policy impact. The people of colour-led environmental justice movement might have only been borne with the local Warren County, North Carolina fight in 1982, but it is beginning to come of age in the new millennium.

Environmental Racism and Unequal Protection: The Civil Rights Movement and Environmental Justice

The legacy of the civil rights movement is one of the most important foundations on which the modern environmental justice movement is predicated. While the quality of life for all US citizens is compromised by a number of environmental and human health problems, not all segments of the citizenry are impacted equally. In contrast to high-income salaried and professional workers, who can often buy themselves access to ecological amenities and a cleaner environment in non-industrial urban, suburban and rural areas, people of colour face a much greater exposure rate to toxic pollution and other environmental hazards. For communities of colour, this takes the form of exposure to: (1) greater concentrations of polluting industrial facilities and power plants; (2) greater concentrations of hazardous waste sites and disposal/treatment facilities, including landfills, incinerators and trash transfer stations; and (3) lower rates of environmental enforcement and clean-up (Faber and Krieg, 2001). Thus, unequal exposure to environmental hazards are experienced by people of colour in terms of where they 'work, live and play' (Alston, 1991).

Hazardous waste sites nationwide are among the more concentrated environmental hazards confronting communities of colour. According to a 1987 report by the United Church of Christ's Commission on Racial Justice, three out of five African-Americans and Latinos nationwide live in communities that have illegal or abandoned toxic dumps. Communities with one hazardous waste facility have twice the percentage of people of colour as those with none, while the percentage triples in communities with two or more waste sites (Chavis and Lee, 1987). A subsequent follow-up study conducted in 1994 has now found the risks for people of colour to be even greater than in 1980, as they are 47 per cent more likely than whites to live near these potentially health-threatening facilities (Goldman and Fitton, 1994). Federal governmental enforcement actions also appear to be uneven with regard to the class and racial composition of the impacted community. According to a 1992 nationwide study in the *National Law Journal*, Superfund toxic waste sites in communities of colour are likely to be cleaned 12 to 42 per cent *later* than sites in white communities. Communities of colour also witness government penalties for violations of hazardous waste laws that are on average only one-sixth (US$55,318) of the average penalty in predominantly white communities (US$335,566). The study also concluded that it takes an average of 20 per cent longer for the government to place toxic waste dumps in minority communities on the National Priorities List (NPL), or Superfund list, for clean-up than sites in white areas (Lavelle and Coyle, 1992, p2–12).

Represented by regional networks such as the Southern Organizing Committee for Economic and Social Justice (SOC) and local and/or state organizations such as People Organized in Defense of Earth and Her Resources (PODER) in East Austin, Texas, this component of the environmental justice movement is committed to battling the disproportionate impacts of pollution in communities of colour, the racial biases in government regulatory practices, the glaring absence of affirmative action and sensitivity to racial issues in the established environmental advocacy organizations and other forms of environmental racism (Bullard, 1994, Bryant and Mohai, 1992). The issue of environmental racism has helped to link issues of civil rights, social justice and environmental protection. It has also inspired investigations into the class, gender and ethnic dimensions of exposure to environmental hazards.

West Harlem Environmental Action (WEACT) was created in 1988, for instance, to educate and organize the predominantly African-American and Latino communities of northern Manhattan in New York City on a broad range of environmental justice issues. These include the use of East, West and Central Harlem and Washington Heights as a dumping ground for noxious facilities and unwanted land uses, including two sewage treatment facilities, six of Manhattan's eight diesel bus depots and a marine garbage collection transfer station. Coupled with the air pollution supplied by three major highways, an Amtrak rail line, the NY/NJ Port Authority and several major diesel truck routes, these facilities gave northern Manhattan an asthma mortality and morbidity rate that is up to five times greater than citywide averages. Through 'The Clean Fuel – Clean Air – Good Health' campaign and other initiatives, these issues are now being addressed. For instance, in December of 1993, efforts to correct problems at the North River Sewage Treatment Plant resulted in settlement with the city for a US$1.1 million community environmental benefits fund and designation of WEACT as a monitor of the city's US$55 million consent agreement to fix the plant.

DYING FOR A LIVING: OCCUPATIONAL HEALTH STRUGGLES AND ENVIRONMENTAL JUSTICE

Another wing of the environmental justice movement is developing out of the struggle for labour rights and better occupational health and safety conditions for vulnerable workers. Spurred by governmental de-regulation and lack of enforcement, neo-liberalism is not only allowing capital to spend less on the prevention of environmental and community health problems outside of the factory, but also to spend less on the prevention of health and safety problems that impact the working class inside the factory. In order to increase the rate of exploitation of labour, business is now reducing and eliminating safety equipment and procedures that lower labour productivity and cut into profits. There are now only 800 inspectors nationwide to cover the 110 million workers in 6.5 million workplaces. As a result, American workers are being exposed to greater hazards at the point of production. Some 16,000 workers are injured on the job *every day*, of which about 17 will die. Another 135 workers die *every day* from diseases caused by exposure to toxins in the workplace (Levenstein and Wooding, 1998). These types of occupational hazards are even more profound for workers lacking the minimal protections afforded by unions or formal rights of citizenship. Over 313,000 of the 2 million farmworkers in the US – of whom

90 per cent are people of colour and undocumented immigrants – suffer from pesticide poisoning each year. Of these victims, between 800 and 1000 die (Perfecto, 1992).

The plight of such vulnerable workers is spurring new coalitions between farm-worker associations such as the United Farm Workers (UFW), immigrant rights groups, consumer and environmental organizations, labour and the environmental justice movement. Recent examples include legislative right-to-know campaigns, farmworkers' struggles against pesticide abuses impacting workers in the field and nearby communities, and campaigns against the reproductive dangers of high-tech industry. At the national level, the constituency-based Farmworker Network for Economic and Environmental Justice (FWNEEJ) has taken the lead in linking labour rights issues with workplace and community hazards. Formed in 1993, the FWNEEJ has six affiliated organizations working on pesticide abuses, EPA Worker Protection Standards and immigrant rights. In addition, two smaller, more regionally based farmworker collaborations around pesticide abuse and advocacy have developed environmental justice training programmes. They are CAMPO (Campesinos a la Mesa Politica/Farmworkers to the Policy Table), linking groups from the Midwest, Texas and the Caribbean, and the Farmworker Training Institute, developed by groups from the East Coast and the Caribbean.

PROTECTING CULTURAL AND BIOLOGICAL DIVERSITY: NATIVE LAND STRUGGLES AND ENVIRONMENTAL JUSTICE

The environmental justice movement also emerges out of struggles by Native Americans, Chicanos, African-Americans and other marginalized indigenous communities to retain and protect their traditional lands (see Peña, Chapter 7). A key component of the neo-liberal offensive in the 1990s–2000s against environmentalism involves efforts to contain and roll back policies establishing national parks, as well as protections for wilderness, forests, wild rivers, wetlands and endangered species. The reason is that capital restructuring is facilitating a much more aggressive and destructive scramble by American business for cheaper sources of renewable and non-renewable natural resources. These include efforts to exploit the majestic old-growth forests in Alaska's Tongass National Forest and ancient redwoods in the Pacific Northwest habitat of the endangered spotted owl; the rich deposits of low-sulphur coal that lie underneath the Black Mesa homelands of the Hopi and Navajo Indians in the Four Corners region of the American Southwest; the vast oil and natural gas reserves that lie in the Arctic National Wildlife Refuge in Alaska; and to open up more wetlands and fragile ecosystems to agricultural, commercial, and residential developers. Much of the land richest in natural resource wealth targeted for acquisition by business interests is home to indigenous communities established long ago by Spanish and Mexican land grants in the 18th–19th centuries, or during Reconstruction following the Civil War, or by treaty with the US government. The Native American land base alone amounts to 100 million acres, and is equivalent in size to all 'wilderness lands' in the National

Wilderness Preservation System. In fact, Native lands in the lower 48 states are larger than all of New England. The Navajo Reservation alone is five times the size of Connecticut, and twice the size of Maryland. In an attempt to gain control over and exploit the low-cost resources on these lands, a nationwide corporate attack on Native Americans has been initiated, including calls for the termination of treaty rights (LaDuke, 1999, Weaver and Means, 1996, Grinde et al, 1998).

New resource wars against indigenous communities are consequently intensifying in every corner of the country. Such schemes to exploit new resource reserves are motivated by landed capital's desire to bring in lower cost (and therefore more profitable) sources of oil, coal, timber and other fuels and raw materials to more effectively compete in the world market, as well as to lower the cost of inputs utilized by American capital as a whole in the production process. The result has been the growth in offshore drilling, strip-mining and destructive timber harvests with all attendant adverse social and environmental consequences, including the contamination of indigenous communities and their environment with toxic chemicals and radioactive waste produced by mining and industrial operations. Native lands, and the tribes which depend upon these lands for survival, have already suffered decades of abuse at the hands of indifferent government agencies and rapacious corporations, resulting in problems of severe poverty and ecological degradation. According to the First Nations Development Institute, about 126 species of plants and animals are listed as threatened or endangered on Indian lands (tribal lands include 49 per cent of all threatened or endangered fish, 26 per cent of birds, and 22 per cent of mammal species).

To tackle the social and ecological crises confronting indigenous communities, the environmental justice movement is linking concerns for natural resource protection and sustainability with issues of land and sovereignty rights, cultural survival, racial and social justice, alternative economic development and religious freedom (see Rixecker and Tipene-Matua, Chapter 12, for examples from the Maori in Aotearoa New Zealand). At the forefront of these struggles is the national constituency-based Indigenous Environmental Network (IEN). Formed in 1992, IEN is a resource network committed to building mutual support strategies by providing technical and organizational assistance to over 600 Native American organizations and activists across North America. Working primarily on reservation-based environmental issues, which include forestry, nuclear weapons and waste, mining, toxic dumping, water quality and water rights, IEN is now moving to create regional inter-tribal networks that build the capacity of local organizations as well as the national structure. Its National Council and annual conference are in themselves important centres for collaboration, advocacy and consensus-building among activists representing indigenous peoples from all over the world.

FIGHTING FOR PEOPLE OVER POISONOUS PROFITS: THE PUBLIC HEALTH MOVEMENT AND ENVIRONMENTAL JUSTICE

The environmental justice movement has also developed out of the community/public health and safety movement in general and the anti-toxics movement in particular. In thousands of communities across the US, billions of gallons of highly toxic chemicals including mercury, dioxin, PCBs, arsenic, lead and heavy metals such as chromium have been dumped in the midst of unsuspecting neighbourhoods. These sites poison the land, contaminate drinking water and potentially cause cancer, birth defects, nerve and liver damage and other health effects. The worst of these are called National Priority List (NPL) or Superfund sites, named after the 1980 law to clean up the nation's most dangerous toxic dumps. In a 1991 study, the National Research Council found that there were over 41 million people who lived within four miles of at least one of the nation's over 1500 dangerous Superfund waste sites (National Research Council, 1991). It is estimated that groundwater contamination is a problem at over 85 per cent of the nation's Superfund sites – a particularly alarming statistic when we realize that over 50 per cent of the American people rely upon groundwater sources for drinking. Although these dumps are the worst of the worst, it has been estimated that there are as many as 439,000 other illegal hazardous waste sites in the country (Environmental Research Foundation, 1993). Public health problems related to lead poisoning, pesticide abuse, dioxin and mercury contamination of the environment by municipal incinerators, power plants and a host of other sources, are also critical.

Coupled with the neo-liberal assault on the regulatory capacities of the state, American business is now externalizing more costs and spending less on prevention of health and safety problems inside and outside the factory, as well as on reducing pollution and the depletion of natural resources. According to EPA's Toxic Release Inventory (TRI) for 1998, some 23,000 industrial facilities reported releasing a total of 7.3 billion pounds of chemical pollutants into the nation's air, water, land and underground. The vast majority of these pollutants – some 93.9 per cent (or 6.9 billion pounds) – were released directly on-site, posing greater risks for nearby communities. As is evident from the growing toxic waste problems, pollution and other social/environmental costs of capitalist production, many neo-liberal policy initiatives directed at these current crises are actually intensifying problems they were designed to cure. Most environmental laws require capital to *contain* pollution sources for proper treatment and disposal (in contrast to the previous practice of dumping on-site or into nearby commons). Once the pollution is 'trapped', the manufacturing industry pays the state or a chemical waste management company for its treatment and disposal. The waste, now commodified, becomes mobile, crossing local, state and even national borders in search of 'efficient' (ie low-cost and politically feasible) areas for treatment, incineration and/or disposal (Field, 1998). Because these communities have less political power to defend themselves, possess lower property values and are more hungry for jobs and tax-generating businesses, more often that not, the waste sites and facilities are themselves hazardous and located in poor working-class neighbourhoods and communities of colour. As stated by one government report, billions of dollars are spent to remove pollutants from the air and water only to dispose of such pollutants on the land, and in an environmentally unsound manner (Regenstein, 1986, p160).

The growth in neo-liberal environmental policy initiatives is fuelling the rapid expansion of the waste circuit of capital (in both legal and illegal forms) that, perhaps more than any other phenomenon, has magnified problems of ecological racism and class-based inequities related to toxic pollution that the environmental justice movement is now challenging. Over the last two decades, thousands of local citizen organizations have been created to fight for the clean-up of toxic waste dumps, the regulation of pollutants from industrial facilities, the enforcement and improvement of federal and state environmental standards and many other issues. Emerging from a diverse array of settings, including poor working class communities, with notably high numbers of women in key activist and leadership positions, these local organizations are increasingly making the links between issues of corporate power, governmental neglect and citizen disenfranchisement. As a result, many of these organizations are working in close collaboration with (or evolving into) environmental justice organizations. At the national level, organizations such as the Center for Health, Environment and Justice (CHEJ) headed by Lois Gibbs (formerly the Citizen's Clearinghouse on Hazardous Waste) have taken a lead role in galvanizing the anti-toxics movement to address issue of political-economic power, although most of their efforts were concentrated on white working- and middle-class communities. However, there were a number of activists of colour who emerged from the white-led anti-toxics and environmental health movements (such as the now-defunct National Toxics Campaign) to take up leadership roles in the environmental justice community.

Today, there are a great variety of community-based and regional networks that are organizing communities of colour to protect the health and environment. For instance, the Environmental Health Coalition (EHC) was founded in 1980 and is a community-based organization in San Diego which combines grass-roots organizing, advocacy, technical assistance, research, education and policy development in its work, helping community members develop solutions to environmental health problems. This approach not only brings about institutional change, it also empowers individuals and communities to demand better working and living conditions. Working primarily with people of colour in the San Diego area and Tijuana, Mexico, EHC's programmes concentrate on problems of toxic contamination of local neighbourhoods, the workplace, San Diego Bay and the border region. EHC won a five-year battle with the San Diego Port District in July of 1997, ending the use of the toxic pesticide methyl bromide. A toxic pesticide, which causes birth defects and other health problems, and is an ozone destroyer, methyl bromide had been used to fumigate imported produce unloaded at the port. The practice posed significant health risks to nearby communities, including Barrio Logan, one of San Diego's poorest neighbourhoods. Surrounded by more than 100 toxic polluting facilities, residents in Barrio Logan had experienced high rates of asthma, headaches, sore throats, rashes, damaged vision and other health problems. This unprecedented local victory resulted in the first policy in the world to prohibit the common practice of using methyl bromide as a port fumigant. In fact, EHC was the only local environmental group to participate with national and international non-governmental organizations (NGOs) in 1997 during discussions on the Montreal Protocol, an international treaty regarding the phasing out of ozone-depleting chemicals. The EHC campaign has become a model, which many other environmental health organizations are now using to pressure ports to reduce the use of dangerous pesticides. Since the victory, the Port District has committed US$20 million for the creation of an important wildlife refuge in the economically depressed South Bay, adopted a plan to reduce pesticide use at all of their facilities and agreed to provide funding for comprehensive community planning and expansion of the redevelopment area in Barrio Logan. Because of EHC's efforts, Barrio Logan was recently chosen by a Federal-State Interagency Committee (which included EPA) as one of 15 national environmental justice pilot projects to address air pollution problems. EHC's Border Environmental Justice Campaign also works with groups on the US–Mexican border.

THE EXPORT OF ECOLOGICAL HAZARDS TO THE NEW GLOBAL DUMPING GROUND: THE SOLIDARITY MOVEMENT AND ENVIRONMENTAL JUSTICE IN THE DEVELOPING WORLD

The environmental justice movement is also predicated on the human rights and anti-imperialism campaigns led by the US solidarity movement, including the South African anti-apartheid and anti-US intervention in Central America struggles in the 1980s. Solidarity movements in support of popular-based environmental movement in the developing world are assuming an ever greater importance in the era of corporate-led globalization. The growing ability of multinational corporations and transnational financial institutions to dismantle unions, evade environmental safeguards and weaken worker/community health and safety regulations in the US is being achieved by crossing national boundaries into politically repressive and economically oppressive countries, such as in Mexico, Indonesia, Nigeria and Central America generally (Faber, 1993). As a result, various nationalities and governments are increasingly pitted against one another in a bid to attract capital investment, leading to one successful assault after another on labour and environmental regulations seen as damaging to profits. Aided by recent 'free trade' initiatives such as the North American Free Trade Agreement (NAFTA) and the Free Trade Area of the Americas (FTAA), and enforced by bodies such as the World Trade Organization (WTO), these processes of ecological imperialism include the export of more profitable yet more dangerous production processes and consumer goods, as well as waste disposal methods, to developing countries where environmental standards are lax, unions are weak and worker health and safety issues ignored (Karliner, 1997, Castleman and Navarro, 1987).

Along the US–Mexico border there are more than 2000 factories or *maquiladoras*, many of them relocated US-based multinational corporations. One study of the border town of Mexicali indicated that stiff environmental regulations in the US and weaker ones in Mexico were either the main factor or a factor of importance in their decision to leave the US (Sanchez, 1990, p163–170). In fact, Lawrence Summers, current President of Harvard University and former Undersecretary of the Treasury of International Affairs and key economic policy-maker under the Clinton administration, is infamous for writing a 12 December 1991 memo as a chief economist at the World Bank that argued that 'the economic logic behind dumping a load of toxic waste in the lowest wage country is impeccable', and that the Bank should be 'encouraging more migration of the dirty industries to the LDCs [less developed countries]'. Forging links with developing world popular movements combating such abuses is yet another profound challenge confronting the US environmental justice movement.

Initially led by organizations such as the Environmental Project On Central America (EPOCA) and Third World Network in the 1980s, a host of environmental justice organizations in the US are now focusing on the interconnections between corporate-led globalization and growing problems of poverty, human rights violations, environmental degradation and the lack of democracy for poor developing world peoples. For instance, affiliates with the Southwest Network for Environmental and Economic Justice (SNEEJ) – a regional, bi-national network founded in 1990 by representatives of 80 grass-roots organizations based throughout the US South-west, California and Northern Mexico – worked on the EPA Accountability Campaign in 1994 to force the EPA to subpoena the records of over 95 US corporations operating in Mexico for their contamination of the New River. This was the first enforcement action that used NAFTA environmental 'side bars' and the Executive Order on Environmental Justice, and became one of the largest single enforcement actions ever taken by EPA. Likewise, EarthRights International (ERI) is launching a promising new 'International Right to Know' campaign, which would extend the existing reporting requirements of domestic environmental, occupational health and safety and labour rights legislation to US corporate activities in other countries. The campaign is being built in coalition with the AFL-CIO, Sierra Club, Center for International Environmental Law, Friends of the Earth, Amnesty International and other organizations

COMMUNITY ORGANIZING FOR SOCIAL CHANGE AND ECONOMIC REFORM: THE EMPOWERMENT OF OPPRESSED PEOPLES AND ENVIRONMENTAL JUSTICE

Finally, a significant element of environmental justice activism has evolved out of community-based movements for social and economic justice, particularly in communities of colour. Emphasizing issues of affordable and safe housing, crime and police conduct (including racial profiling and police brutality), un/under-employment and a living wage, accessible public transportation, city services, redlining and discriminatory lending practices by banks, affordable daycare, deteriorating schools and inferior educational systems, job training and welfare reform, and a host of other issues, many of these organizations have expanded their political horizons to incorporate issues such as lead poisoning, abandoned toxic waste dumps, the lack of parks and green spaces, poor air quality and other issues of environmental justice into their agenda for community empowerment. Although many organizations are not strictly self-defined as 'environmental' per se, they may devote considerable attention to environmental issues in their own communities. In fact, in recent years some of the most impressive environmental victories at the local level have been achieved by multi-issue-oriented economic justice organizations.

Direct Action For Rights and Equality (DARE), for instance, was established in 1986 to bring together low-income families in communities of colour within Rhode Island to work for social, economic and environmental justice. In this multi-issue, multi-racial dues-paying membership-based organization made up of 900 low-income families, members are organized into block clubs (similar to chapters), identify issues of common concern at regular organizational meetings and develop a strategy to address the problem. Since its establishment, DARE has successfully campaigned for the clean-up of over 100 polluted vacant lots and improved neighbourhood playgrounds and parks throughout Providence. One of DARE's most significant victories was recently achieved when Rhode Island became the first state in the nation to guarantee health care coverage for day care providers. Through this agreement with DARE, Rhode Island has set a new standard for other states to follow and implement. DARE is beginning work on campaigns to win jobs and career training from local companies for young people and is implementing further strategies to reduce pollution in low-income neighbourhoods.

Also included in this corner of environmental justice activism are the contributions of social justice-oriented religious groups and alliances, particularly those located in disenfranchised communities of colour. For instance, the St Paul Ecumenical Alliance of Congregations (SPEAC) began faith-based organizing in 1990 through a wide variety of civic and religious-based institutions within St Paul, Minnesota's lowest-income census tracts. Today, SPEAC's 19 low-income congregations and congregations of colour have strategically expanded their alliances at the neighbourhood, metropolitan and regional levels to impact St Paul's core city issues of reclaiming metro-polluted land for living wage job creation, as well as related issues of regional tax base sharing and reinvestment, public finance reform, affordable home ownership and fair welfare reform. Working in close collaboration with ageing inner ring suburban municipalities, SPEAC and the Interfaith Action (IA) of Minneapolis recently won a total of US$68 million in state funds which is being utilized to turn polluted dirt into pay dirt, by redirecting funds from outer ring suburban development on agricultural land (green fields) into the reclamation of abandoned, polluted industrial land in the inner cities (brown fields). This funding, when fully spent and matched by private investment over the next six years, will yield about 2000 permanent, good wage industrial jobs which will be easily accessible to people who need them most, rather than promoting urban sprawl. This campaign has become a model for metropolitan stability throughout the country.

THE STRUGGLE FOR ECOLOGICAL DEMOCRACY: LINKING SUSTAINABILITY AND ENVIRONMENTAL JUSTICE

As we move further into the new millennium, the mainstream US ecology movement is confronting an immense paradox. On the one hand, over the last three decades environmentalists have built one of the more broadly based and politically powerful new social movements in this country's history. As a result, US governmental policies for protecting the environment and human health are among the most stringent in the world. On the other hand, despite having won many important battles, it is becoming increasing apparent that the traditional environmental movement is losing the war for a healthy planet. With the ascendancy of neo-liberalism, globalization and the growing concentration of corporate power over all spheres of life, the ability of the movement to solve the ecological crisis is undermined. While there is no doubt that ecological problems would be much worse without the mainstream environmental movement and current system of regulation, it is also clear that the traditional strategies and policy solutions being employed are proving to be increasingly impaired. Most existing environmental laws are poorly enforced and overly limited in prescription, emphasizing, for instance, ineffectual *pollution control* measures which aim to limit public exposure to 'tolerable levels' of industrial toxins rather than promoting *pollution prevention* measures which prohibit whole families of dangerous pollutants from being produced in the first place. In addition, other problems such as the acceleration of sprawl and the growth in US emissions of greenhouse gases continue to worsen. The US system of environmental regulation may be among the best in the world, but it is grossly inadequate for safeguarding human health and the integrity of nature.

Perhaps the most critical factor for explaining the hegemony of neo-liberalism and the growing incapacity of the state to adequately address the ecological crisis is what Robert Putnam has termed the decline in *social capital* – those social networks and assets that facilitate the education, coordination and cooperation of citizens for mutual benefit (Putnam, 2000). Over the past generation, the social networks that integrate citizens into environmental organizations and other civic institutions have seriously deteriorated in communities across the country. The resulting decline in social capital inhibits genuine citizen participation in the affairs of civil society and engagement in the realm of politics, including the ability to tackle environmental problems in an equitable and effective fashion (Borgos and Douglas, 1996). With interactions that build mutual trust eroded, greater sectors of the populace become increasingly cynical of their ability to collectively effect meaningful ecological and social changes. Instead, a growing number of people retreat into what Jurgen Habermas (1975) terms *civil privatism*, with an emphasis on personal lifestyle issues such as career advancement, social mobility and conspicuous consumption. When social and environmental problems are confronted, increasingly individualized or 'privatized' solutions become the favoured response. As a result, the various racial, ethnic, class and religious divides in American society become accentuated, as the 'haves' increasingly disregard the needs of the 'have nots': witness the attack on affirmative action, the social safety net, labour rights and ecological protection in favour of reduced taxes, fiscal conservatism and increasingly harsh punishments for criminal misconduct.

Unfortunately, too many mainstream environmental organizations adapt corporate-like organizational models that further inhibit broad-based citizen involvement in environmental problem-solving. For these groups, citizen engagement means simply sending in membership dues, signing a petition and writing the occasional letter to a government official. As stated by William Shutkin (2000, pp1–20), there is a 'tendency for many non-profit environmental organizations to treat members as clients and consumers of services, or volunteers who help the needy, rather than as participants in the evolution of ideas and projects that forge our common life'. In the effort to conduct studies, draft legislation and organize constituencies to support passage of environment-friendly initiatives, the mainstream movement has gravitated toward a greater reliance on law and science conducted by professional experts. The aim of this move towards increased professionalization is to regain legitimacy and expert status in increasingly hostile neo-liberal policy circles. The effect, however, is to reduce internal democratic practices within some environmental organizations and state regulatory agencies. The focus on technical-rational questions, solutions and compromises, rather than issues of political power and democratic decision-making, is causing a decline in public interest and participation in national environmental politics (Faber and O'Connor, 1993).

To overcome this crisis of democracy and the corporate assault upon nature requires the reinvigoration of an *active environmental citizenship* committed to the principles of *ecological democracy*. These principles include a commitment to: (1) grass-roots democracy and inclusiveness – the vigorous participation of people from all walks of life in the decision-making processes of capital, the state and social institutions that regulate their lives, as well as civic organizations and social movements which represent their interests; (2) social and economic justice – meeting all basic human needs and ensuring fundamental human rights for all members of society; and (3) sustainability and environmental protection – ensuring that the integrity of nature is preserved for both present and future generations. These three pillars on which the concept of ecological democracy rests provide a meaningful vision for building a more just and ecologically sound American society.

Fortunately, there are signs that a powerfully new active environmental citizenship committed to the principles of ecological democracy is beginning to emerge in America and throughout the world. The revitalization of grass-roots environmental organizations committed to genuine base-building and political-economic reform is a reaction to the new challenges posed by neo-liberalism and globalization, and includes the use of direct action against timber companies, polluters, the World Trade Organization (as seen in the 'Battle in

Seattle'), the World Bank and others (as well as criticism toward the 'corporatist' and exclusionary approaches of mainstream environmental organizations). Pressing for greater economic equality, greater corporate and government accountability (such as the 'right to know' about hazards facing the community) and more comprehensive approaches to environmental problem-solving (such as adoption of the precautionary principle over risk-assessment, source reduction and pollution *prevention* over pollution *control* strategies, 'Just Transition' for workers out of polluting industries over job blackmail, etc), the struggle for ecological democracy represents the birth of a *transformative* environmental politics (Faber and O'Connor, 1993, Dowie, 1995).

At the forefront of the struggle for ecological democracy and a new active environmental citizenship is the environmental justice movement. No other force within the broader context of grass-roots environmentalism offers the same potential as the environmental justice movement for: (1) bringing new constituencies into environmental activism, particularly in terms of oppressed peoples of colour, the working poor and other populations who bear the greatest ecological burden; (2) broadening and deepening our understanding of ecological impacts, particularly in terms of linking issues to larger structures of corporate power; (3) constructing and implementing new grass-roots organizing and base-building strategies over traditional forms of advocacy, as well as developing new organizational models which rebuild social capital and maximize democratic participation by community residents in decision-making processes; (4) connecting grass-roots and national layers of environmental activism; (5) creating new pressure points for policy change; (6) building coalitions and coordinated strategies with other progressive social movements, including the labour movement and (7) bringing more innovative and comprehensive approaches to environmental problem-solving, particularly in terms of linking sustainability with issues of social justice.

Environmental justice activists clearly recognize the importance of community building, promoting active forms of citizen participation in decision-making processes and forging stronger partnerships with other community organizations in order to build a more vibrant and democratic civil society. As stated by Mark Gerzon (1995, pp188–95), '... strengthening the capacity of communities for self-governance – that is, making the crucial choices and decisions that affect their lives', is the most critical task confronting the environmental movement in rebuilding social capital and a vibrant ecological democracy. Because the environmental justice activists emphasize base-building strategies that take a multi-issue approach, they function as *community capacity builders* to organize campaigns that address the common links between various social and environmental problems (in contrast to isolated single-issue-oriented groups, which treat problems as distinct). In this respect, the movement has done an outstanding job of *enlarging the constituency* of the environmental movement as a whole by incorporating poorer communities and oppressed peoples of colour into strong, independent organizational structures insulated from colonization and co-optation by white-led, mainstream environmental organizations and government bodies. Although an identity-based politics focused on environmental racism poses some limitations to coalition building with white working- and middle-class families, the movement has done important work in helping to *span community boundaries* by crossing difficult racial, class, gender-based and ideological divides which weaken and fragment communities (Mathews, 1997, pp275–280).

Finally, the movement is facilitating *community empowerment* by emphasizing *grass-roots organizing and base-building* over traditional forms of environmental advocacy. Under the traditional advocacy model, professional activists create organizations that speak and act on behalf of a community. In contrast, the grass-roots organizing approach by the environmental justice movement emphasizes the mobilization of community residents to push through the systemic barriers that bar citizens from directly participating in the identification of problems and solutions so that they may *speak and act for themselves* (Alston, 1990). Base-building implies creating accountable, democratic organizational structures and institutional procedures which facilitate inclusion by ordinary citizens, and especially dispossessed people of colour and low-income families, in the public and private decision-making practices affecting their communities.

If the environmental justice movement continues to build upon the already impressive successes it has established in these areas, and find ways to collaborate with the broad array of grass-roots citizens groups representative of the white middle class, we may finally witness the creation of a truly broad-based ecology movement, inclusive of all races, the working poor and women, that is finally capable of implementing a national and international strategy to end the abuses of nature wrought by corporate America. In short, the environmental justice movement is critical to the larger effort to build a more inclusive, democratic and effective ecology movement in the US – one which can challenge and transform structures of power and profit which lie at the root of the ecological crisis.

CONCLUSION

It is now clear that the traditional environmental movement has become so fragmented, parochial and dominated by single-issue approaches that its capacity to champion fundamental social and institutional changes needed to address America's ecological crisis is greatly diminished. As stated by Pablo Eisenberg (1997, pp331–341), 'although we know that our socio-economic, ecological and political problems are interrelated, a growing portion of our nonprofit world nevertheless continues to operate in a way that fails to reflect this complexity and connectedness'. In this respect, if the traditional environmental movement continues to conceive of the ecological crisis as a collection of unrelated problems, and if the reigning paradigms are defined in the neo-liberalist terms of a minimally regulated capitalist economy, then it is possible that some combination of regulations, incentives and technical innovations can keep pollution at tolerable levels for many people of higher socio-economic status. Poorer working-class communities and people of colour who lack the political-economic resources to defend themselves will continue to suffer the worst abuses. If, however, the interdependency of issues is emphasized, so that environmental devastation, ecological racism, poverty, crime and social despair are all seen as aspects of a multi-dimensional web of a larger structure crisis, then a transformative ecology movement can begin to be invented (Rodman, 1980).

It is precisely this single-issue orientation that the environmental justice movement is coming to challenge by developing broad-based coalitions that are pushing for comprehensive approaches to community, national and global problems. The struggle for environmental justice is not just about distributing environmental risks equally (ie distributive environmental justice) but about preventing them from being produced in the first place so that no one is harmed at all (ie productive environmental justice). The struggle for environmental justice must be about the politics of corporate power and capitalist production per se and the elimination of the ecological threat, not just the 'fair' distribution of ecological hazards via better government regulation of inequities in the marketplace. And while increased participatory democracy by popular forces in governmental decision-making and community planning is desirable (if not essential), and should be supported, it is, in and of itself, insufficient for achieving true sustainability and environmental justice. What is needed is a richer conception of ecological democracy.

From this perspective, organizing efforts against procedures that result in an unequal distribution of environmental problems (distribution inequity) cannot ultimately succeed unless environmental justice activists continue to address the procedures by which the environmental problems are *produced* in the first place (procedural inequity) (Lake, 1996, p169). Any effort to rectify distributional inequities without attacking the fundamental processes that produced the problems in the first place focuses on symptoms rather than causes and is therefore only a partial, temporary, and necessarily incomplete and insufficient solution. What is needed is an environmental justice politics for procedural equity that emphasizes democratic participation in the capital investment decisions through which environmental burdens are *produced* then distributed. As Michael Heiman (1996, p120) has observed, 'If we settle for liberal procedural and distributional equity, relying upon negotiation, mitigation and fair-share allocation to address some sort of disproportional impact, we merely perpetuate the current production system that by its very structure is discriminatory and non-sustainable'. Productive environmental justice can only be achieved in a sustainable economic system – a post-capitalist society in which material production and distribution is democratically planned and equitably administered according to the needs of both present and future members of society.

Rather than existing as a collection of isolated organizations fighting defensive 'not-in-my-backyard' battles (as important as they may be), the environmental justice movement must continue to evolve into a political force capable of challenging the systemic causes of social and ecological injustices as they exist 'in everyone's backyard'. It is precisely this distinction between *distributional environmental justice* versus *productive environmental justice* that many in the movement are now beginning to address in a more systematic fashion. Only by bringing about what Barry Commoner (1990) calls 'the social governance of the means of production' – a radical democratization of all major political, social and economic institutions – can humanity begin to gain control over the course of its relationship with nature. Such a programme for social governance would require that the institutions of workplace and local direct democracy, liberal democratic procedures and constitutional guarantees, state planning and the initiatives of popular-based social and environmental movements be sublated into a genuine ecological democracy (O'Connor, 1992, p1–5).

The challenge confronting the environmental justice movement is to help forge a truly broad-based political movement for ecological democracy. While the traditional environmental movement has played a critical and progressive role in stemming many of the worst threats posed to the health of the planet and its inhabitants, the movement is now proving increasingly unable to institute more sustainable and socially just models of development in the face of neo-liberalism, globalization and the economic restructuring of US and international capitalism. And as unsustainable practices and environmental injustices intensify across the globe, the need for a mass-based international movement committed to the principles of ecological democracy will become more pressing. Just as in the 1930s, when the labour movement was forced to change from craft to industrial unionism, so today does it appear to many that labour needs to transform itself from industrial unionism into an international conglomerate union, inclusive of women and all racial/ethnic peoples, just to keep pace with the restructuring of international capital. And just as in the 1960s, when the environmental movement changed from a narrowly based conservation/preservation movement to include the middle class (and some sectors of the white working class), so today does it seem to many that it needs to change from single-issue local and national struggles to a broad-based multi-racial international environmental justice movement. We must work in solidarity to promote strong unions, environmental justice movements and worker health and safety standards throughout the rest of the world in order to protect local initiatives and gains. This historic task now confronts the environmental justice movement.

ACKNOWLEDGEMENTS

Funding for this research paper was provided by the Nonprofit Sector Research Fund of the Aspen Institute, Washington, DC; the Marion and Jasper Whiting Foundation; as well as the Research, Scholarship and Development Fund (RSDF) of the Provost Office at Northeastern University.

REFERENCES

Alston, D (1990) *We Speak for Ourselves: Social Justice, Race, and Environment*, Panos Institute, Washington, DC
Alston, D (1992) 'Transforming a movement: People of colour united at summit against environmental racism', *Sojourner,* vol 21, pp30–31
Bastian, A and Alston, D (1993) 'An open letter to funding colleagues: New developments in the environmental justice movement', New World Foundation & Public Welfare Foundation, pp1–4

Borgos, S and Douglas, S (1996) 'Community organizing and civil renewal: A view from the south', *Social Policy,* vol 1, pp18–28

Bryant, B (1995) *Environmental Justice: Issues, Policies, and Solutions,* Island Press, Washington, DC

Bryant, B and Mohai, P (1992) *Race and the Incidence of Environmental Hazards: A Time for Discourse,* Westview, Boulder

Bullard, R (1990) *Dumping in Dixie: Race, Class, and Environmental Quality,* Westview, Boulder

Bullard, R (1994) *Unequal Protection: Environmental Justice and Communities of Colour,* Sierra Club Books, San Francisco

Castleman, B and Navarro, V (1987) 'International mobility of hazardous products, industries, and wastes', *Annual Review of Public Health,* vol 8, pp1–19

Chavis, B and Lee, C (1987) *Toxic Wastes and Race in the United States: A National Report on the Racial and Socioeconomic Characteristics of Communities Surrounding Hazardous Waste Sites,* United Church of Christ Commission for Racial Justice, New York

Commoner, B (1990) *Making Peace with the Planet,* Pantheon, New York

Dicken, P (1992) *Global Shift: The Internationalization of Economic Activity,* 2nd edn, Guilford, New York

Dowie, M (1995) *Losing Ground: American Environmentalism at the Close of the Twentieth Century,* MIT Press, Cambridge, MA

Eisenberg, P (1997) 'A crisis in the nonprofit sector', *National Civil Review,* vol 86, pp331–41

Environmental Research Foundation (1993) *Rachel's Hazardous Waste News,* no 332, pp1–2

Faber, D (1993) 'Environment under fire: Imperialism and the ecological crisis in Central America', *Monthly Review,* New York

Faber, D (1998) *The Struggle for Ecological Democracy: Environmental Justice Movements in the United States,* Guilford, New York

Faber, D and Krieg, E (2001) *Unequal Exposure to Ecological Hazards: Environmental Injustices in the Commonwealth of Massachusetts,* Northeastern University, Boston

Faber, D and O'Connor, J (1993) 'Capitalism and the crisis of environmentalism' in R Hofrichter (ed), *Toxic Struggles: The Theory and Practice of Environmental Justice,* New Society, Philadelphia

Field, R (1998) 'Risk and justice: Capitalist production and the environment' in D Faber (ed), *The Struggle for Ecological Democracy,* Guilford, New York

Gerzon, M (1995) 'Reinventing philanthropy: foundations and the renewal of civil society', *National Civil Review,* vol 84, pp188–95

Goldman, B and Fitton, L (1994) *Toxic Waste and Race Revisited: An Update of the 1987 Report on the Racial and Socioeconomic Characteristics of Communities with Hazardous Waste Sites,* United Church of Christ Commission for Racial Justice, New York

Gordon, D (1996) *The Fat and the Mean: The Corporate Squeeze of Working Americans and the Myth of Managerial Downsizing,* Free Press, New York

Grinde, D, Zinn, H, and Johansen, B (1998) *Ecocide of Native America: Environmental Destruction of Indian Lands and Peoples,* Clear Light, New York

Habermas, J (1975) *Legitimation Crisis,* Beacon Press, Boston

Heiman, M (1996) 'Race, waste, and class: New perspectives on environmental justice', *Antipode,* vol 28, pp111–21

Johnston, BR (1994) *Who Pays the Price?: The Sociocultural Context of Environmental Crisis,* Island Press, Washington, DC

Karliner, J (1997) *The Corporate Planet: Ecology and Politics in the Age of Globalization,* Sierra Club Books, San Francisco

LaDuke, W (1999) *All Our Relations: Native Struggles for Land and Life*, South End, Boston

Lake, R (1996) 'Volunteers, NIMBYs, and environmental justice: Dilemmas of democratic practice', *Antipode,* vol 28, pp160–74

Lavelle, M and Coyle, M (1992) 'Unequal protection: The racial divide in environmental law', *National Law Journal,* vol 21, pp2–12

Lee, C (1992) *Proceedings: The National People of Colour Environmental Leadership Summit*, United Church of Christ Commission for Racial Justice, New York

Levenstein, C and Wooding, J (1998) 'Dying for a living: Workers, production, and the environment' in D Faber (ed) *The Struggle for Ecological Democracy*, Guilford, New York

Mathews, D (1997) 'Changing times in the foundation world', *National Civic Review,* vol 86, pp275–80

National Research Council (1991) *Environmental Epidemiology: Public Health and Hazardous Wastes*, National Academy Press, Washington, DC

O'Connor, J (1992) 'A political strategy for ecology movements', *Capitalism, Nature, Socialism,* vol 3, pp1–5

Perfecto, I (1992) 'Farm workers, pesticides and the international connection' in B Bryant and P Mohai (eds) *Race and the Incidence of Environmental Hazards*, Westview, Boulder

Putnam, R (2000) *Bowling Alone: The Collapse and Revival of American Community*, Simon & Schuster, New York

Regenstein, L (1986) *How to Survive in America the Poisoned*, Acropolis, Washington, DC

Rodman, J (1980) 'Paradigm change in political science: An ecological perspective', *American Behavioral Scientist,* vol 24, pp49–78

Roque, J (1993) 'Environmental equity: reducing risk for all communities', Review of EPA report, *Environment*, vol 35, pp25–28

Ross, R and Trachte, K (1990) *Global Capitalism: The New Leviathan*, State University of New York Press, Albany

Sanchez, RA (1990) 'Health and environmental risks of the maquiladora in Mexicali', *Natural Resources Journal,* vol 30, pp163–70

Schwab, J (1994) *Deeper Shades of Green: The Rise of Blue-Collar and Minority Environmentalism in America*, Sierra Club, San Francisco

Selcraig, B (1997) 'What you don't know can hurt you', *Sierra*, Jan/Feb, pp38–43

Shutkin, W (2000) *The Land That Could Be: Environmentalism and Democracy in the Twenty-First Century*, MIT Press, Cambridge, MA

Tokar, B (1996) 'Trading away the earth: Pollution credits and the perils of free market environmentalism', *Dollars and Sense,* vol 204, pp24–29

Weaver, J and Means, R (1996) *Defending Mother Earth: Native American Perspectives on Environmental Justice*, Orbis, New York

Chapter 3

Inequality and Community and the Challenge to Modernization: Evidence from the Nuclear Oases

Andrew Blowers

THE DISCOURSE OF SUSTAINABLE DEVELOPMENT

Broadly speaking, there are two distinctive strands in discourses seeking to explain the relationship between environmental and social change. One stresses the inherent compatibility between economic development and the conservation of the environment. The other emphasizes the inevitable deterioration of the environment through the development of modern forms of technological development. In theoretical terms these contrasting positions are represented by, on the one hand, ecological modernization, and on the other, by the 'risk society' thesis. I shall examine these in a moment. But first I wish to note that these very different positions both neglect the analytical and practical importance of two components in the relationship between society and the environment. These components are inequality and community, each of which must be central to an analysis and understanding of the possibilities and prospects for sustainable development.

By contrast, the environmental justice movement with its origins in the US, both in principle and in terms of practical political action, emphasizes the distributional implications of environmental change. Environmental justice as a theoretical and moral issue focuses on the uneven impacts over space and time imposed by modern technologies. In spatial terms there is the disadvantage experienced by local communities exposed to environmental degradation or risk. In terms of intergenerational impacts there is a concern for the risks to the future resulting from the actions of the present. These distributional outcomes are explained in terms of the power relations which routinely secure environmental advantages for the rich while consigning the disadvantages to the poor. The focus of attention is on inequality as it is expressed in communities at the local level. Such an approach reveals, as I shall show, some interesting and surprising possibilities for empowering hitherto disadvantaged communities. At a theoretical level, this analysis provides an interesting challenge to some of the assumptions of contemporary discourses of modernization.

My purpose in what follows is, first, to demonstrate the limitations of contemporary discourses of ecological modernization and 'risk society', notably their neglect of inequality and community and the implications for changing power relations. I shall then identify how these deficiencies are addressed by alternative approaches. This abstract discussion is then given some empirical focus by drawing on evidence from so-called 'peripheral communities' to demonstrate how power relations may shift over time to mitigate disadvantage. This critique of modernization theory leads into a discussion of the principles that must inform society's efforts to shift towards a more sustainable environment and the implications for policy-making.

THE ENVIRONMENTAL DIMENSIONS OF MODERNIZATION

Over the past two decades in many Western countries there has been a transformation in governance in response to the increasing invasion of market processes into the various domains of social and political life. This process has been termed 'political modernization' (Leroy and van Tatenhove, 2000, van Tatenhove et al, 2000) and has been particularly characteristic, with national variants, of trends in north-western European countries. These changes have been encouraged by the ideological shift to the Right but they must also be seen as connected to developing economic globalization. The rhetoric of the transformation embraces notions of consensus, cooperation and collaboration, presenting a view of social development which is apolitical, emphasizing cohesion and thereby reducing the focus on conflict. The reality has been a fundamental change in the relationship between the state and society, a shift from a welfare state to an enabling state, which has had profound impacts at all levels of policy-making including the environment.

Ecological modernization is the environmental variant of this political modernization. Indeed, Janicke claims that 'the ecology question has been a strong motor of political modernization' (1996, p77). Its advocates argue that with ecological modernization, 'economic growth and the resolution of ecological problems can, in principle, be reconciled' (Hajer, 1995, p248). Herein lies its attractiveness, for ecological modernization neither fundamentally threatens business-as-usual nor requires a different form of political-economic system. Ecological modernization is seductive and persuasive, 'a doctrine of reassurance, at least for residents of relatively prosperous developed societies. It assures us that no tough choices need to be made between economic growth and environmental protection, or between the present and the long term future' (Dryzeck, 1997, p146).

Ecological modernization has been exposed to a range of interpretations, claims, criticisms, reviews and applications (see for example special issues of *Environmental Politics*, vol 9 (1), 2000 and *Geoforum*, vol 31 (1)). It is discussed here in terms of the basic components identified in its initial formulation:

1   One is the necessity of incorporating environmental issues into the production and consumption processes – the so-called 'refinement of production' (Mol, 1995). Among the examples of this are recycling, waste minimization, pollution abatement, use of renewable resources, conservation of energy and life cycle analysis.

2   The market is regarded as flexible, efficient, innovative and responsive. The encouragement of competition and wealth creation has been accompanied by the dismantling of regulations and its concomitant, the recognition of self-regulation by industry as the normal way of dealing with environmental problems. Moreover, the private sector has taken on the responsibility for delivering a range of formerly public (including environmental) services.

3   The state maintains its regulatory function though in a more residual form providing a framework of standards and incentives for environmental performance. The engagement of the state and business as partners in environmental policy-making and planning is an application of a central tenet of political modernization.

4   The state/market nexus presents a form of *governance* that is distinctly different from the public policy-led, state-regulated forms that hitherto existed. There is an emphasis on collaboration and cooperation to be achieved through *negotiative forms of decision-making* (described in various ways as 'collaborative management', 'communicative governance' or 'cooperative management regimes' (Glasbergen, 1998)). The *policy network* in which these various actors interact has come to be seen 'as *the* organizational instrument to improve public policy-making' (Leroy, 2000, p4). Its features are a shared definition of problems, a common desire to tackle them and knowledge of the contribution each partner can make to solutions. Indeed, *partnership* has become the *idée fixe* of political modernization in the UK. The partnership that really matters is that between business and government although, in recent years, the emphasis on these key partners has been tempered by the incorporation of other so-called 'stakeholders'. In the environmental field these include local government, voluntary organizations and community-based groups encouraged to cooperate in a variety of programmes such as Local Agenda 21 and partnerships for the regeneration of disadvantaged areas.

5   As a result, environmental movements become involved in collaborative relationships with business and government in the delineation of environmental policy. As Harvey observes, 'The thesis of ecological modernization has now become deeply entrenched within many segments of the environmental movement' (1996, p378). For some, there is evidence of an equality of partnership, what Janicke (1996) terms an 'iron triangle', a kind of 'ecocorporatism' between business, government and environmental movements. However, it can also be argued that the role of environmental movements is one of supporting and legitimizing the dominant partnership of state and business.

6   The influence of environmental movements derives from the decline of formal representative politics, another feature of political modernization but which is supposedly compensated for by a highly developed civil society.

In this political space, it is argued, political interests and movements are able to influence policy through consciousness raising, persuasion, lobbying, mobilization of opinion or direct action. Civil society is the arena in which various social movements, including environmental movements, seek to make their presence felt. Many commentators perceive civil society as somehow neutral ground providing opportunity for progressive and democratic influences to flourish and counter the dominant forces in society. This view is flawed for three reasons. First, social movements frequently become incorporated into the decision-making structures operated by dominant interests. Second, civil society also offers fertile ground for reactionary forces to flourish on the basis of a politics of resentment. Third, social movements themselves are not particularly democratic. They may claim to represent a wider public interest but, in their organizational structures and specific objectives, they are often quite exclusive and unrepresentative (Potter, 1996).

There are some specific limitations to the application of ecological modernization. It is limited in its application to specific industries. While it is possible to introduce pollution controls, resource efficiency and conservation in a wide array of processes, certain industries cannot be modernized to the extent of removing threats to the environment. This is especially so with industries such as nuclear energy which pose high consequence, long term risks which can never be eliminated. In its application to specific processes in specific countries ecological modernization diverts attention from the problem of ecosystem complexity and interrelationships which may (as global warming or ozone depletion illustrate) spring surprises or create intractable problems that require global action. Furthermore, it ignores the impact of externalities on other places and in other parts of the world. It is specific communities, often in relatively remote locations, which bear the burden of long term risk imposed by certain industries (nuclear reprocessing being a typical example).

Political modernization and its ecological variant are, in many ways, even less democratic than the system they have supplanted. In theory, political modernization urges the doctrine of consensus and purports to include a variety of interests; in reality it supports a system that is exclusive, elitist and unrepresentative, a condition described by Crouch (2000) as 'post-democracy'. It has been accompanied by two processes that are a barrier to broader, more socially inclusive forms of cooperation. One is the persistence of inequality which is itself a barrier to cooperation. The other is the weakening of those institutions (such as community) which provide the integration necessary for cooperation to develop (though, as I shall show, such institutions have by no means been eliminated).

These limitations of ecological modernization have led some commentators to broaden the concept beyond its state/market focus. Christoff (1996) tries to take account of globalizing processes, the advent of megarisks, the problem of inequality and the emphasis on modernization to propose a form of 'strong' ecological modernization that emphasizes local cultural and environmental conditions and the need for broad social and institutional change. Dryzeck argues that melding capitalism with more enlightened environmental values 'offers a plausible strategy for transforming industrial society into a radically different and more environmentally defensible (but still capitalist) alternative' (1997, p143). More recently, the arch exponents of ecological modernization have confronted their critics, describing it as a 'theory in the making' (Mol and Sonnenfeld, 2000, p12) able to link societal change to a relevant political programme of environmental improvement (Mol and Spaargaren, 2000). Ecological modernization has even been enthusiastically elevated as a critique and counterweight to the processes at work in the global economy (Mol, 2001). It remains at heart a reformist, even conservative thesis. It is argued that environmental concerns should not be privileged over other social issues and that social practices and institutions 'are already transforming to a major extent' (Mol and Spaargaren, 2000, p35) such that the need for radical social change can at least be questioned.

Such arguments, while seeking to accommodate criticisms, appear to stretch the idea of ecological modernization beyond credibility, to render it so inclusive as to be vague, tendentious and analytically useless. Ultimately, ecological modernization does not challenge capitalism; if anything it celebrates it. It identifies a society where pluralism, openness, civil society and consensus all supposedly flourish. It tends to deny the realities of elitism, conflict and inequality which generate a social context in which it becomes impossible to realize sustainable forms of development. In short, ecological modernization portrays a partial, not to say ideological, perspective on the relationships between society and the environment.

ALTERNATIVE APPROACHES

'Risk society'

The comfortable conclusions of ecological modernization are confronted by the altogether more pessimistic outlook propounded by Ulrich Beck's analysis of the implications of modernization propounded in his 'risk society' thesis (1992, 1995). Risk society emphasizes the awesome destructive power inherent in some modern industries and the control exercized by scientific experts in the management of risk. Far from being involved in decision-making, environmental movements utter their warnings and seek to prevent the development of risky technologies in a series of confrontations. While ecological modernization suggests a steady progress towards sustainable development, risk society reveals a world in which science and technology are directed towards the perpetuation of risk.

Unlike ecological modernization, Beck's analysis is not confined to the relationship between the economy and the environment; it encompasses social change as well. One of the features of modern technology is that as individual consumers we contribute to the production of risk but, since risks can only be controlled by collective action, as individuals we cannot prevent risks occurring and therefore continue to act irresponsibly. This creates a fatalistic acceptance of risks over which we have no control. Beck graphically describes the tendency as Murphy's Law, 'Whatever can go wrong will go wrong, and even what can't go wrong will do so' (1995, p174). This feeling of vulnerability to technological risk has been exacerbated by consciousness of personal insecurity, described by Beck as 'individualization', created by the conditions of living in modern society with its threat of unemployment, reduction of welfare and the dislocation of personal life.

Beck sees two avenues of escape. One is through 'sub-politics' where an apparently apolitical disavowal of conventional party politics is replaced by commitment to various causes, campaigns and ideas such as those promoted through environmental movements (Beck, 1992, 1998). The other potential escape is through 'reflexivity' by which individuals, confronted by the reality of their condition, are able to undergo reflection and self-criticism which leads on to self-transformation. In this lies the hope that society as a whole will be able to effect a transformation of industrial processes and political institutions. It must be said that this reflexivity has a wistful air about it and seems to be offered more in hope than expectation. But it does at least offer the prospect of change to be brought about by a collective determination to draw back from the abyss.

Although risk society draws a broad theoretical perspective, it is limited in its empirical scope. Like ecological modernization it, too, draws its inspiration from Western societies (and especially those of north-western Europe) so it is difficult to apply it more universally. It focuses on high-risk technologies (incidentally the very ones that are not captured in the ecological modernization analysis) as the creators of all pervading environmental risk. Although the risks to which we are exposed may ultimately engulf us all, the impacts of risky technologies are present here and now and are experienced more by some than others.

Nor is the condition of individualization universal – some individuals and groups are more vulnerable than others and it is often the most vulnerable who, struggling to survive, are least able to contemplate their condition in reflexive fashion. Beck's escape routes do not necessarily lead to benign outcomes. He reposes a somewhat naïve faith in the potential of grass-roots activism, the idea that 'politics may peter out as sub-politics is activated' (1998, p38). In the zone of sub-politics it is just as likely that reactionary forces may flourish and reflexivity may easily lead to self-interested rather than altruistic reflections. It is, therefore, conceivable that the social consequences of risk society may be an increase in inequality and a loss of collective purpose to defend against the consequences of ecological deterioration. Although Beck recognizes the importance of differences based on gender, ethnicity and so on, he tends to deal with them at a rather abstract and general level. He does not, I think, see inequality as a key factor in dealing with environmental risk. Instead, he assumes that modern technological risk is pervasive and thereby 'democratic' in that all are affected and none can escape. As with ecological modernization, risk society concentrates attention on supposedly encompassing processes thereby neglecting the significance of differentiation that characterizes both society and the environment.

Counter-modernist perspectives

Ecological modernization puts the case for the continuation of an environmentally sensitive form of modernization; risk society confronts the necessity for change. But, in both accounts, the relevance of inequality in dealing with the processes of environmental change is analytically neglected. In asserting the prevalence of modernization the two perspectives also fail to identify the importance of those features, notably community and locality, which give expression to the diversity of social responses to environmental change.

Inequality and community are central concerns of a range of alternative approaches and movements, including environmental justice, which, in very different ways, provide a challenge to the dominant discourse of modernity. Anti-modernist perspectives were relatively common a generation ago, coinciding with the intellectual paradigms then current (eg Stretton, 1976, Gorz, 1980, Sandbach, 1980, Schnaiberg, 1980). Subsequently they tended to become eclipsed as attention was focused on the neo-liberal agenda that accelerated during the 1980s. More recently there has been a revival of a very diverse set of arguments coming from different ideological and theoretical positions (for example, Dobson, 1990, Jacobs, 1991, Eckersley, 1992, Pepper, 1993, Harvey, 1996, Gould et al, 1996, Benton, 1996) which identify some key elements missing from the dominant discourse:

1   Local/global. It is in the locality that environmental problems are generated and it is the locality that experiences the effects. With the increase in transboundary problems there is no longer necessarily a coincidence between local cause and effect. As Gould et al put it, 'Local problems are more and more often generated by geographically distant producers and consumers. The socio-economic causes of environmental problems are therefore organized at a different level than are the social responses to these problems' (1996, p34). It is clear that certain places, communities and countries suffer disproportionately from the pollution, resource depletion and environmental degradation that result from the exploitative processes of the modern economy. 'The impacts of environmental degradation are always socially and spatially differentiated. They may end up affecting the global environment, but first they damage small parts of it' (Low and Gleeson, 1998, p19).

2   Intergenerational effects. According to current predictions global warming will present serious impacts within one or two generations; the depletion of biodiversity removes existing or potential assets from the future; and the problems of nuclear waste will persist indefinitely. The transmission of environmental effects down the generations means that the future is already compromised by the actions of the present.

3   Empowerment. Radical critics point to the imbalance of power reflected in the powerlessness of disadvantaged communities. Citing many examples, including such celebrated cases as Love Canal, Bhopal and Chernobyl, they illustrate the problems of ill health, danger, dereliction and even displacement experienced by poor communities. Decisions affecting their

environment are often taken far away and their dependence on employment renders them passive in the face of environmental risk (though, as we shall see later, passivity is not an inevitable outcome).

4   Environmental movements. Some radical theorists (and in this instance ecological modernists too) take a more critical stance on the role of environmental movements. A central criticism is that environmental movements, by focusing on ecological issues, tend to lose sight of the social implications. They point out that many, and often the most powerful, groups are composed of middle-class people intent on preventing damage to their own amenity, that such opposition often leads to the diversion of polluting projects to disadvantaged communities, and that such groups are unrepresentative and elitist in their organization and decision-making.

It is the focus on social context, on local impacts, empowerment and disadvantage which distinguishes the environmental justice movement from those movements primarily concerned with environmental issues. Environmental justice seeks to root out the inequalities which structure environmental disadvantage. Although the environmental justice movement has tended to become associated with specific struggles in the US, my intention is to broaden the concept to include the myriad of citizens-based organizations in many parts of the world which mobilize to resist dangerous activities (such as nuclear facilities or toxic waste dumps) or the destruction of communities through infrastructure projects (such as roads or airports).

Ecological modernization and the risk society thesis have relatively little to say about political change. They are apolitical theories. Ecological modernization presumes a consensus in which, by definition, political conflict is largely absent. Risk society exposes the inadequacies of the present system to deal with global risks but, in terms of political solutions, can only offer 'a sigh to the end of the world' (Beck, 1996, p38). Environmental justice, with its concern for real people in real places, gives a moral, practical and empirical context to the more abstract theoretical discussions. Inequality, and especially the problem of localized disadvantage, is exposed as a stumbling block to the cooperative relationships promoted by ecological modernization or the need for collective reflexivity espoused in the risk society analysis. Environmental justice, in its broader definition, provides a compelling empirical and theoretical counter-perspective to modernization theories. In the next section I shall use the example of a specific context to shed some empirical light, drawing on the social consequences of environmental inequality experienced in nuclear communities.

NUCLEAR OASES: INEQUALITY AND THE PERIPHERY

Nuclear oases as peripheral communities

The nuclear industry is dying. Although global warming and concerns about energy security hold out the possibility of some new nuclear build in the US and some other Western countries, nuclear is not, at present, a realistic option to replace fossil fuels. Nuclear energy is no longer competitive in most Western countries, if it ever was. Commercial production of power reactors has ceased in all but a few countries in the Far East. The military need for plutonium is much diminished and surplus stocks of the material have eliminated the need for further production. The radioactive emissions from nuclear activities, both routine and more dramatic and catastrophic releases such as at Chernobyl have engendered disillusion and anxiety among the public at large. Yet the accumulated legacy of radioactive contamination remains scattered across thousands of sites and the problem of managing radioactive wastes from past and present production is a major problem with consequences extending into the far future.

Although the risks presented by nuclear operations are widespread (and, in the case of a major accident or conflict, potentially global in reach) the problem of radioactive wastes is geographically concentrated. Ultimately these wastes, presently stored at power plants or other facilities, will be assembled at sites for long term storage or deep disposal. Already there are some sites where past and present military and civil reprocessing operations have created large volumes of wastes which pose risks to local communities. These are the 'nuclear oases' (Blowers et al, 1991), places where substantial volumes of wastes, emissions of radioactivity and, in some cases, contaminated land, pose risks to the local communities dependent on the industry. Examples of such nuclear oases are: Hanford in Washington state in north-west US where plutonium was produced for the atomic bomb during World War Two; La Hague, a reprocessing complex on the tip of the Cotentin peninsula in Normandy, France; Sellafield in north-west England, the centre for civil and military nuclear facilities; and Gorleben in Germany, the site of an interim waste store and of a potential disposal repository (Blowers, 1999, 2000). Elsewhere, notably in the US, Russia and Japan, there are other sites where past and present nuclear activities and dependent communities are concentrated.

These nuclear oases appear to convey the characteristics of 'peripheral communities' (Blowers and Leroy, 1994) – places which experience the disadvantage of risk and environmental degradation. They are peripheral in the sense that they are geographically remote, relatively inaccessible and isolated. Dependence on a dominant employer renders them monocultural, subject to economic risk as well as relatively powerless, their fortunes controlled by external influences. As a consequence they exhibit an inward-looking, defensive and acquiescent culture. They are the outcome of a process of 'peripheralization' whereby a coalition of economic interests (of labour and capital) reinforces the presence of an activity that is resisted by a coalition of interests (environmental and economic) defending the intrusion of the unwanted industry elsewhere. The recent history of conflicts over nuclear waste amply confirms this tendency for the nuclear industry to retreat to its established oases, where jobs and familiarity conspire to provide an acceptable location. Thus, the geography of nuclear waste appears to be set in aspic in disadvantaged communities, the victims of power relations which ensure that the industry has nowhere else to go.

Nuclear oases and modernization

Viewed in this way, these communities appear to express the social consequences of modernization and reflect aspects of both ecological modernization and the risk society. In terms of ecological modernization, the nuclear industry exemplifies the state/market nexus for, although the industry is substantially state controlled (and in some cases state-owned), reprocessing, clean-up and waste management on these nuclear sites is undertaken by various forms of partnership with private sector companies and contractors while the state provides financial guarantees. The regulatory framework has secured tighter environmental controls (the ecologization of production) through emissions limitation, decontamination and waste minimization through packaging, vitrification and encapsulation of complex waste streams. And there is increasing evidence of the participation of environmental groups in the development of waste management policies and in developing a culture of openness. Despite these features of ecological modernization the nuclear industry can never be fully modernized ecologically in that risk, however minimized, will always persist; the market is ultimately dependent on state support and guarantees for the long term; and participation by environmental movements is always likely to be constrained by the dominance of expert systems of control which are inherently obscure or secretive. Ecological modernization, as interpreted in this essay, has little to say about the social, cultural and political processes which shape communities of disadvantage such as nuclear oases.

By contrast, the nuclear industry appears to be the archetype of the risk society, an activity dominated by experts, where risk is pervasive and enduring, and which, through reflexive recognition of powerlessness, creates a culture of fatalistic acceptance. Nuclear oases confront both the technological and the social attributes of the risk society. The thesis fails to identify in any empirical way the social importance of the inequality that arises from the uneven distribution of risk which exposes communities to localized environmental degradation. However, at a more abstract level risk society's concern with the notion of reflexivity does suggest possibilities for social change which can be identified more precisely in the nuclear oases.

The problem with both ecological modernization and the risk society is the emphasis on processes of modernization and the assumptions about social change which flow from it. While the nuclear industry is undoubtedly an exemplar of former technological modernization it is an increasingly outmoded and bygone technology which, like other industries such as coal, steel and heavy engineering, has become itself traditional in contrast with the newer electronic and biotechnological industries which are the dynamic expressions of the contemporary age. This traditionalism is also present in a social sense. Within these communities forms of traditional integrating institutions persist whether in the form of the mechanistic solidarity of community or the organic solidarity created by the institutions of the welfare state or trade unions. These integrating institutions co-exist alongside the so-called individualization which Beck identifies as a consequence of modernization. Thus, the social conditions in nuclear communities are complex and have interesting implications for social and environmental change which escape the modernization theories.

Changing power relations

The nuclear oases are the product of a particular set of power relations. The industry's strength in these communities lies in the support it can expect from a dependent workforce as well as from the investment supplied by governments that recognize their responsibility for environmental clean-up. Typically, the economic leverage exerted by the industry combined with the threat of investment withdrawal has led governments to provide continuing support. For example, the need to support the West Cumbrian economy was a major reason for the decision, in 1994, to approve the commissioning of the new reprocessing facility (THORP) at Sellafield at a time when the justification on commercial grounds was being questioned (Sadnicki et al, 1999). Economic considerations as well as environmental justification are responsible for heavy investment (of around US$6 billion annually) into clean-up of the Hanford site where the nuclear industry supports the remote Tri-Cities communities of around 100,000 people. Conversely, in areas where the nuclear industry is not fully established the local community may possess the resources to delay or prevent the intrusion of an activity that is perceived as a threat to the environment and to the existing economy. This explains the success of communities at greenfield sites in resisting the nuclear industry, for example at Gorleben where proposals for a reprocessing plant were withdrawn and those for a repository deferred and where the interim storage facility has been the focus of widespread and sustained protest (Blowers and Lowry, 1997).

The power relations within these communities and between them and the outside world are changing. There is a defensive coalition between capital and labour whose interests converge in maintaining employment and investment. But the support of workers is not unconditional and may be ambiguous since these communities also live with the potential risks and unpredictable hazards associated with the industry. Loeb (1986, p253), describing Hanford, comments that 'those most immersed in the atomic world are most insistent about denying its implications'. Zonabend (1993, p120) witnessed at La Hague a 'selective blindness' which 'helps the nuclear worker of today to resign himself to a fate that has always existed'. At Sellafield a similar pragmatic passivity was encountered where self-deprecation 'at being the only place in Britain prepared to accept nuclear waste was mitigated by a countervailing sense of pride' (Wynne et al, 1993, p58).

These analyses strongly suggest that such communities are not wholly individualized nor fragmented as modernization theories might have us believe. There are conflicts within them, to be sure, notably between those dependent on the nuclear industry and either pre-existing, traditional communities (of farmers and fishermen at La Hague or Sellafield and a strong rural culture in Gorleben) or adventitious groups developing their own sense of local territorial identity (for example, incomers seeking the attractions of the English Lake District in which Sellafield is situated). These add to the social complexities which have to be understood. But, taken together, these traditional integrating elements based on employment and location create resources of influence and power which, when exerted, can have a determining influence on the local economy and environment.

The decline of the industry, paradoxically, is accompanied by a shift in power relations, which ensure its continued existence in the nuclear oases. It is not simply a case of the community depending on the industry but of the industry becoming increasingly dependent on its surviving oases. This interdependence is buttressed by external power relations which increase the difficulty for the industry of finding sites elsewhere and which raise the awareness that nuclear communities bearing disproportionate risk for general benefit are entitled to support by society at large. We do not need the opaque notion of reflexivity to tell us this basically self-evident feature of power. As circumstances change so opportunities open up and will be seized upon.

The example of nuclear oases draws attention to features of social change in relation to the environment which are absent in the grand narratives of ecological modernization and risk society. First, it shows that environmental risk is unevenly distributed and that the adverse consequences of technological modernization are spatially concentrated. This feature of environmental inequality tends to be overlooked by ecological modernization which is primarily concerned with environmental improvement rather than degradation, and by risk society which emphasizes the so-called 'democratic' nature of risk. Second, the theoretical focus on modernity understates the persistence of traditional forms of integration which constitute the survival of community. Ecological modernization is largely silent on this social question while risk society focuses on individualization as the prevailing condition and reflexivity as its consequence. Third, the nuclear oases indicate the changing nature of power relations both within communities and in relation to external forces. Ecological modernization tends to take a pluralistic view of power as shared between participants (market, state and environmental actors) in partnership while risk society emphasizes the controlling power of experts. Fourth, the complexity and interconnectedness of social relations is underlined in the nuclear oases and suggest that specific circumstances in specific places must be understood to explain outcomes. Again, these social matters are not an integral element in theories which seek to provide more general explanations of societal change in response to environmental problems.

These features – inequality, community, local power relations and interconnectedness – are a necessary complement to broader theoretical ideas. They provide a local focus with wider implications and are the basis for understanding the problem of environmental justice. In the concluding section I shall look at the social and political implications of the analysis and the questions they pose for contemporary theories of modernisation.

INEQUALITY, COMMUNITY AND THE PRINCIPLES FOR ENVIRONMENTAL JUSTICE

Power and the periphery

The example of nuclear oases makes it clear that inequality must be a key consideration in society's search for solutions to its environmental problems.

Inequality needs to be tackled at various levels. At the political level the contemporary emphasis on ecological modernization has tended to encourage the access of particular groups or 'stakeholders' while others, notably already disadvantaged communities, tend to be politically excluded. These communities may lack the capacity, motivation or resources to defend against environmental deterioration; in any case the environment is likely to be lower on their list of priorities than the need for economic security. As a result such communities may manifest the condition of 'negative fatalism' described by Beck. On the other hand, as our examples indicate, passive acceptance may be transformed into a more active resistance against environmental risks for two reasons. One is that growing environmental awareness influences the disadvantaged as well as the population in general. Second, high-risk and polluting industries such as the nuclear industry need the support of local communities. Consequently, these communities can exert greater leverage in their demands for environmental improvements or for compensation for bearing the blight and anxiety of environmental risk.

The power available to these communities comes both from within and without. Internally, their growing (reflexive) awareness of their social and environmental condition emanates from the integration arising from shared experience of economic vulnerability and, at the same time, a consciousness that they are bearing the burden of risk on behalf of society. Externally, they may be supported by environmental movements, which, while opposing any further development of the nuclear industry, recognize the need to clean up its legacy. Together communities and movements combine using the mechanisms available in civil society (lobbying, protest, participation, referendum, voting and so on) to give visibility to the cause of environmental justice. The complexity of the interactions between the local interests of community and industry and the wider public interest in environmental protection have produced a shift in power relations.

This change has been accompanied by a lessening of the peripheral characteristics of these communities. While they remain geographically isolated, environmentally degraded and economically dependent on a single source of employment, they are changing socially. No longer culturally defensive and acquiescent, communities such as Hanford and Sellafield assert their right to environmental as well as economic justice while Gorleben demonstrates the cultural condition of active resistance. These communities remain peripheral and disadvantaged but they are not ignored.

Nuclear oases are, in certain respects, a special case. Nuclear workers are relatively well paid, well housed and enjoy a range of facilities in health, education, infrastructure and welfare that contrast with the poverty, dereliction, neglect and disadvantage that is the common experience of communities suffering environmental injustice. Their spacious and often prosperous surroundings seemingly contradict their hazardous occupation. They are the modern examples of the 'company towns' of the early industrial age. But, they are also places whose future is uncertain, where risk is a permanent condition, and which are responsible for managing a problem unwanted elsewhere. Society at large is thereby indebted to the nuclear oases and they are, therefore, society's problem. The uneven distribution of nuclear risk has become not simply a moral problem, but a practical issue that must be tackled by seeking equitable solutions.

Equity as a moral and practical issue

The principle of equity applied to these communities has social, political and environmental dimensions. At the *social* level equity can be applied by means of compensation. The intention is to 'offset any stigma, perceived or actual' (Richardson, 1999, p4) associated with the activity. This involves a range of economic measures such as relief of taxation, regeneration, diversification and infrastructure provision as well as the provision of community facilities. Compensation is being actively pursued in many countries though mainly as an incentive to attract volunteers for siting of future disposal facilities (Richardson, 1997, 1998, van den Berg and Damveld, 2000). In practice, it applies also to existing sites and places such as Hanford and La Hague have benefited from a range of measures.

In its *political* sense, equity involves a commitment to openness, the provision of information, participation in decision-making and, in some instances, the provision for a veto over decisions affecting the community. These various measures of community empowerment are necessary if trust, the essential ingredient for any agreement between a community and the industry, is to be secured. The application of a right of veto implies that agreement may be forfeited which creates uncertainty in decision-making. But, as Richardson observes, 'to withdraw that right, or threaten to suspend it in the "national interest", will only serve to destroy any trust and mutual respect that may develop between the involved stakeholders' (1997, p26). Again, these principles are being introduced as inducements to volunteers but, in the established nuclear oases, the veil of secrecy has been, to varying degrees, removed and the local community and environmental groups are routinely consulted over decisions though rarely have any veto over them.

Environmental equity, applied to nuclear oases, covers the minimization of risk through such matters as site remediation and decontamination, the decommissioning of redundant facilities, monitoring and the management of wastes through improved storage, packaging and immobilization measures. In addition it also covers more cosmetic measures such as landscaping, the provision of recreational areas and nature reserves and other environmental enhancements. While increasing attention is paid to the environment, the technical and scientific problems facing clean-up are formidable. At Hanford the problem of leakage from tanks carrying liquid high-level wastes potentially threatens the catchment of the Columbia River and at Sellafield the management of the accumulated historic wastes on the site is the most difficult clean-up task. The cost of sustaining programmes of environmental protection and enhancement over long timescales constitutes a challenge for intergenerational equity which is the key issue in the sustainable management of radioactive wastes.

The case of the nuclear oases indicates that sustainable development is not simply an environmental problem but also a social and political one. Environmental inequality has a social context and mitigation of risk and pollution is a moral issue. But, more than that, it creates communities which are increasingly conscious of the harmful consequences of uneven development. The nuclear oases, albeit a special case, suggest that the risks they experience cannot continue to be borne without recognition by the wider society. The dependence of the nuclear industry and society at large on these communities provides them with the power to demand and receive compensation. These social conflicts and shifts in power relations at the local level are missed by discourses which, like ecological modernization, herald an age of consensus or, like risk society, concentrate on the global consequences of modern megarisks. While the nuclear oases remain disadvantaged, at least their claims to environmental justice are being acknowledged. And, although their experience is unlikely to be matched in the poverty stricken and polluted communities in the developed and developing world, they provide a glimpse of the potential for environmental and social change that derives from shifting power relationships. It is this evidence of power derived from community integration and emanating from a sense of injustice that provides a compelling theoretical and practical critique of the dominant discourses of modernization.

REFERENCES

Beck, U (1992) *Risk Society: Towards a New Modernity,* Sage, London

Beck, U (1995) *Ecological Politics in an Age of Risk,* Polity Press, Cambridge

Beck, U (1996) 'Risk society and the provident state' in S Lash, B Szerszynski and B Wynne (eds) *Risk, Environment and Modernity: Towards a New Ecology,* Sage, London

Beck, U (1998) *Democracy Without Enemies,* Polity Press, Cambridge

Benton, T (ed) (1996) *The Greening of Marxism,* Guilford, New York

Blowers, A (1999) 'Nuclear waste and landscapes of risk', *Landscape Research,* vol 24 (3), pp241–64

Blowers, A (2000) 'Landscapes of risk: Conflict and change in nuclear oases' in J Gold and G Revill (eds) *Landscapes of Defence,* Routledge, London (forthcoming)

Blowers, A, Lowry, D and Solomon, B (1991) *The International Politics of Nuclear Waste,* Macmillan, London

Blowers, A and Leroy, P (1994) 'Power, politics and environmental inequality: a theoretical and empirical analysis of the process of "peripheralisation"', *Environmental Politics,* vol. 3 (2), summer, pp197–228

Blowers, A and Lowry, D (1997) 'Nuclear conflict in Germany: The wider context', *Environmental Politics,* vol 6 (3), autumn, pp148–55

Christoff, P (1996) 'Ecological modernization, ecological modernities', *Environmental Politics,* vol 5 (3), pp476–500

Crouch, C (2000) *Coping with Post-democracy,* Fabian Society, London

Dobson, A (1990) *Green Political Thought,* Unwin Hyman, London

Dryzeck, J (1997) *The Politics of the Earth: Environmental Discourses,* Oxford University Press, Oxford

Eckersley, R (1992) *Environmentalism and Political Theory: Towards an Ecocentric Approach,* UCL Press, London

Glasbergen, P (1998) 'The question of environmental governance' in P Glasbergen (ed) *Co-operative Environmental Governance: Public-Private Agreements as a Policy Strategy,* Kluwer Academic Publishers, Dordrecht

Gould, K, Scnaiberg, A and Weinberg, A (1996) *Local Environmental Struggles: Citizen Activism in the Treadmill of Production,* Cambridge University Press, Cambridge

Gorz, A (1980) *Ecology as Politics,* South End Press, Boston

Hajer, M (1995) *The Politics of Environmental Discourse: Ecological Modernization and the Policy Process,* Oxford University Press, Oxford

Harvey, D (1996) *Justice, Nature and the Geography of Difference,* Basil Blackwell, Oxford

Jacobs, M (1991) *The Green Economy,* Pluto Press, London

Janicke, M (1996) 'Democracy as a condition for environmental policy success: The importance of non-institutional factors' in W Lafferty and J Meadowcroft (eds) *Democracy and the Environment,* Edward Elgar, Cheltenham

Leroy, P (2000) 'Political modernisation and the renewal of environmental policy arrangements', unpublished paper

Leroy, P and van Tatenhove, J (2000) 'Political modernization theory and environmental politics' in F Buttel and G Spaargaren (eds) *Environment and Global Modernity,* Sage, London

Loeb, P (1986) *Nuclear Culture: Living and Working in the World's Largest Atomic Complex,* New Society Publishers, Philadelphia

Low, N and Gleeson, B (1998) *Justice, Nature and Society: an Exploration of Political Ecology,* Routledge, London

Mol, A (1995) *The Refinement of Production,* van Arkel, Utrecht

Mol, A (2001) 'Ecological modernisation and the global economy', paper presented at the 5th Nordic Environmental Research Conference on The Ecological Modernisation of Society, Aarhus

Mol, A and Sonnenfeld, D (2000) 'Ecological modernisation around the world: An introduction', *Environmental Politics,* vol 9 (1), spring, pp3–14

Mol, A and Spaargaren, G (2000) 'Ecological modernisation theory in debate', *Environmental Politics,* vol 9 (1), spring, pp17–49

Pepper, D (1993) *Eco-Socialism: From Deep Ecology to Social Justice,* Routledge, London

Potter, D (1996) 'Non-governmental organisations and environmental policies' in A Blowers and P Glasbergen (eds) *Environmental Policy in an International Context: Prospects,* Arnold, London

Richardson, P (1997) *Public Involvement in the Siting of Contentious Facilities: Lessons from the radioactive waste repository siting programmes in Canada and the United States, with special reference to the Swedish Repository Siting Process,* Swedish Radiation Protection Institute, Stockholm

Richardson, P (1998) 'A review of benefits offered to volunteer communities for siting nuclear waste facilities', paper prepared for Swedish National Co-ordinator for Nuclear Waste Disposal

Richardson, P (1999) 'Development of retrievability plans', prepared for Swedish National Co-ordinator for Nuclear Waste Disposal

Sadnicki, M, Barker, F and MacKerron, G (1999) *THORP: the Case for Contract Renegotiation,* Friends of the Earth, London

Sandbach, F (1980) *Environment, Ideology and Policy,* Basil Blackwell, Oxford

Schnaiberg A (1980) *The Environment: From Surplus to Scarcity,* Oxford University Press, Oxford

Stretton, H (1976) *Capitalism, Socialism and the Environment,* Cambridge University Press, Cambridge

van den Berg, R and Damveld, H (2000) *Discussions on Nuclear Waste: A Survey on Public Participation, Decision-making and Discussions in Eight Countries,* Laka Foundation, Amsterdam

van Tatenhove, J, Arts, B and Leroy, P (2000) *Political Modernisation and the Environment: The Renewal of Environmental Policy Arrangements,* Kluwer Academic Publishers, Dordrecht
Wynne, B, Waterton, C and Grove-White, R (1993) *Public Perception and the Nuclear Industry in West Cumbria,* Centre for the Study of Environmental Change, Lancaster University
Zonabend, F (1993) *The Nuclear Peninsula,* Cambridge University Press, Cambridge

Part 2

Challenges

For reasons stated in the Introduction, and those outlined by other authors in this book, and finally, those in our Conclusion, we are under no illusions as to the magnitude of the task inherent in identifying, scoping, and finally colonizing the common ground between sustainability and environmental justice. Not only are most of the actors different, but so are the language(s) spoken, the change strategies and the timescales. However, the goal of a more dynamic, inclusive, diverse and effective movement is surely worth pursuing.

Dobson, in Chapter 4, comes to 'the conclusion that social justice and environmental sustainability are not always compatible objectives'. He argues that academics, policy-makers, practitioners and activists who think they are, base their assertions on a desire for neat solutions, rather than on hard analysis or empirical evidence. From a political point standpoint, he sees great benefits in firmly linking environmental sustainability and social justice, which he sees as a replay of the old question: 'can there be a rapprochement between 'red' and 'green'?' where 'social justice' stands for 'red' and 'environmental sustainability' stands for 'green'".

* * *

Chapter 4

Social Justice and Environmental Sustainability: Ne'er the Twain Shall Meet?

Andrew Dobson

I have come to the reluctant conclusion that social justice and environmental sustainability are not always compatible objectives. Assertions to the contrary are more often, I think, based on wishful thinking than on clear-sighted analysis or hard empirical evidence. This conclusion is 'reluctant' because from a political point of view I can see tremendous benefits in marrying environmental sustainability and social justice. These terms act as ciphers for a debate that is as old as the contemporary environmental movement itself: can there be a rapprochement between 'red' and 'green'? In the terms of this chapter 'social justice' stands for 'red' and 'environmental sustainability' stands for 'green'. Ever since the collapse of 'actually existing socialism' and the practically concomitant rise of the contemporary environmental movement, activists and theorists on the progressive left have been assessing the possibility of green and red coming together in a way that might challenge the global hegemony of liberal capitalism. My own view, based both on theoretical considerations and on observation of real-life attempts at coalition initiatives – both in the highest reaches of government as in contemporary Germany and in the movement politics of civil society – is that rapprochements will only ever be temporary and transient. This is simply, yet crucially, because reds and greens have fundamentally different objectives. Indeed I think that to expect socialists and environmentalists to form a common cause is as unrealistic as to expect liberals and socialists to make a common cause: the differences between reds and greens are of the same structural ideological order as those between liberals and socialists. This is to say that the differences between them are not merely tactical, but strategic: their objectives differ in fundamental ways. It is against this broad background that the debate regarding the compatibility of social justice and environmental sustainability must be seen.

I suggested above that this debate has empirical and theoretical dimensions. Most of my work on this topic has been theoretical, and most of what follows is theoretical too. One thing that struck me very forcibly during my research, though, was how very little empirical work had been done on the relationship between policies for social justice and environmental sustainability. I take it that the starting point for those who regard them as compatible objectives is either that social justice is a precondition for environmental sustainability or – more strongly – that social justice *produces* environmental sustainability. In the first place, though, the proposition might look intuitively stronger the other way round – that environmental sustainability is a precondition for social justice. (But then this is surely a truism: environmental sustainability is a precondition for anything and everything – including, sadly, for social injustice). Second, and more germane to my point here, what is striking is how little empirical work has been done to test the above-mentioned proposition. In all of the research I did in the two years leading up to publication of *Justice and the Environment* never once did I come across a piece of empirical research aimed at confirming or denying the proposition that social justice is a precondition for environmental sustainability. Instead, the proposition was – and is – simply assumed to be true. I found no *theoretically informed* empirical research on this topic. By this I mean research that does not take for granted particular understandings of what social justice and environmental sustainability might mean, but takes as its starting point the fact that these terms are *contested* and have, therefore, no determinate meanings.

Effective empirical research would have to examine the compatibility between social justice and environmental sustainability across the full range of their meanings rather than just one or two of them. It is instructive to remember, for example, that for many people – liberals especially – a socially just situation is one in which we expect to find significant inequalities in the distribution of society's 'goods and bads', so long as these unequal distributions are fairly arrived at – for example through some people 'deserving' to have more goods than others. Is this the kind of social justice that compatibility enthusiasts have in mind when they argue that social justice leads to environmental sustainability? Would they really want to endorse inequality as a route to sustainability? I think not. Yet unequal shares – 'deserved' – is one version of what social justice might entail. So it would at least be incumbent on our putative empirical researchers to be clear about what type of social justice they were referring to in testing the proposition that social justice is a precondition for – or leads to – environmental sustainability. This, though, as I say, is food for future thought. At present there seems to be no empirical research of any kind, let alone theoretically informed research, designed to confirm or deny the compatibility thesis.

This is an extraordinary thing when one thinks, for example, of the sustainable development movement, based as it is on the notion that social justice (of a broadly egalitarian kind, it should be noted) is a precondition for environmental sustainability. Might we not have expected such a widely endorsed movement to be more firmly grounded in theoretically informed empirical evidence? If such evidence exists I have not come across it in great quantities. Likewise the 'environmental justice' movement which is of such moment in the

US, and of which much is made in the rest of this book. As is well known, the environmental justice movement has proved conclusively that the distribution of environmental 'bads' in the US (landfill sites, toxic waste dumps and the like) are disproportionately visited upon communities of poor people and/or people of colour. The empirical evidence to back up this conclusion is sophisticated, overwhelming and vast. The objective of the environmental justice movement is to have environmental 'bads' more fairly distributed across the country, so that wealthier communities have their fair share of landfill sites. As far as justice goes this is a reasonable agenda, but given the terms of reference of this chapter and the book as a whole we are also interested in knowing whether the agenda will also 'produce' environmental sustainability. Is the environmental justice movement also a movement for environmental sustainability? If it is, then arguments for the compatibility of justice and sustainability are strengthened. If not… well, we need to know that too.

The working hypothesis – to be confirmed or denied – might be that the environmental justice movement is not a movement for environmental sustainability. Let us assume, here, that 'environmental sustainability' refers to a reduction in the aggregate tonnage of waste consigned to landfill sites. On the face of it the environmental justice movement's objective is not to reduce aggregate tonnage but to have existing tonnage more fairly distributed across wealth and racial cleavages. This suggests that the sustainability and justice movements have rather different objectives – hence the working hypothesis referred to at the beginning of this paragraph.

There are, though, at least three ways in which this hypothesis might be resisted. First, it might be claimed that my characterization of the environmental justice movement is not a characterization but a caricature, and that environmental justice campaigners are, actually and already, campaigners for environmental sustainability objectives. At the very least, though, this would be a disputed counter-claim. Two contributors to a recent book on the subject, for example, came to radically different conclusions regarding the nature of the environmental justice campaign. On the one hand David Camacho writes that 'environmental justice advocates are not saying, "Take the poisons out of our community and put them in another community". They are saying that NO community should have to live with these poisons' (Camacho, 1998, p12). This suggests that environmental justice campaigners have, simultaneously, a sustainability agenda. On the other hand, Stephen Sandweiss argues that, 'For most environmental justice activists, the problem is one of distributional inequity' (Sandweiss, 1998, p35). Who is right? Neither, or both, perhaps. My point is simply that we cannot call between them until we have a set of systematic empirical studies on which to base our conclusions. In the meantime, the following parenthetical comment from David Schlosberg, and that fact that it is in a parenthesis, is very significant: 'Through public participation, [environmental justice] activists and communities may accomplish both more equitable distribution of environmental risks (or, more ideally, a decrease in toxic exposure and environmental risks for all) and the recognition of various communities, cultures, and understandings of environmental health and sustainability' (Schlosberg, 1999, p13).

The second counter-hypothesis is that it might be argued that individuals who participate in the environmental justice movement become politicized in a sustainability direction – that is, that their involvement has the transformative effect of shifting their objective from sharing out landfill sites more fairly to campaigning for a reduction in the aggregate production of waste. Is this what actually happens? I don't know – but I believe that no one else does either. I have been unable to find any empirical studies of the political development of individual environmental justice campaigners or communities that might prove or disprove this counter-hypothesis – and others much more versed than me in this literature tell me that none exists as far as they are aware.[1]

The third counter-hypothesis is that a fairer distribution of environmental 'bads' to wealthier communities would encourage those communities to take more robust action to reduce aggregate levels of waste. As long as they are able to pass off the toxic by-products of the production process to disadvantaged communities, the story goes, they have no incentive to campaign for reduced waste levels. This counter-hypothesis must remain just that – a hypothesis – unless and until there is a fairer distribution of environmental 'bads'. At that point we would hope for the empirical studies that would enable us to reach authoritative conclusions regarding the compatibility between the justice and sustainability movements. Judging by past and present experience, however, the omens regarding the possibility of such studies do not look good.

The US environmental justice movement is, therefore, simultaneously a site for extravagant claims regarding the compatibility of the justice and environmental agendas and a black hole as far as empirical studies designed to substantiate those claims are concerned. My view is that the credibility of those claims is at stake and that the onus is now on the claims' sponsors to provide some foundations for their brave-sounding hopes and aspirations. Might I say once again that I would be delighted to see these hopes substantiated? My present scepticism is based largely on a systematic analysis of the terms 'environmental sustainability' and 'social justice' and my conclusions regarding their respective objectives. It is to this I now turn.

Let me begin this section by reiterating the point made earlier that 'social justice' and 'environmental sustainability' are contested terms, each of which has multiple meanings. It is useful here to deploy the standard distinction between a 'concept' and a 'conception'. Social justice and environmental sustainability are both 'concepts', and at this level there is likely to be relatively little disagreement as to their meaning. Let us say, for the moment, that environmental sustainability is about sustaining some feature(s) of the 'natural' environment into an indeterminate future, and that social justice is about the fair distribution of benefits and burdens in a community of justice. It is immediately clear that these conceptual 'definitions' conceal a minefield of indeterminacies. As for environmental sustainability, which features of the 'natural' environment are to be sustained into the future? For how long? For whose benefit? In the context of social justice, what does 'fair' mean? How do we decide? What constitutes a 'benefit' and a 'burden'? Who or what should be included in the 'community of justice'? Different answers to these questions produce different conceptions of environmental sustainability and social justice, therefore it is at this level, rather than at the level of the concepts themselves, that contestation takes place. Even at this early stage of the argument it should be clear how complex the question of whether social justice and environmental sustainability are compatible objectives is. We must ask ourselves *which* social justice? And *which* environmental sustainability? Given this multiplicity of meanings of both terms, it might not be unreasonable to assume that *some* conceptions of social justice will be compatible with *some* conceptions of environmental sustainability, but not all. Once again, this kind of hypothesis can only be tested effectively in the crucible of empirical practice.

So the beginnings of an authoritative empirical examination into the relationship between social justice and environmental sustainability lie in recognizing the multiplicity of their meanings. In one sense, however, this seems a disabling thought. How could the empirical researcher ever be sure that all the meanings had been captured in her or his research? Fortunately, the theorist can be of further assistance. Having 'deconstructed', so to speak, the terms environmental sustainability and social justice, it is possible to 'reconstruct' them in a manner that produces a relatively small yet reasonably comprehensive number of conceptions of both terms which the empirical investigator can use. Let me explain how this works.

The crucial insight is that there are some questions to which any conception of environmental sustainability or social justice must have an answer. These questions will give us the 'scope' of each term, the conceptual territory in which they move, their 'grammar'. I hinted at these questions earlier. For example, in regard to environmental sustainability, which features of the 'natural' environment are to be sustained into the future? And for social justice, what does a 'fair' distribution consist in? The next step is to see that there is a determinate number of answers to these questions, and that each answer gives us a particular conception of justice or sustainability. I have followed through this analytical agenda in detail in *Justice and the Environment*, but it is worth outlining here the broad conclusions I reach so that I can be more explicit as to the use to which they might be put by the empirical researcher.

It turns out that the most important question for environmental sustainability is 'what is to be sustained?' Three radically different types of answers can be given to this question, each of which puts a very different spin on our organizing question of whether social justice and environmental sustainability are compatible. For convenience I present this question and three family-related answers to it in a table below (see Figure 4.1). These answers 'emerge' from the vast and ever-expanding literature on environmental sustainability and sustainable development that I surveyed.

Our question of whether social justice is compatible with environmental sustainability now becomes three questions:

1   Is social justice compatible with the sustaining of critical natural capital?
2   Is social justice compatible with the sustaining of biodiversity?
3   Is social justice compatible with the sustaining of the value of natural objects?

	Answer 1	Answer 2	Answer 3
What is to be sustained?	Critical natural capital[2]	Biodiversity	The value of natural objects[3]

Figure 4.1 *Types of environmental sustainability*

Without going into too much detail, the answer to each of these three questions is likely to be rather different – giving the lie to the notion that there is a determinate answer to the question of whether social justice and environmental sustainability are compatible objectives. In the case of the first question, our intuition is that not only is social justice compatible with the sustaining of critical natural capital, but that social justice *demands* the sustaining of critical natural capital and its fair sharing around the community of justice. Put differently we might say that fair shares in the parts and processes of nature that are critical for the maintenance of human life is actually an aspect of what social justice is about. In this context, then, there seems to be a rather precise relationship between justice and sustainability.

This is less clear, though, in the context of the second question – whether social justice is compatible with the sustaining of biodiversity. Social justice certainly does not seem to *demand* the sustaining of biodiversity, except of course where biodiversity is critical for the maintenance of human life, but this merely takes us back to the first question. It cannot be argued that *all* biodiversity is critical for the maintenance of human life since this would mean suggesting that the survival of the human species depended on maintaining the present number of all species. This cannot be right: we all know that species numbers are in decline, but this has had no discernible effect so far on the survival of the human race.

But even if social justice does not demand the sustaining of biodiversity, might it not at least be compatible with it? Perhaps, but perhaps not. There seems no reason why policies designed to increase fairness in the human community would result in the maintenance of present levels of biodiversity. Indeed, there seems here not only to be no policy connection, but no conceptual connection. Policies for social justice influence levels of biodiversity, of course, but not necessarily in such a way that more of the former means more of the latter. This is not to say that social justice is incompatible with biodiversity, either; merely that they are in a relationship of trade-off rather than mutual advantage.

The same is true with regard to the last of the three questions, whether social justice is compatible with the sustaining into the future of the value of natural objects. Imagine a venerable oak tree, stuffed full with 'natural value'. Will its chances of survival be enhanced, decreased or unaffected by policies for social justice? Surely all three are possibilities: it depends on the particular circumstances and the particular policy. Once again we see that there is no determinate relationship between justice and this conception, at least, of sustainability: the relationship between them is contingent rather than necessary, and while compatibility is possible it cannot be guaranteed.

We are now in a position to say a little more about the 'theoretically informed' empirical research on the relationship between social justice and environmental sustainability that I was calling for earlier. At the very least, we should expect our empirical researcher to test for compatibility, bearing in mind the three possible understandings of environmental sustainability found in Figure 4.1. So if, for example, and returning to an earlier topic in this chapter, the researcher's question was: 'is the US environmental justice movement also a movement for environmental sustainability?', we would expect the examination to embrace the environmental justice movement's relationship with the maintenance of critical natural capital, of biodiversity, and of 'natural value'. I have no idea what the answers would be, but I very much hope that one day someone will find out.

But of course even this is only half the story. Social justice is as contested a concept as environmental sustainability, and while we know what it is 'about' – the fair distribution of benefits and burdens – there are multiple answers to the question of what it 'is'. I suggested that one way into this multiplicity in the case of environmental sustainability was to ask: 'what are the questions to which *any* conception of environmental sustainability would have to have an answer?,' and to offer a series of answers drawn from the environmental sustainability literature. This gave us the three conceptions of environmental sustainability outlined in Figure 4.1, based on the common question 'what is to be sustained?'

We can perform the same operation on the concept of social justice. There are a number of questions to which any theory of social justice would have to have an answer, such as: 'who or what belongs to the community of justice?', 'should the justice of a particular distributive outcome be judged by the fairness of the procedures that led to it, or by the consequences of those procedures?', and 'should a theory of distributive justice be impartial in respect of views of the "good life" or should it act in the service of such views?' The different answers to these questions lead to different conceptions or theories of social justice, just as they do in the case of environmental sustainability. This evidently introduces a further degree of complexity into our original question of whether social justice is compatible with – or, even, is functional for – environmental sustainability. Now we would need to ask, for example, which of procedural or consequentialist theories of justice is the more compatible with environmental sustainability – that is, all three conceptions of environmental sustainability outlined in Figure 4.1. To illustrate the task that confronts our theoretically informed empirical researcher, let us assume that there are three identifiable conceptions of social justice (SJ^1, SJ^2, SJ^3), just as there are three identifiable conceptions of environmental sustainability (ES^1, ES^2, ES^3). Now the question of whether social justice is compatible with environmental sustainability is in effect *nine* questions (is SJ^1 compatible with ES^1 and ES^2 and ES^3, and SJ^2 with ES^1 and ES^2 and ES^3?, and so on).

There is no space here to see how this might work out in detail, although I attempted to do so in *Justice and the Environment*. It is, though, possible to illustrate the likely outcome of such an investigation by taking just one of the questions to which any theory of social justice must have an answer, and offering three alternative answers. The question is: 'what should the principle of distribution be?' Three venerable answers emerge from the history of social justice theory and practice: need (you get what you need), desert (you get what you deserve), and historical entitlement (eg you get what you inherit). In tabular form the picture looks like this:

	Answer 1	Answer 2	Answer 3
What is the principle of distribution?	Need	Desert	Entitlement

Figure 4.2 *Types of social justice*

In a manner similar to that following Figure 4.1, our question of whether social justice is compatible with environmental sustainability is now three questions:

1   Is distribution according to need compatible with environmental sustainability?
2   Is distribution according to desert compatible with environmental sustainability?
3   Is distribution according to entitlement compatible with environmental sustainability?

But we know that there are three conceptions of environmental sustainability too, so for completeness' sake the nine resulting questions – amounting to a semi-comprehensive research agenda – are collected in an endnote.[4] Once again there is no space here to discuss these nine possible relationships, but I can easily illustrate the kind of effect that embracing the multiplicity of meanings of social justice can have on our compatibility question.

Let me recall, then, a remark I made in commenting on Figure 4.1. There I said that in the context of environmental sustainability understood as the sustaining of critical natural capital, 'our intuition is that not only is social justice compatible with the sustaining of critical natural capital, but that social justice *demands* the sustaining of critical natural capital and its fair sharing around the community of justice'. *Now it should be clear that this bold conclusion is reached at the cost of assuming that the principle of social justice in question is 'need'*. By its very nature critical natural capital is a fundamental human need, and therefore 'distribution according to need' suggests that this basic need be equally satisfied. But what if the principle of social justice in question is not need but, say, desert? Can we say that critical natural capital is equally *deserved*? This seems counter-intuitive, in part because the principle of desert is usually used to justify inequalities of distribution, but also because the term has no purchase on the thing being distributed. This is what philosophers call a 'category mistake', and I shall follow up the important implications of this point in the paragraph after next.

For the moment, though, it should be clear that our apparently one-dimensional, if still complex, question of whether social justice is compatible with environmental sustainability has many more ramifications than appear at first sight. This is much more than a merely theoretical interest. The implications for policy-makers are profound. The view in government in the UK at present, for instance, seems to be that policies for social justice, in the international arena, will result automatically in greater environmental sustainability. This belief is exemplified in the activities of the Department for International Development which has plenty to say about international justice but is virtually silent on environmental sustainability, in the mistaken belief that more of the former will necessarily produce more of the latter. Similarly, a more nuanced understanding of the relationship between social justice and environmental sustainability would make for more realistic coalition-building in 'civil society'. Spats between the Sierra Club and the environmental justice movement in the US would seem less surprising, for example, and disagreements could be more carefully negotiated with a view to building longer lasting, more secure and more realistic coalitions.

These coalitions are difficult to build because, I think, social justice and environmental sustainability speak different languages and have different objectives. There are three ways in which the languages of social justice and environmental sustainability might be related: (a) the environment as something to be distributed, (b) justice as functional for sustainability, and (c) 'justice to the environment'. The first of these discourses is that of the environmental justice movement. We noted above that all theories of justice have to have answers to certain common questions, and the one we focused on above was that of the principle of distribution (according to need, desert, or entitlement). Another question to which all theories of social justice must have an answer is: what is to be distributed? There are all sorts of possible answers: money, calories, political posts, for example. The answer that environmental justice gives to this question of what is to be distributed is: landfill sites and toxic waste dumps. Landfill sites and toxic waste dumps are distributed around the country in ways which the environmental justice movement considers to be unfair, and the movement's political project is organized around making people aware of this and asking that wealthier communities take their fair share of the rubbish.

As I argued above, this project – on its own and in the absence of other measures – will not necessarily result in environmental sustainability (although I also suggested three ways in which this assertion might be resisted). There is no guarantee that dividing up landfill sites more fairly between rich and poor communities will result in an overall decrease of the tonnage of waste consigned to such sites. The environmental justice project will not necessarily, in other words, merge seamlessly with the second discourse to which I referred just now, the one in which 'justice is functional for sustainability'. This is just one example of the general status of this putative functional relationship, which is that the jury is out. The answer to the question of whether justice is functional for sustainability is: it depends. It depends – as I explained above – on the type of justice we are talking about and on the conception of sustainability at stake. I also suspect that it depends on quite specific empirical conditions too, but I am not qualified to comment knowledgeably on that, especially in the absence of the raft of empirical studies on which we might base our conclusions. Let me put this another way: I have no doubt that under certain conditions, and given particular understandings of what 'justice' and 'sustainability' might mean, justice will be functional for sustainability. But I am equally certain that this will neither hold for all conditions, nor for all understandings of justice and of sustainability.

This leaves us with the third possible relationship between the discourses of justice and sustainability: the idea of 'justice to the environment'. According to this idea, the sustainability movement can piggyback on the powerful legitimating function that the discourse of 'justice' possesses to make its sustainability point. If, for example, we take sustainability to be about the sustaining of biodiversity (the second conception in Figure 4.1, above), then sustainability proponents would say that it is 'unjust' to reduce levels of biodiversity. And they would say that this not because such reduction is unjust for members of the human community, but because it is unjust to those species that are lost. From this point of view, other species (or some of them at least) are regarded as members of the community of justice, and it is as unjust to deprive them of the means of survival as it is to do so for human beings.

This is, to say the least, a controversial position. Paradigmatically, justice is predicated only of human beings, which is to say that only human beings can properly be regarded as distributors and recipients of justice. This is because, so supporters of this view say, only human beings possess the necessary qualities of sentience and rationality that can lead to a sense of justice in the first place, and because only human beings can act with reciprocity towards one another – a further condition for 'justice' to be predicated of the participating parties.

Some sustainability proponents have called into question the 'rationality, sentience and reciprocity' preconditions for entry into the community of justice. They will accept that only human beings can properly be regarded as *distributors* of justice, for obvious reasons, but will wonder whether the community of *recipients* might not be widened to include at least some non-human animals and even perhaps the non-human natural world more broadly. Are factory-farmed animals victims of injustice? And if so, what about deforested trees?

One way to admit beings other than human beings to the community of justice is to ascribe them rights, on the basis that justice is a matter of rights recognition (Feinberg, 1981). Then the question becomes one of whether we can ascribe rights to animals – and beyond. These debates nearly always involve developing a list of characteristics for membership, and those who would deny animals the possession rights or membership of the community of justice will say that animals do not possess those characteristics. Supporters of the idea of rights or justice for animals, though, will employ the metaphor of the continuum rather than the boundary, and point to the 'limit cases' where some human beings, through illness or other incapacity, find themselves possessing fewer of the characteristics that would admit them to the community of justice than is 'normally' the case. If, despite this, we grant them justice, it would be 'speciesist' to deny it to similarly equipped non-human animals.

Most of those who deny that justice can be predicated of the non-human world do not conclude that 'anything goes' as far as our relationship with it is concerned, but they argue that this relationship should be regulated by broader moral considerations, rather than by the more specific rules and conditions associated with justice. Thus one can oppose cruelty to animals, or deforestation, or species loss, without admitting animals, trees, or species other than the human one to the community of justice.

At the very least, then, we can see that the idea of 'justice to the environment' is fiercely contested. This in itself is no reason to abandon the idea. The environmental agenda in its radical guise is, after all, a transformative agenda – transformative of hearts and minds, and of the terms of political discourse. But even those who would argue strenuously for the notion of justice to the environment struggle to do so effectively. Nick Low and Brendan Gleeson, for example, argue that 'non-human nature, or at least aspects of nature, [does] have a claim on justice' (Low and Gleeson, 1998, p21). But their attempts to generate principles for what they call 'ecological justice' are more a series of moral statements than arguments written in the grammar and language of justice. Their first principle is that, '*every natural entity is entitled to enjoy the fullness of its own form of life.* Non-human nature is entitled to moral consideration.' (Low and Gleeson, 1998, p155; emphasis in the original). This is not a principle of distributive justice since it says nothing about the distribution of goods and bads within the community of justice. And their second principle of ecological justice is a descriptive rather than a prescriptive statement, and one from which no determinate justice principles can be drawn uncontroversially: '*The second principle is that all life forms are mutually dependent and dependent on non-life forms*' (Low and Gleeson, 1998, p155; emphasis in the original). The inverted commas in a subsequent comment seem to indicate the authors' own doubts about what they're doing: 'The ideas discussed above seem to us to indicate valid paths for opening up "justice" to non-human nature' (Low and Gleeson, 1998, p157).

The difficulties associated with working up the idea of 'justice to the environment' are based, I think, on the fact that while the sustainability and environmental justice movements are both concerned with the environment, they have very different objectives in respect of it. The sustainability movement is concerned to preserve and/or to conserve the non-human natural environment, while the environmental justice movement aims to divide up the worked environment (particularly the bad bits of it) more fairly. Sustainability is about preservation and/or conservation; justice is about distribution. While the former might be a precondition for the latter, this is only so at such a high level of generality as to be meaningless in respect of drawing any determinate conclusions as to the *kind* of justice for which sustainability is a precondition (indeed, sustainability is also a precondition for *in*justice, as I pointed out earlier). And I have said enough about the putative functionality of justice for sustainability to indicate that I regard that relationship to be under-explored and unproven and, in any case, only likely to be true under quite circumscribed empirical and theoretical conditions.

I suggested above that there are three ways in which the languages of justice and sustainability might be brought together: (a) the environment as something to be distributed, (b) justice as functional for sustainability and (c) 'justice to the environment'. I have argued that in none of these cases is it evident that the songs of justice and sustainability come from the same hymn sheet, and that we should therefore expect movements for justice and sustainability to bicker at least as often as they bond. In conclusion, though, I do think that there is one area where the languages convincingly converge – and this is the area of intergenerational justice. In discussing the prospects for the idea of 'justice to the environment' above, I gave the example of biodiversity and suggested that, for proponents of justice to the environment, species loss is unjust not because it is unfair to humans but because it is unfair to the species that are lost. This only works as an argument from justice, though, if we can convincingly say that species other than the human species can be regarded as legitimate recipients of justice. As we saw, this view is a contested one.

But while species loss may not be unjust in respect of the present generation of humans (since we have the theoretical opportunity to engage with them before they go), it might be regarded as unjust in respect of *future* generations of humans who will be deprived of such an opportunity. If the idea of intergenerational justice makes sense (and this is less contested than the idea of justice to other species, but contested nevertheless), then one understanding of what it might involve is the fair distribution of opportunities across the generations. Opportunities are, in turn, a function of the availability of 'stuff' through which opportunities might be realized. This is why in the context of non-renewable resources the intergenerationally just strategy is to invest, now, in alternative resources that will do the same job – and therefore provide the same opportunities – as those that are running down. The argument for conserving biodiversity is the same. Species loss in the present deprives future generations of the opportunity to 'enjoy' the species that are lost. Species loss amounts, therefore, to an intergenerationally unjust distribution of opportunity. By these lights the objectives of justice (as equal distribution of opportunities) and sustainability (as the preservation of biodiversity) are the same.

Notes

1   Sylvia Tesh, personal communication, 3 September 2000
2   That is those parts of 'nature' and those natural processes that are critical for the maintenance of human life.
3   That is the value that objects have by virtue of their history as *natural* objects. From this point of view a restored landscape is not as valuable as the 'original' one, however visually identical it might appear.
4   1   Is distribution according to need compatible with the sustaining of critical natural capital?
    2   Is distribution according to need compatible with the sustaining of biodiversity?
    3   Is distribution according to need compatible with the sustaining of the value of natural objects?
    4   Is distribution according to desert compatible with the sustaining of critical natural capital?
    5   Is distribution according to desert compatible with the sustaining of biodiversity?
    6   Is distribution according to desert compatible with the sustaining of the value of natural objects?

7   Is distribution according to entitlement compatible with the sustaining of critical natural capital?
8   Is distribution according to entitlement compatible with the sustaining of biodiversity?
9   Is distribution according to entitlement compatible with the sustaining of the value of natural objects?

REFERENCES

Camacho, D (1998) 'The environmental justice movement: A political framework' in D Camacho (ed), *Environmental Injustices, Political Struggles: Race, Class and the Environment*, Duke University Press, Durham and London

Dobson, A (1998) *Justice and the Environment: Conceptions of Environmental Sustainability and Dimensions of Social Justice*, Oxford University Press, Oxford

Feinberg, J (1981) 'The rights of animals and unborn generations' in E Partridge (ed), *Responsibilities to Future Generations*, Prometheus Books, New York

Low, N and Gleeson, B (1998) *Justice, Society and Nature: an Exploration of Political Ecology*, Routledge, London

Sandweiss, S. (1998) 'The social construction of environmental justice' in D Camacho (ed), *Environmental Injustices, Political Struggles: Race, Class and the Environment*, Duke University Press, Durham and London

Schlosberg, D (1999) *Environmental Justice and the New Pluralism: The Challenge of Difference for Environmentalism*, Oxford University Press, Oxford

Part 3

Cities, Communities and Social and Environmental Justice

As urbanization continues apace, many authors prioritize the recognition and achievement of greater urban sustainability as the challenge of this century. Still others see depopulated rural areas and communities as being in need of strategic interventions to both lessen ecological degradation and maintain the ecological services provided. Others are working at the scale of community represented by the state. At whatever scale of problem, creative approaches and solutions are being developed in response to social and environmental justice, and sustainability issues. Part 3 explores these problems, approaches and solutions in four very different contexts.

In Chapter 5, Rees and Westra examine the relationship between environmental justice and prospects for urban sustainability in the 21st century. They begin with the premise that prevailing economic structures and international relationships, which are required to maintain current levels of material consumption, mainly by the wealthiest fifth of the world's population, are already causing harm to impoverished populations and ultimately threaten geopolitical security. The chapter examines the causes and effects of environmental injustice in cities and explores the implications of increasing resource scarcity (ie water, food and energy) for both urban vulnerability and prospects for sustainability. Taking this environmental security argument further, they contend that if the tensions caused by growing inequity between the world's rich and poor, both within and between countries, cannot be addressed through means other than material growth, the prospects for urban (and ultimately global) sustainability are fatally compromised.

The convergence of race, politics and pollution in the Mississippi Chemical Corridor is explored by Wright in Chapter 6. She investigates environmental equity in air toxin releases to determine the extent of disproportionate assumption of risk burdens by racial communities in Louisiana parishes along what has been termed the Mississippi River 'chemical corridor'. She finds clearly discernable patterns in the placement of toxic facilities near African-American communities. The chapter further examines the extent to which race and politics interact in the development of governmental policies and cultural practices that result in racially discriminatory siting patterns. Wright concludes with a discussion of how these patterns of pollution have affected the lives of citizens in communities in both Louisiana and Nigeria, both totally dependent on oil, and gives some attention to the global implications of pollution injustice.

Drawing on a study of land-based Chicano farming communities in the American south-west, Peña in Chapter 7 explores the relationship between environmental justice and sustainable development. The chapter examines the emergence of the environmental justice movement in the south-west and the strategies to protect Chicano farmlands from damage and expropriation. He asks how the concepts of sustainability, ecosystem management and social justice are framed within the context of discourses about place, identity and legal rights to contested space. How do differently positioned actors seek to attain legitimacy? Which forms of spatial logic do they invoke and seek to legitimize? What are the local, regional, national and global conditions that influence the process of mobilization of resistance? How do contrasting views of place and identity in relation to land, water and legal rights influence different organizational forms? What are the constraints and contradictions facing different actors?

In Chapter 8 Eady examines the role of state government in formulating environmental justice policy. She outlines how the US Environmental Protection Agency (EPA) created a framework for addressing environmental justice through public policy decisions. States have slowly begun to follow EPA's lead by modifying their own policy-making and regulatory infrastructures for the purpose of promoting environmental justice, or in several states, equity. The growth of state interest in environmental justice policy is critical for several reasons. Siting, zoning, transportation and growth management are at the centre of most environmental justice controversies. These issues, for the most part, have traditionally been reserved to state and local governments. This chapter examines the potential for state public policy to level the environmental playing field for communities of colour and low-income communities. It also identifies challenges in environmental justice policy and decision-making that is unique and distinct from federal policy decisions.

*   *   *

Chapter 5

When Consumption Does Violence: Can There be Sustainability and Environmental Justice in a Resource-limited World?

William E Rees and Laura Westra

INTRODUCTION: THE GLOBALIZATION OF ENVIRONMENTAL JUSTICE

The 21st century arrived on the blustery winds of accelerating global change. Optimists were buoyed up by reports that the world economy was surging ahead on an expanding wave of unprecedented duration and creating heretofore undreamed of money wealth; technological innovations in fields ranging from microbiology and genetic engineering, through microelectronics and communications, to robotics and nano-technology promised a future more exciting than even science fiction had led us to expect; the spread of industrial agriculture and genetically modified crops (the latest extension of the green revolution) seemed to promise that famine would soon be abolished forever; and with the end of the Cold War and the great political standoffs of the 20th century, it looked like democracy might yet become the political *modus operandi* for most of the world's people.

At the same time, darkening clouds blotted the dawning horizon, calling to question the sustainability of the entire human enterprise. Unresolved environmental issues from climate change to biodiversity loss symbolized an ecosphere under greater stress from human economic pressure than from any 'natural' force; the history of pesticide use and ozone depletion were telling us that there is as much reason to regard human technological prowess with anxious dread as with joyful anticipation;[1] production agriculture and cropland consolidation together with the lure of better economic opportunities in the city were driving or enticing millions of rural poor from the land throughout the developing world into already overcrowded mega-cities ill-prepared to cope with the influx; meanwhile, the widening economic gap between wealthy elites and the materially dispossessed threatened to undermine civil society and geopolitical stability. Little wonder that informed citizens everywhere greeted the new century with an uneasy mixture of hope and dismay.

The terrorist attacks on New York and Washington on 11 September 2001 finally tipped the balance – any remaining pockets of unbridled optimism vanished with the surreally violent implosion of the twin towers of New York's World Trade Center. The uneasy sense of security the world had enjoyed since the end of the Cold War period dissipated over the city in a massive cloud of dust and ash.

This chapter focuses on one of the most troublesome intersections of the conflicting political, economic, ecological and social forces facing humanity in the 21st century. Environmental decay is arguably the inevitable product of the post-World War Two explosion of economic production that has so far gone mainly to satisfy the material demands of the wealthy 20 per cent of the human family. While the benefits of material growth accrue to the rich and powerful, the burden of resource depletion, land degradation and pollution, increasingly falls on the weak and poor. This burden weighs particularly heavily on the urban poor in the burgeoning cities of low-income countries but also on the poor and on minority races in high-income countries. The growing ecological segregation ('eco-apartheid') of the world's people along economic and racial lines has led to charges of environmental racism and calls for environmental justice in countries as economically and socially disparate as Angola and the US. It is also part of the historical backdrop against which global terrorism must be assessed.

Our purpose in this chapter is therefore fourfold: first, we characterize eco-apartheid and expose its sociocultural and economic roots; second, we show how continued unequal growth and developing world urbanization can only exacerbate the problem; third, we develop the case that human-induced global ecological change and resource scarcity increasingly constrain conventional solutions; and, finally, we suggest some basic ethical and legal principles that should be reflected in international agreements for sustainability with justice. We start from the premise that ecological sustainability and equitable socio-economic development go hand in hand, that each is prerequisite for the other. (Society and the economy cannot exist without a functional ecosphere but events following 11 September show that we will not be able to focus on stabilizing the ecosphere without greater equity and social justice). In short, our analysis attempts to answer a single overriding question: just what are the current prospects for simultaneously achieving ecological sustainability and environmental justice in a rapidly urbanizing world driven by an economic development model devoid of either moral or biophysical reference points?

We are motivated, in part, by concern that the environmental burden on marginalized groups is so iniquitous, and the assault on human dignity so egregious, that calls for social reform, per se, tend to draw attention away from the structural economic forces, the ecological dysfunction, and, most important, the cultural values at *cause* of the problem. When intentional, this is a tactical error. The socio-economic, ecological and cultural problems of global change can neither be analyzed nor solved in isolation from one another. Environmental decay and attendant social injustice are intertwined symptoms of a misguided global economic development paradigm that is itself a product of a much deeper cultural malaise. In effect, we argue that the fundamental flaw lies in the techno-industrial worldview from which the much heralded economic 'new world order' is being constructed. More specifically, we identify contemporary 'Western' values and consumer lifestyles as penultimate causes of both ecological decay and eco-apartheid.[2] Acknowledging that overconsumption by the rich visits ecological violence on the poor raises moral, ethical and legal questions that go considerably beyond existing concerns about inequity and charges of environmental racism, however serious the latter might be.

DEFINING THE PROBLEM: PAST IS PROLOGUE

Worldwide, the urban poor tend to live in neglected neighbourhoods, enduring pollution, waste dumping and ill health, but lacking the political influence to effect improvements. (UNEP, 2000)

There can be little doubt that the relatively impoverished have always suffered the greatest consequences of local environmental decay. Certainly since the beginning of the industrial revolution, the urban poor, racial and ethnic minorities, and the otherwise marginalized – those who have neither the power to control nor the resources to avoid noxious hazards in the workplace or in their homes – have borne the greatest ecological costs of material growth (Robins and Kumar, 1999). The general pattern persists today all over the world. As McGranahan et al (1996) observe, extreme examples of city-level environmental distress are found in the industrial cities of the former socialist and communist economies and in middle- and low-income mega-cities in the developing world. Certainly the urban environmental hazards causing the most ill health are those found in the impoverished homes, neighbourhoods and workplaces located principally in the poorer countries of the South (Hardoy et al, 1992, McGranahan et al, 1996).

But the problem is hardly confined to former Eastern bloc and developing world cities. Even in the US the need for environmental justice 'has been raised to widespread national consciousness [in the past few decades] through the awareness, anger and passion of people speaking for environmentally beleaguered communities' (English, 2000). In that country, the geographic distribution of air pollution, contaminated waters and fish, toxic waste sites, landfills, etc, correlates strongly with the distribution of both racial minorities and poverty. However, Bullard (1995) emphasizes that the correlation between chronic exposure to ecological hazards and race is much stronger than that between exposure and income poverty. A National Wildlife Federation review of 64 studies of environmental inequity found 63 cases of disparity by race or income. However, race proved to be the more important factor in 22 of 30 tests (Goldman, 1994, cited in Bullard, 1995). Similarly, the Argonne National Laboratory found that: 'Out of the whole population ... 33 per cent of whites,

50 per cent of African-Americans, and 60 per cent of Hispanics live in the 136 counties in which two or more air pollutants exceed standards' (Wernette and Nieves, 1992, cited in Bullard, 1995).

To make matters worse, the evidence is clear that even in these allegedly enlightened times, rich neighbourhoods are better served by environmental law and regulatory agencies than are less advantaged ones. As Bullard notes, 'if a community is poor or inhabited largely by people of colour, there is a good chance that it receives less protection that a community that is affluent or white'. He argues that:

> *the current environmental protection paradigm has institutionalized unequal enforcement, traded human health for profit, placed the burden of proof on the 'victims' rather than on the pollution industry, legitimated human exposure to harmful substances, promoted 'risky' technologies such as incinerators, exploited the vulnerability of economically and politically disenfranchized communities, subsidized ecological destruction, created an industry around risk assessment, delayed clean-up actions, and failed to develop pollution prevention as the overarching and dominant strategy.* (Bullard, 1995)

It seems that in the US economic privilege and power not only insulate the wealthy from the worst effects of ecological degradation but also confer additional protection under the law.

Growing the problem: A failed development model[3]

Those suffering from eco-apartheid in poorer countries can expect little succour from the prevailing international development model. Indeed, the case can be made that globalization and market-based export-led development are causally linked to both poverty and eco-violence. Certainly the explosive growth of the human enterprise, the widespread degradation of local ecosystems, the destruction of community, the erasure of traditional knowledge-systems and the inequitable distribution of impacts – good and bad – are accelerated or enhanced by contemporary globalization, expanding material trade and economic integration.

It needn't be this way. One can imagine a form of enhanced internationalization (not homogenizing globalization) based on national integrity, respect for local community and managed fair trade that could bring greater ecological stability and economic security to a larger proportion – and certainly to the largest absolute number – of human beings than any previous arrangement in history. Why, then, is globalization today more associated with the world's most pressing crises than with possible solutions in the minds of so many concerned citizens? We believe the answer to this question lies in sociopolitical *context*, the particular set of facts, values, beliefs and assumptions upon which contemporary globalization is founded. Globalization today is very much a *social construct*, but it is one that undermines democracy and civil society while favouring the interests of those imposing it by further concentrating wealth and power in their hands.

The myth of competitive equilibrium

Contemporary globalization is based on the near-universal acceptance of neo-liberal (neo-classical) market economics as the basis for international 'development'. However, as economists themselves increasingly acknowledge, this model is flawed to its core. To begin, neo-liberal economics is founded on a model of 'general competitive equilibrium', a prominent distinguishing feature of which is that it bears little relationship to the real economy (Ormerod, 1997).

In theory, a free-market competitive equilibrium is optimally efficient – demand equals supply in every market and all resources are fully utilized. At equilibrium, no individual or firm can be made better off by altering the allocation of resources in any way without making someone worse off. (This is a state know as 'Pareto optimality'. Note that by definition, it means any government intervention in the marketplace in defence of the public interest would introduce inefficiencies.) However, all this depends on the following critical assumptions:

- diminishing marginal returns in consumption and production;
- perfect competition among a hyper-infinite continuum of traders (buyers and sellers) none of whom can individually influence prices;
- all traders have perfect knowledge of all present and future markets; and
- there are an infinite number of future markets.

Clearly, none of these necessary conditions for competitive equilibrium obtain in the real world. In fact, increasing returns to production are often a fact of modern economies and, far from hyper-infinity, the number of corporate traders in many sectors is steadily decreasing toward oligarchy as firms attempt to hone their competitive edges through mergers and acquisitions.[4] This places smaller locally based producers everywhere at a distinct disadvantage, subverts competition, diminishes consumer choice and ultimately allows powerful firms to control prices. Finally, no participating 'trader' can have perfect knowledge of even a single local market, let alone of a theoretical hyper-infinity of present and future markets. Ormerod concludes that '… there appear to be so many violations of the conditions under which competitive equilibrium exists that it is hard to see why the concept survives, except for the vested interests of the economics profession and the link between prevailing political ideology and the conclusions which the theory of general equilibrium provides' (Ormerod, 1997, p66).

James K Galbraith makes a similar point in his devastating condemnation of the year 2000 meeting of the American Economics Association (AEA) (Galbraith, 2000). Despite the fact that the empirical evidence 'flatly contradicts' each of the five leading ideas of modern economics, discussion of the 'great issues of economic policy' were missing from the program of the AEA. Given its evident dissociation from the real world, Galbraith argues that 'modern economics … seems to be, mainly, about *itself*' (Galbraith, 2000, p1, original emphasis). He goes on: 'But self-absorption and consistent policy error are just two of the endemic problems of the leading American economists. The deeper problem is the nearly complete collapse of the prevailing economic theory – of the structure of thought that supports their policy ideas. It is a collapse so complete, so pervasive, that the profession can only deny it by refusing to discuss theoretical questions in the first place' (Galbraith, 2000, p4).

Untrammelled trade: Corroding communities and ecosystems

In a particularly pernicious sleight of hand, today's political economy implicitly (sometimes explicitly) equates human welfare with income growth, and concurrently advances the untrammelled marketplace as the well-spring and arbiter of social values. Indeed, in globalist circles sustainable development is essentially equated with sustained growth of per capita world product (gross domestic product at the national level). Globalists go so far as to argue that chronic poverty in the developing world is a primary cause of ecological degradation and that the only sure way to eliminate poverty (and 'fix' the environment) is through growth (Beckerman, 1992).

Since trade can relieve local shortages (thus seeming to increase local carrying capacity) and catalyze growth, more liberal or 'free' trade is a mainstay of contemporary globalization. In theory, if each country specializes in those few goods or commodities in which it has a comparative advantage and trades for everything else, the world should be able to maximize gross material efficiency and therefore total output. With this singular goal in mind, globalization creates an increasingly prominent role for transnational corporations, encourages the transportation of resources and manufactured goods all over the planet, facilitates the instantaneous opportunistic movement of finance capital across national boundaries in search of the highest returns and generally encourages the integration of regional and national economies (Korten, 1995). These trends represent a threat to national sovereignty, to accountable democracy and to economic stability, even as they undermine options for community economic development. Meanwhile, corporate agglomeration fosters today's characteristic trickle-up (or flood) of wealth to the top.

It should be apparent that the competitive scramble to raise gross domestic product (GDP) relegates non-market and essential common-pool values to a decidedly secondary place in international development. The modern market model eschews moral and ethical considerations, ignores distributive inequity, abolishes 'the common good' and undermines intangible values such as loyalty to person and place, community, self-reliance and local cultural mores. Similarly, qualities and functions of nature that have no ready market value become invisible to the decision-making process (effectively pricing them at zero). In these circumstances, it is hardly surprising that the model is not adequately delivering, even on its own terms. Despite increasing GDP growth, chronic poverty prevails in many developing countries and the income gap between high-income Organisation for Economic Co-operation and Development (OECD) countries and the South is growing. The absolute gap is widening everywhere and even the relative income gap is increasing for most regions. (East Asia is the major exception – per capita incomes have gone from one tenth to almost one fifth of those in the high-income OECD countries since 1960). In 1970 the richest 10 per cent of the world's citizens earned 19 times as much as the poorest 10 per cent. By 1997 the ratio had increased to 27:1. By that time, the wealthiest 1 per cent of the world's people commanded the same income as the poorest 57 per cent. (Income ratios reflect purchasing power parity [data from UNDP, 2001]). Far from raising all boats, the rising economic tide is stranding the flimsiest craft on the reefs of despair, deepening the misery of millions of impoverished people. Here are the economic roots of eco-apartheid.

We should also note that contrary to conventional theory (and common understanding), balanced trade to the mutual benefit of both partners is actually no longer the objective. This is because much of the globally competitive scramble for international markets is driven by debt. The compulsive drive 'to maximize exports, minimize imports and *create a trade imbalance* … represents a financial struggle between [firms and] nations; a struggle which is entirely the result of the debt-financed financial system and the fact that all nations trade from a position of gross insolvency' (Rowbotham 1998, p88, emphasis added). Because so much 'income' at all levels of society goes into debt servicing, almost all countries suffer from chronically inadequate purchasing power to absorb the output of their domestic economies. In today's world, therefore, firms and countries attempt to maximize the net inflow of debt-free money by selling abroad. Since trade is a 'zero-sum game' there will be winners and losers – for every net exporter of debt-free cash (or material) there must be a net importer.

As similar enterprises invade each others' markets, the result is a global trading system in which 'goods that could easily be produced locally flow backwards and forwards across the country … and across the whole world' at great ecological and social cost to most trading partners and the world at large (Rowbotham 1998, p89). The intense competition bids down prices, encourages overproduction and consumption, undermines local/regional firms and economies, and eliminates surpluses needed for sound resource management. Among other things, these forces have generated a chronic global farm income crisis which extends even to rich farming nations like the US and Canada.[5] Meanwhile, the exploding demand for transportation, much of it non-essential, burns up a third of the world's precious oil supplies and contributes to climate change. In short, the rhetorical veil of efficiency actually conceals one of the most wasteful and destructive economic systems imaginable.[6]

The contemporary economy presses particularly hard on developing countries. For example, the need to meet international debt charges explains much of the cash-cropping for export that dominates agriculture in many developing world countries. Meanwhile, the land reforms, the introduction of intensive cropping methods, and the economic 'structural adjustments' required as a condition for the development loan in the first place, often have devastating impacts on the local environment, on subsistence production and on local community integrity (see Box 5.1). These factors also help drive urbanization in much of the developing world as 'inefficient' traditional farmers and field hands are driven from the land or have their markets undercut by corporate producers or imports.[7] Despite these costs and sacrifices, commodity exports under prevailing terms of trade generally fail to generate sufficient income to dissolve the debt. Forty-seven nations still have a *per capita* GDP of less than US$855 and remain heavily indebted, their governments owing foreigners the equivalent

BOX 5.1 TRADE AS THERMODYNAMIC IMPERIALISM

The structure of trade, as we know it at present, is a curse from the perspective of sustainable development. (Haavelmo and Hansen, 1991, p46)

Our disturbing assessment of the prospects for global development based on trade is rooted in modern interpretations of the second law of thermodynamics. The uncontrolled growth and development of complex industrial society *necessarily* disrupts the structural order and integrity of the ecologicla and other social systems upon which it is dependent.

This is because the techno-economic growth and high material standards of developed countries require net transfers of available energy and material (negentropy or exergy) from the periphery to the industrial centre. Indeed, 'industiral production and world trade are ... mutually reinforcing modes of exergy apporpriation. To export industrial goods would be meaningless if the money gained could not be used to purchase new resources with *higher exergy content*. The 'net order' thus appropriated by industry is as fundamental to its reproduction (as a synthetic biomass) as food is to an organism' (Hornborg, 1992a, original emphasis). At the same time, consistent with the second law of thermodynamics, less developed regions and countries 'must experience a net increase in entropy [disorder] as natural resoureces and traditional social structures are dismembered' (Hornborg, 1992b).

These effects are not all confined to developing countries. The depletion of the nothern cod stocks of Atlantic Canada and the temperate rain-forests of British Columbia has been driven largely by exports and is producing unprecedented socio-economic stress in the human communities that have historically depended on these resources. Similarly, North America's prairie soils have lost over half their original organic matter and natural nutrients as a result of intensive agricultural practices and decades of grain exports to the world. All this while the globally competitive economics of production agriculture depopulates the countryside and rural communities evaporate.

Internationally, to the extent that trade and the restructuring of rural economies displaces people from productive landscapes to overcrowded cities, particularly in the South, to supply urban industrial regions, mainly in the North, it contributes to impoverishment, urban migration and local ecolgcial decay. Moreover, to the extent that the current international development model and debt-financed development favours the depletion of natural capital in the South and net transfers of wealth to the North, this poverty and ecological decline are chronic conditions. The people of less developed countries cannot live on either potential carrying capacity (exergy) or money wealth exported back to the industrial heartland – and they increasingly suffer the direct physical consequences of local ecological depletion.

The editors of *The Ecologist* (1992) develop a similar argument, that the prevailing international development model represents the effective 'enclosure' of the global commons:

> *The market economy has expanded primarily by enabling state and commercial interests to gain control of territory that has traditionally been used and cherished by others, and by transforming that territory – together with the people themselves – into expendable 'resources' for exploitation.*
> (The Ecologist, 1992, p131)

This supports both the present analysis of eco-violence and Hornborg's view that 'the ecological and socio-economic impoverishment of the periphery are two sides of the same coin ...' (1992b).

of at least 18 months of export earnings. Many debtor nations spend more of their income servicing debts to the world's richest nations than providing social services to their own impoverished citizens (Roodman, 2001). The international debt crisis leaves 'emerging economies' unable to cope with problems of social inequity, uncontrolled urbanization and eco-injustice.

Perverse economics erodes human welfare

We can hardly overstate the extent to which today's extreme 'free-market' thinking actually perverts sound economics. Good economic theory would indeed have us maximize welfare, but recognizes that production/consumption is only one factor in the equation. A healthy environment, natural beauty, stable communities, safe neighbourhoods, economic security, social justice, a sense of belonging and countless other life-qualities contribute to human well-being. Thus, to the extent that people value any of these non-market and *public* goods more than they might value a little more material consumption, forgoing additional production/income growth to obtain these goods (ie through taxation) is actually sound economics – it would increase net social welfare (Heuting, 1996).

The point is that with a more comprehensive and equitable agenda, the technologies and tendencies to globalization could be turned to enhancing not only income security, but also many other values that make life worthwhile. As it stands, available data suggest that the prevailing international development paradigm may actually be destroying more unmeasured yet real wealth, much of it in the common pool, than is accruing to private interests. This is gross market failure – in a total social cost/benefit framework, it is clearly uneconomic to allow the destruction of two dollars' worth of resources or the global commons so that some individual or firm can realize one more dollar of profit. Sound economics gives governments a legitimate role in protecting and enhancing the public interest whenever the market fails to do so. Yet, in today's world, government intervention in the economy is reviled – globalists all sing in the deregulation choir.

To summarize, the dominant 'development' paradigm has brought us to the point where sustained material growth destroys ecosystems, impoverishes the planet, diminishes the human spirit and visits violence upon whole poor communities. But it also further enriches a powerful minority, mainly the already wealthy – and this poses a major barrier to change. Those who benefit most from the prevailing development model are isolated by distance and wealth from its negative consequences. Their feedback from 'development' is all positive. Since in the real world wealth is very much equated with power, those who would 'save the world' will have to overcome the resistance of those who perceive that their interests lie with the status quo. The increasing pace and scale of human-induced ecological change underscores the urgency of this task.

Upping the ante: Global ecological change

> *Neoclassical economics lacks any representation of the materials, energy sources, physical structures and time-dependent processes that are basic to an ecological approach. Worse, it is inconsistent with the physical connectivity and*

positive feedback dynamics of energy and information systems. (Christensen, 1991)

Conventional economics and globalization show no more sensitivity to ecosystems' properties and dynamics than they do to human communities. The resultant deteriorating state of the ecosphere was brought into sharp relief in November 1992 when some 1700 of the world's leading scientists, including the majority of Nobel laureates in the sciences, issued the following warning concerning humankind–environment relationships: 'We the under-signed, senior members of the world's scientific community, hereby warn all humanity of what lies ahead. A great change in our stewardship of the earth and the life on it is required if vast human misery is to be avoided and our global home on this planet is not to be irretrievably mutilated'. The reasons for the scientists' warning seem clear enough. Empowered by fossil energy, human exploitation of the ecosphere has increased dramatically since the beginning of the industrial revolution – and the pace is still accelerating. The global economy has expanded threefold and the human population has ballooned 30 per cent to over 6 billion in the 22 years since 1980 alone.

The sheer scale of human activity now ensures that the environmental impacts are global in scope. Of the world's major fish stocks, 70 per cent are being fished at or beyond their sustainable limits[8]; logging and land conversion to accommodate human demand has shrunk the world's forests by half[9] and so-called 'development' claimed half the world's wetlands in the 20th century. Overall, half the world's land-mass has already been transformed for human purposes and more than half of the planet's accessible fresh water is being used by people. One result of the shrinking area of 'natural' habitats is a biodiversity loss rate 1000 times greater than the 'background' rate. Meanwhile, more atmospheric nitrogen is fixed and injected into terrestrial ecosystems by humans than by all natural terrestrial processes combined; stratospheric ozone depletion now affects both the Southern and Northern hemispheres; atmospheric carbon dioxide has increased by 30 per cent in the industrial era and is now higher than at any time in at least the past 160,000 years;[10] mean global temperature has reached a similar record high and the world seems to be plagued by increasingly variable climate and more frequent and violent extreme weather events (Lubchenco, 1998, Tuxill, 1998, WRI/UNEP, 2001, Vitousek et al, 1997). Significantly, the WRI/UNEP report also makes the case that it is the world's poor – those most directly dependent on local ecosystems for their livelihoods – who suffer the most when ecosystems are degraded or collapse.

It is precisely these ecological trends that pose the threat of 'vast human misery' and the 'irretrievable [mutilation]' of our planetary home. Moreover, their proximate causes are well known to science – the clearing of forests, the conversion of natural ecosystems to high-input 'production' agriculture, over-fishing, the combustion of fossil fuels, the excessive discharge of biophysically active chemicals, etc – all in the service of economic and population growth. One might reasonably expect, therefore, that the global political process and policy-makers everywhere would be acting affirmatively to relieve the pressure under the banner of sustainable development.

There has, in fact, been a great increase in high-sounding rhetoric and a flurry of environmental legislation in various countries around the world. However, economic growth remains the focal item on the political agenda. Even Principle 2 of the 1992 'Rio Declaration on Environment and Development', while recognizing the need for humans to live in harmony with nature, emphasizes that, 'States have, in accordance with the Charter of the United Nations and the principles of international law, the sovereign right to exploit their own resources pursuant to their own environmental and developmental policies…' provided that domestic economic activities don't damage other states or areas beyond the limits of national jurisdiction (UNCED, 1992). Such assertions of the inalienable right to develop, however qualified, may be politically necessary to achieve international agreement on soft-law affecting environment and development. However, they remain ecologically naïve and functionally ineffective in protecting either local resource or global life support systems.

On the one hand then, the world appears to be wakening to the reality of the ecological crisis. On the other, we remain in deep denial of the extent of the value shift and behavioural transition needed to avoid disaster. Conventional 'sustainable development' doctrine insists that there is no inherent conflict between the economy and the environment – with improved management and technology, the world should be able to eat its economic cake and have the environment too (see also Blowers' critique of ecological modernization in Chapter 3).

Such optimism might even be justified under strictly specified conditions but these conditions cannot emerge from contemporary development models. In present circumstances, demand for nature's 'goods and services' will continue to climb exponentially. For the first time, however, we can anticipate significant shortfalls in supply. There simply isn't enough 'nature' to go around under prevailing growth-bound 'development' assumptions.

The ecological (and social) footprints of consumption

The size of the potential shortfall is clearly revealed by ecological footprint analysis. Ecological footprinting estimates human demand (or 'load') on the Earth in terms of the ecosystem area required to provide basic material support for any defined population (Rees, 1992, 1996). Thus, the ecological footprint of a specified population is the area of land and water ecosystems required on a continuous basis to produce the resources that the population consumes, and to assimilate the wastes that the population produces, wherever on Earth the relevant land/water may be located. A complete eco-footprint analysis includes both the area the population 'appropriates' through commercial trade and the area required to provide certain free land and water-based services of nature (ie the carbon sink function) (Rees, 2001, Wackernagel and Rees, 1996).

As noted above, population ecological footprints are based on material consumption where consumption is defined in ecological and physical rather than in economic (dollar) terms (see NRC, 1997 for various definitions). Indeed, ecologists explicitly classify humans as *consumer* organisms since virtually everything we do – including all economic activity – involves the consumptive use of so-called 'resources' first produced by ecosystems or through other natural processes. (The simplest example is our dependence on green plants (primary *producers*) for all our food.) In physical terms, consumption involves the irreversible transformation of available energy and material partly into useful products, but mainly into waste (and even the useful products eventually become waste), in accordance with the second law of thermodynamics. The second law dictates that the economic use of resources invariably results in an increase in net entropy – the dissipation of available energy and the degradation of useful materials. In our view, human societies exceed the carrying capacity of their supportive ecosystems when the growing economy-as-dissipative-structure results in resource depletion and the pollution of air, water and land.

This conceptualization is compatible with Stern's (1997) working definition of consumption as 'human or human-induced transformations of materials and energy'. So defined, consumption is ecologically significant to the extent that 'it makes materials or energy less available for future use, moves a biophysical system toward a different state or, through its effects on those systems, threatens human health, welfare or other things people value' (Stern, 1997).

Eco-footprint analysis takes the additional step of converting the material and energy flows associated with consumption into a corresponding ecosystem area – the land and water area a study population effectively 'appropriates' from nature to produce its resources and to assimilate (at least some of) its wastes. And a revealing 'additional step' it is. Eco-footprinting shows that the residents of the US, Canada, many Western European and other high-income countries each require five to ten hectares (12–25 acres) of productive land/water to support their consumer lifestyles. By contrast, the citizens of the world's poorest countries have average eco-footprints of less than one hectare. Even burgeoning China's per capita eco-footprint is only 1.2 hectares (Wackernagel et al, 1999).[11]

Given the size of their citizens' per capita eco-footprints, it should come as no surprise that many high-income countries exceed their domestic bio-productivity by 100 per cent or more. Indeed, many industrial nations impose ecological footprints on the Earth several times larger than their political territories. In effect, the enormous purchasing power of the world's richest countries enables them to finance massive 'ecological deficits' by appropriating through commercial trade or natural flows the unused productive capacity of other nations and the global commons. Even the apparent biophysical surpluses of large seemingly under-populated countries like Canada and Australia are taken up by the ecological deficits of other countries.

Most importantly, quantifying the ecological deficits of rich countries should force recognition of the fact that much of the ecological damage afflicting developing countries and their peoples is caused not by consumption to satisfy local needs, but rather by intensive export-oriented production to satisfy developed world demand. Consumption by the world's wealthy causes much ecological destruction around the world, but, as noted, distance and wealth insulate the rich from the negative consequences of their consumer lifestyles (see Box 5.1).

Eco-footprinting thus underscores the gross inequity characterizing use of the Earth's productive capacity. Figure 5.1 allocates the biophysical output of the

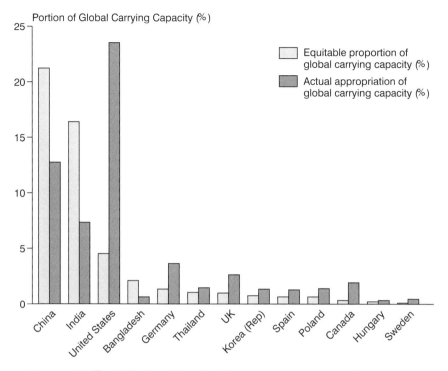

Figure 5.1 *Equitable (population-based) vs actual appropriations of global carrying capacity by selected countries*

planet to a selection of countries in proportion to their 1977 populations (based on data from Wackernagel et al, 1999). The figure also shows each country's eco-footprint as a proportion of the Earth's total productive land/water area.

This comparison reveals that wealthy market economies like those of the US, Canada and Western Europe appropriate two to five times their equitable share of the planet's productive land/water, much of it through trade. By contrast, low-income countries like India, Bangladesh and even China use only a fraction of their equitable population-based allocation. The prevailing forces of globalization tend to exacerbate rather than level these gross eco-economic inequities.

Eco-footprinting also shows that the world economy has already exceeded global ecological limits. There are only about 9 billion hectares of productive cropland, pasture and forest on Earth and perhaps 3 billion hectares of equivalent shallow ocean – this marine area produces about 96 per cent of the global fish catch – for a total of 12 billion productive hectares. However, with an estimated average eco-footprint of 2.8 hectare per capita (Wackernagel et al, 1999), the present human population already has a total eco-footprint of almost 17 billion hectares.

These data suggest we have overshot the Earth's long term human carrying capacity by as much as 40 per cent. How can this be? A population can live in overshoot – ie, beyond its ecological means – for a limited period by depleting vital ecosystems and drawing down non-renewable resource stocks. (This 'capital liquidation' permanently reduces *future* carrying capacity.) The empirical proof of human overshoot is the stuff of daily headlines – ozone depletion, atmospheric and climate change, land degradation, fisheries collapse, biodiversity losses, etc.

Since the wealthy fifth or so of humanity consumes 80+ per cent of global economic output, the rich alone effectively 'appropriate' the entire capacity of Earth in important dimensions. The ecosphere itself is now a scarce resource. This dooms to failure any efforts to secure sustainability with justice through sheer material growth. There simply isn't sufficient natural income – the 'goods and services' of nature – to support even the present world population at Northern material standards while simultaneously maintaining the functional integrity of the ecosphere.[12] Developing countries cannot follow today's high-income countries along their historical path to material excess, using prevailing technologies, without undermining global sustainability (see also Daly, 1991). Herein lies the ecological root of international eco-apartheid.

Meanwhile, 2 or 3 billion additional people are expected to arrive at the feast by 2050 (and we still have to account for the independent habitat needs of the millions of other species with whom we share the planet). If the world cannot safely expand its way to sustainability (see also Haavelmo and Hansen, 1991, Goodland, 1991, Goodland and Daly, 1993), we will have to discover other ways of relieving the material impoverishment of half of humanity. For example, developed world consumers might start to think seriously of ways of reducing their bloated ecological footprints to create the ecological space required for needed growth in the developing world.

To summarize, eco-footprinting underscores the fact that consumption, particularly over-consumption by the rich, is the principal driver of ecosystems degradation, frequently even in developing countries. This contrasts with the popular view that subsistence activities by the poor are the most important cause of ecological decay in the South and that '… the surest way to improve [the] environment is to become rich' (Beckerman, 1992). Eroding ecosystems and changing climate may directly harm innocent, economically marginalized peoples, but the distal cause is often the unsustainable lifestyles of the world's wealthy.

Urbanizing the crisis

Prospects for eco-justice are complicated by urbanization in the developing world. The reorganization of the rural landscape for production agriculture and forestry, and the promise of a better life in the city, have displaced millions of rural poor from the land in the past few decades (Rees, 1999, UNCHS, 2001). While the total population growth rate has fallen to about 1.3 per cent annually, the rate of urbanization in the developing world is over 3 per cent. The UN estimates that urban populations, particularly in the South, grew by about 50 per cent in the 1990s to just under 3 billion (UN, 1995, UNDP, 1998). At present, migration and natural increase contribute equally to this growth. The UN expects the world's urban population to reach 4.9 billion by 2030, an increase of 72 per cent in the first three decades of the century (UNCHS, 2001). This means

Table 5.1 *Third-World cities expected to attain populations in the 15–25 million range by 2015*

Mega-City	Population (millions)	
	1995	2015
Mexico City	16.6	19.2
São Paulo, Brazil	16.5	20.3
Bombay, India	15.1	26.2*
Shanghai, China	13.6	18.0
Calcutta, India	11.9	17.3
Beijing, China	11.3	15.6
Lagos, Nigeria	10.3	24.6*
Delhi, India	10.0	16.9*
Karachi, Pakistan	9.7	19.4*
Dahka, Bangladesh	8.6	19.5*

* indicates growth in 50–100+ per cent range by 2015.
Source: Smith and Naím, 2000

that over this 30-year period, the world's urban population alone is expected to grow by the equivalent of the total human population in the early 1930s.

In 2015 almost 17 per cent of urban dwellers will live in large cities of 5 million or more and the UN expects there to be 71 such mega-cities. As much as 95 per cent of the anticipated urban growth will occur in the developing South. This is a major shift. In 1950 the only city to exceed 10 million people (New York) was in the 'North' (London – also in a high-income country – had only 8 million). However, should present trends continue, 23 of the 27 cities expected to reach 10 million by 2015 will be in the developing world, 15 of these in Asia. Similarly, of the 44 cities of between 5 and 10 million inhabitants in 2015, some 36–39 will be in the developing world (UN, 1995, UNCHS, 2001). Several developing world cities will have populations in the 15–25 million range by 2015 (Table 5.1).[13]

Material constraints in the midst of plenty

We have already noted that environmental oppression is greatest in cities and that the most egregious examples are found in the mega-cities of the developing world. It is particularly foreboding therefore that low-income cities are woefully ill-prepared to accommodate the anticipated newcomers. In the mid-1990s as many as 25 per cent of developing world city dwellers did not have access to safe potable water (rising to 66 per cent of urban households in Africa) and 50 per cent lacked adequate sewage facilities (only 38 per cent of households in Asia-Pacific are connected to sanitary sewage). Even in 2000, more than 600 million of the urban poor lack adequate sanitation and 450 million suffer from unsafe drinking water (NRTEE, 1998, UNCHS, 2001). Accordingly, the World Bank estimates that developing countries alone will need to invest $US200 billion a year in basic infrastructure in the period to 2005, most of it for urban regions. Given anticipated urban growth and material demand to 2025, '… it would be reasonable to expect the total volume of investment [in infrastructure] to reach US$6 trillion by that time' (NRTEE, 1998, p11).

Keep in mind that we must now contemplate constructing adequate new physical plant to support an urban population *increase* in just the next three decades equivalent to the total accumulation of people in all of history up to the 1930s! In effect, we will be doubling the 1970s' urban presence on the planet. If one assumes that the pace of GDP growth in the developing world can be maintained and that the international political will exists to make the necessary investments, this would mean millions of new dwelling units, stores and offices; thousands of new schools, hospitals and water and waste-water treatment plants; countless square kilometers of new roads and parking facilities for tens of millions of additional motor vehicles and all manner of supportive transportation, communications, and related urban infrastructure. On the face of these data, it is little wonder that analysts expect cities to be 'using much more energy, [materials,] water, and land than ever before and doing so in more concentrated, land-, capital-, knowledge-, and technology-intensive ways' (NRTEE, 1998, p2).

But can we reasonably assume adequate economic growth and that the necessary investments will in fact be made? There is little in the historical record or contemporary geopolitics to encourage a sanguine perspective on developing world urbanization. The international community has not been particularly responsive to the relatively modest requirements of developing cities for such basics as clean water and sanitary sewers during a period of unprecedented money wealth-creation in the high-income North. On the contrary, at a time when the ethics of the market have come to dominate international relations, many rich countries have failed to meet even their existing international development commitments and development-aid budgets are actually falling. Are the rich likely to be any more forthcoming in response to vastly larger demand during a period of increasing uncertainty about the domestic effects of global change as the world breaches ecological limits? Meanwhile, many developing countries, constrained as they are by their existing debt burdens, will be unable to find the necessary investment capital domestically. Conditions are therefore ripe for the many symptoms of peri-urban ecological decay and ecological injustice to spread unabated for the foreseeable future.

WHEN CONSUMPTION DOES VIOLENCE: FROM CONSCIOUSNESS TO RESPONSIBILITY

Early in July 2000 a mountain of rubbish, swollen and lubricated by recent heavy rains, collapsed in Payatas (Manila), the Philippines. The resultant avalanche of slimy debris engulfed many adjacent shacks and shanties killing outright 218 of the desperate people who lived in and on the refuse. An equivalent number were listed as missing under the garbage. A horrified world watched on television as dozens of people stumbled around and choked on the sludge that threatened to bury them. For many viewers who had never heard the term 'environmental injustice', the concept would never be mere metaphor or intellectual wordplay. Real human beings who had been reduced to scavenging in garbage, were ultimately killed by it.

This local example of eco-injustice is a graphic symbol of the environmental violence that strikes hardest against those whom Vandana Shiva terms the 'absolutely poor' (Shiva, 1989). These most vulnerable of the human family already suffer the worst effects of degraded landscapes and polluted air, water and crops, and are also the most exposed to the ravages of climate change and the potential spread of (particularly tropical) disease. Ten thousand people died when hurricane Mitch, the deadliest Atlantic storm in 200 years, slammed into Central America in 1998. Worldwide, more than 120,000 people were killed in 1998–99 (including 50,000 when a supercyclone struck Orissa, in India) and millions more lost their homes (Abramovitz, 2001). According to the International Red Cross World Disasters Report, singular events in 1998, such as Hurricane Mitch and the El Niño weather phenomenon, plus declining soil fertility and deforestation, killed thousands and drove a record 25 million people from the countryside into crowded, under-serviced shanty towns around the developing world's fast-growing cities (IRC, 1999). This represents 58 per cent of the world's refugees. For the first time, people fleeing violent weather events and ecological decay outnumbered political refugees. While 'the 1990s set a new record for disasters worldwide' (Abramovitz, 2001, p123), it is unlikely that the world has yet seen the worst. The IRC report predicts that developing countries in particular will continue to be hit by super-disasters driven by human-induced atmospheric and climatic change, ecological degradation, and rising population pressures.

Significantly, it is becoming increasingly difficult to distinguish environmental from sociopolitical refugees. Resource scarcity and the degradation of ecosystems can exacerbate the chronic inequity among social groups, thus heightening the threat of both civic strife and international conflict. Various authors have documented the violent tensions that emerge from ecological decay as desperate people compete for shrinking land and resource supplies, particularly in developing countries (Gurr, 1985, Homer-Dixon, 1991, 1994, Homer-Dixon et al, 1993, Homer-Dixon and Blitt, 1998).

Framing sustainability with justice

We have argued, that given the dominant values of 'Western' industrial society and the structure of the prevailing economic development model, the incidence and intensity of both ecological and social violence against the defenceless are likely to increase, possibly dramatically. With the expansion of humanity's ecological footprint further beyond carrying capacity, we can expect global ecological change to accelerate. Hundreds of millions of additional desperately impoverished people will be forced from the land only to be exposed to the appalling living conditions of the barrios and favelas surrounding many cities in the developing world in coming decades. Thus, *even as things stand*, the social and ecological vulnerability that comes with dismal economic prospects virtually guarantees that gross ecological injustice will be a defining characteristic of world development in the 21st century. The situation will deteriorate further if fallout from the events of 11 September leads to extended war or other manifestations of geopolitical chaos.

We have also made the case that material consumption, particularly consumption by the economically privileged, is the 'forcing mechanism' for global ecological change. Incipient resource scarcities represented by everything from the collapse of fish stocks, through the decline of water tables, to the depletion of readily accessible petroleum reserves can be traced mainly to the overwhelming demand of Northern consumers. Similarly, 'industrial countries are responsible for more than 90 per cent of the 350 million metric tons of hazardous waste produced globally each year' (Sachs, 1995, pp36–37). Wealthy cities are, of course, the principal loci of both production and consumption. As McGranahan et al (1996, p109) put it, the largest contributors to global environmental problems are the affluent, 'living preponderantly in the urban areas of the North'. Consumption in developed world cities alone accounts for 65 per cent of the world's resource use and waste production (Rees, 1999).

These findings are ethically and morally charged. In effect, they mean that at the limits of biophysical carrying capacity, routine acts of non-essential consumption ultimately translate into acts of violent harm against the poor and racial minorities. There is, of course, no intent to harm and wealthy consumers who are ignorant of the distant systemic consequences of their material habits might be excused their seeming moral inertia. But once we raise to collective consciousness the link between consumption/pollution and eco-violence, society has an obligation to view such violence in the appropriate light. Not acting to reduce or prevent eco-injustice would convert erstwhile blameless consumer choices into acts of aggression.

In this light we need to move beyond analyzing what might constitute 'just garbage' (Wenz, 1995) on a case by case basis; we must instead acknowledge that unnecessary waste produced in affluent countries has the capacity, directly or indirectly, to harm the innocent. Careless consumption and the negligent disposal of the resultant garbage is *fundamentally* unjust and increasingly tends to blur the fine line between the trash and the trashed (Mills, 2000). In short, global society has a moral imperative '... to devise systems which ensure that those responsible for making environmental demands assume the main responsibility for the consequences of their activity – they should not expect other people, other species and other places to absorb the associated costs of environmental and social breakdown' (Haughton, 1999, p65). Over-consuming nations (and individual over-consumers) must come to terms with the fact that the ancient concept of gluttony-as-deadly-sin has acquired new modern meaning.

The Charter of the United Nations (UN, 1945) provides the moral authority to act in the interests of justice. Chapter 1 (Article 1, Purposes and Principles) states that the Purposes of the United Nations are to maintain international peace and security, and to that end: 'to take effective collective measures for the prevention and the removal of threats to the peace, and for the suppression of acts of aggression'. The Executive Order to establish an 'Office of Environmental Justice' (1 Feb 1994) in the US, and the provisions of the European Convention on Human Rights (Article 14 on discrimination) also move us in the right direction. But the increasing gravity of the problem and its indisputable global reach suggest that we will need no less than the strongest powers in international law.

Towards this end, consider the five basic principles originally proposed by Robert Bullard (1995) as the basis of a *national* legal framework for environmental justice:

- The right to protection – no individual should need to fear exposure to extreme personal risk from human-induced environmental degradation.
- A strategy of prevention – prevention (the elimination of risk before harm occurs) should be the preferred strategy of governments.
- Shift the burden of proof – environmental justice requires that entities applying for operating permits or undertaking activities potentially damaging to public health and others' property be required to prove the safety of their operations.
- Obviate proof of intent – the law should allow differential impact (on the poor or minorities) as shown by statistical analysis to infer discrimination 'because proving intentional or purposeful discrimination [in court] is next to impossible' (Bullard, 1995, p18). Do not actions that continue to manifest gross negligence in regard to the poor or minority races constitute de facto eco-injustice (or eco-racism)?
- Redress existing inequities – targets should be set and resources made available to compensate individuals and mitigate damages where environmental and related health problems are greatest.

Since eco-justice for all is clearly a desirable goal of international development, these principles could also help frame *international* law governing relationships among countries and controlling the behaviour of transnational corporations. The latter are increasingly being granted the rights of persons in international treaties such as the NAFTA.[14] It is time they were also required to assume moral responsibility for their actions.

The formal logic of common law throws additional useful light on the subject. In Canada, for example, 'the most important [common law] tort action today, both in terms of number of claims made and its theoretical importance, is the negligence cause of action' (Charles and VanderZwaag, 1998, p80). Negligence law focuses on compensation for losses caused by unintentional but unreasonable conduct that harms legally protected interests. Unreasonable conduct is taken to mean 'the omission to do something which a reasonable [person] guided by those considerations which ordinarily regulate human affairs, would do, or doing something that a prudent, reasonable [person] would not do' (cited in Charles and VanderZwaag, 1998, p80).

Here is clear recognition that fault may be found even in the case of *unintended* harm if the latter results from careless or *unreasonable* conduct.[15] Moreover, a negligence action may be launched in Canada in the event of environmental assault. The plaintiff must establish, on balance, five key elements of the tort – legal duty, breach of the standard of care, cause in fact, proximate cause and damage to the plaintiff (Charles and VanderZwaag, 1998). How might this work in the international arena? There is no doubt that eco-violence damages the plaintiff. The causal links between careless consumption/disposal and eco-violence are also becoming better established.

In this light, failure to act responsibly on the part of offending nations would seem to breach a reasonable standard of care. What is missing in international law is acknowledgment of the offense and the capacity to create and enforce a legal duty to act.[16]

The Criminal Code of Canada (Section 219) is even clearer that lack of intent to harm is no defence if damage results from knowing acts performed in careless disregard for others: 'Everyone is criminally negligent who (a) in doing anything, or (b) in omitting to do anything that it is his duty to do, shows wanton or reckless disregard for the lives or safety of other persons' (where 'duty' means a duty imposed by law). Every person who causes the death of another person through criminal negligence is guilty of an indictable offence and liable to imprisonment for life. Significantly, Section 222(5)(b) of the code states that 'a person commits homicide when, *directly or indirectly, by any means*, he causes the death of a human being, by being negligent' (emphasis added).

There is no prima facie moral reason why the behavioural standards imposed by international law should not be as rigorous as those required by domestic law. If human-induced climate change can be shown to be a cause of death and destruction, then are not countries like Canada and the US guilty of 'wanton or reckless disregard for the lives or safety of other persons' in their failure to act to reduce their profligate fossil fuel consumption and carbon dioxide emissions, particularly given the proliferation of options for conservation and the increasing availability of alternatives? (These countries have among the highest levels of CO_2 emissions per capita.)

In fact, however, the international community has been slow to acknowledge possible fault on the part of over-consuming nations even in the face of growing evidence of death and damage from eco-violence. Significantly in the present context, consumption per se was kept off the table at the United Nations Conference on Environment and Development (the Rio 'Earth Summit') in 1992.[17] On the legal side, 'the International Law Commission has been working for decades on separate draft articles on state responsibility for wrongful acts under international law and state liability for acts not prohibited by international law'. As yet, however, '… no comprehensive treaty on state responsibility/ liability exists … and academics continue to argue over the what the "customary law" basis of state responsibility is'. A major issue is 'whether the basis of state liability should be fault-based or strict (just based on damage occurring)' (VanderZwaag, 2001, pers comm). In short, the question of state responsibility/liability internationally for negligent damage to persons, property, communities or the global commons remains unresolved and mired in controversy. Must we wait for famine, resource wars, or the displacement of 40 million people from their villages by sea-level rise before the international community has an appropriate institutional and legal framework from which to address such problems?

EPILOGUE

So long as [ecological] decline is seen as temporary, advantaged groups are likely to accept policies of relief and redistribution as the price of order and the resumption of growth. Once it is accepted as a persisting condition, however, they will increasingly exert economic and political power to regain their absolute and relative advantages. (Gurr, 1985)

As discouraging as it may seem, the behaviour described in Gurr's bleak prognosis may well reflect some fundamental human trait. Earlier in human biological and social evolution, selfish retrenchment during periods of relative scarcity may have conferred survival value on individuals and groups practising such behaviour. Simple behaviours such as hoarding and suspicion of unlike peoples, as well as more complex human traits such as discrimination and outright racism (including eco-racism), may have their roots in the deep shadows of human bioevolution. In the modern world, however, any advantage of such crudely self-interested behaviour would likely be dissipated. Today, the sheer scale of human activity renders us all dependent on the continued integrity of 'common-pool' resources and global life-support systems. The fates of the world's disparate peoples and nations are becoming inextricably bound together.

In effect, then, global sustainability represents the ultimate public good–common property problem, a problem that can be solved only through mutual constraint practised out of mutual self-interest. This, in turn, requires an unprecedented level of international agreement on instruments to restrain actions of individuals and nations that would threaten the global commons. The simple fact is that no country can achieve sustainability alone, no matter how just its social policies or how much it manages to shrink its domestic ecological footprint. Indeed, under the remorseless logic of the commons, individual self-restraint is 'unprofitable and ultimately futile unless one can be certain of universal concurrence' (Ophuls and Boyan, 1992, pp198–99).

Thus we seem to have reached a stage in human evolution when survival value may well accrue more to cooperative than to competitive behaviour. It is therefore more than a little dispiriting that the prevailing development paradigm encourages both individuals and nations to behave as self-interested utility maximizers with insatiable material demands. The inevitable result on a finite planet is competitive over-exploitation, growing inequity and looming ecological chaos that threatens us all. If the world community cannot resolve the intra- and interregional tensions caused by ecological decay and social injustice through some form of extended compassion, reason or at least more enlightened self-interest – by any means other than through sheer material growth – the prospects for sustainability are indeed fatally compromised.

Ironically, perhaps for the first time, no person today need be deprived of the necessities of life and no one should feel threatened by the material needs of others. We have the technology to provide for everyone's economic security in the short term while global society organizes to seek a more sustainable long term development path. Of course, technology alone is not enough – sustainability *will* require significant changes in prevailing values and behaviour in support of mechanisms to redistribute the world's eco-economic output. The question is whether we humans are sufficiently self-aware and adequately in control of our destiny to allow reason finally to dominate passion in the determination of our mutual fate. Can we convince ourselves that we have more to gain by shifting paths than by staying our present course? The answer is by no means certain. As Lynton Caldwell observes: 'The prospect of worldwide cooperation to forestall a disaster … seems far less likely where deeply entrenched economic and political interests are involved' (Caldwell, 1990).

Humanity is thus caught up in a conundrum. We are coming to understand that ecological sustainability is primary, indeed, foundational for a sustainable society. But to achieve ecological sustainability will require geopolitical stability founded on greater international material equity and environmental justice. At present, all these goals are threatened by a morally bankrupt and ecologically naïve economic growth model that actually increases inequity and spawns eco-violence. Little wonder that sustainability still recedes from our grasp.

Fortunately the conundrum is of our own making and can therefore be undone – at least in theory. What we need is sufficient motivation. The threat of 'mutually assured destruction' is sometimes credited with staving off nuclear annihilation during the Cold War. Perhaps the magnitude of the present threat is enough to focus our collective will on achieving 'mutually assured survival'. In the face of the evidence, the unthinking continuation of today's unsustainable practices would be an act of colossal collective stupidity.[18] By contrast, positive action to achieve a just sustainability would be the ultimate triumph of intellect and reason over unconscious fate. Success in this enterprise would certainly go a long way toward substantiating humanity's claim that there is intelligent life on Earth.

ACKNOWLEDGEMENTS

We would like to thank John Cobb and David VanderZwaag for their helpful comments on sections of an earlier draft of this chapter.

NOTES

1   And this doesn't even take into account the growing concern that people are increasingly more the servants than the masters of their technologies (see Postman, 1992).
2   The ultimate cause may be deeply rooted in human behavioural evolution.
3   Portions of this section also appear in slightly different form in Rees (2002).
4   Larger integrated companies benefit from intra-firm trading, economies of scale and increased market clout.
5   Despite rising food prices, farmgate receipts in Canada are lower in real terms than during the depression (NFU, 2000).
6   This is further highlighted when we consider the role of the advertising industry in creating demand for an increasing array of products of questionable value in a world where so many essential needs go unsatisfied.

7   These circumstances show that so-called 'development' is often actually a cause of both urban and rural poverty.
8   Already 20 per cent of the world's freshwater fish are extinct, endangered or threatened.
9   Tropical deforestation may now exceed 130,000 km^2 per year.
10 More recent reports put carbon dioxide at its highest levels in 20 million years (Radford, 2000).
11 For an update on the eco-footprints of nations, see WWF (2000).
12 We would require two or three additional Earth-like planets to support just the present human family of 6 billion at average Canadian material standards. Of course, the entire population might be supportable indefinitely at lower per capita consumption levels. This by no means implies lower quality of life.
13 However, Tokyo will be the world's largest city with about 28.9 million inhabitants.
14 Often at the expense of national sovereignty, democracy, local desires and, possibly, eco-justice. For example, in August 2000, a NAFTA tribunal ordered Mexico to pay California-based Metalclad Corp $US16.7 million as compensation for a Mexican municipality's refusal to allow the company to run a hazardous waste dump located in the community (G&M, 2000).
15 According to Spears (1989), negligence on the part of a [property owner] can 'trigger a finding of fault...' 'Fault based on negligence is assessed by considering the reasonable expectations and standards of conduct that society has set for [a landowner] ... failure to meet this objective standard – his negligence – constitutes a quasi-delictual offence and fault is ... legally presumed'. Re-read this passage substituting 'signatory nation' for 'property owner'.
16 Note, however, that there is no such reluctance to act in the creation of treaties pertaining to trade such as the NAFTA or the GATT/WTO or when alleged offences are committed against such treaties (see note 141).
17 Former US President George Bush, for example, insisted that the American way of life was not up for negotiation.
18 Must humans suffer the same ignominious collapse that invariably befalls other species that overrun and deplete their habitats?

REFERENCES

Abramovitz, J (2001) 'Averting natural disasters' in L Brown, C Flavin and H French (eds) *State of the World 2001*, W W Norton, New York (for the Worldwatch Institute, Washington, DC)
Beckerman, W (1992) 'Economic growth and the environment: Whose growth? Whose environment?', *World Development,* vol 20 (4), pp481–96
Bullard, R (1995) 'Decision making' in L Westra and P Wenz (eds), *Faces of Environmental Racism*, Rowman and Littlefield, Lanham MD (reprinted from *Environment,* vol 36(4), pp39-44)
Caldwell, L (1990) *Between Two Worlds: Science, the Environmental Movement, and Policy Choice*, Cambridge University Press, Cambridge
Charles, W and VanderZwaag, D (1998) 'Common law and environmental protection: Legal realities and judicial challenges' in E Hughes, A Lucas and W Tilleman (eds) *Environmental Law and Policy* (2nd edn), Emond Montgomery Publications, Toronto
Christensen, P (1991) 'Driving forces, increasing returns, and ecological sustainability' in R Costanza (ed), *Ecological Economics: The Science and Management of Sustainability*, Columbia University Press, New York

Daly, H (1991) 'From empty world economics to full world economics: Recognizing an historic turning point in economic development' in R Goodland, H Daly, S El Serafy and B von Droste (eds), *Environmentally Sustainable Economic Development: Building on Brundtland*, UNESCO, Paris

English, M (2000) *Environmental Risk and Justice*, unpublished MS, University of Tennessee, Energy, Environment and Resources Center

G&M (2000) 'NAFTA ruling raises environmental questions', *Globe and Mail* (National Edition, 1 September, pB5), Toronto

Galbraith, J (2000) 'How the economists got it wrong', *The American Prospect*, vol 11 (7), 14 February

Goldman, B (1994) *Not Just Prosperity: Achieving Sustainability with Environmental Justice*, National Wildlife Federation Corporate Conservation Council, Washington

Goodland, R (1991) 'The case that the world has reached limits' in R Goodland, H Daly, S El Serafy and B von Droste (eds), *Environmentally Sustainable Economic Development: Building on Brundtland*, UNESCO, Paris

Goodland, R and Daly, H (1993) 'Why northern income growth is not the solution to southern poverty', *Ecological Economics,* vol 8, pp85–101

Gurr, T (1985) 'On the political consequences of scarcity and economic decline', *International Studies Quarterly*, vol 29, pp51–75

Haavelmo, T and Hansen, S (1991) 'On the strategy of trying to reduce economic inequality by expanding the scale of human activity' in R Goodland, H Daly, S El Serafy and B von Droste (eds), *Environmentally Sustainable Economic Development: Building on Brundtland*, UNESCO, Paris

Hardoy, J, Mitlin, D, Satterthwaite, D (1992) *Environmental Problems in Third World Cities*, Earthscan Publications, London

Haughton, G (1999) 'Environmental justice and the sustainable city' in D Satterthwaite (ed) *The Earthscan Reader in Sustainable Cities*, Earthscan Publications, London (reprinted from *Journal of Planning Education and Research*, vol 18 (3), pp233–43 [1999])

Heuting, R (1996) 'Three persistent myths in the environment debate', *Ecological Economics* vol 18, pp81–88

Homer-Dixon, T (1991) 'On the threshold: Environmental changes as causes of acute conflict', *International Security*, vol 19, no 2, pp76–116

Homer-Dixon, T (1994) 'Environmental scarcity and violent conflict: Evidence from cases', *International Security*, vol 19 (1), pp5–40

Homer-Dixon, T and Blitt, J (eds) (1998) *Ecoviolence: Links among Environment, Population and Security*, Roman and Littlefield, Lanham, MD

Homer-Dixon, T, Boutwell, J and Rathjens, G (1993) 'Environmental change and violent conflict', *Scientific American*, February

Hornborg, A (1992a) 'Machine fetishism, value and the image of unlimited goods: Toward a thermodynamics of imperialism', *Man*, vol 27 (1), pp1–18

Hornborg, A (1992b) 'Codifying complexity: Towards an economy of incommensurable values', paper presented to the Second Meeting of the International Society for Ecological Economics (Investing in Natural Capital), Stockholm, 3–6 August

IRC (1999) *World Disasters Report*, International Red Cross

Korten, D (1995) *When Corporations Rule the World*, Earthscan Publications, London

Lubchenco, J (1998) 'Entering the century of the environment: A new social contract for science', *Science*, vol 297, pp491–97

McGranahan, G, Songsore, J and Kjellén, M (1996) 'Sustainability, poverty, and urban environmental transitions' in D Satterthwaite (ed) (1999) *The Earthscan Reader in Sustainable Cities*, Earthscan Publications, London (originally published in C Pugh (ed), *Sustainability, the Environment and Urbanization*, Earthscan Publications, London [1996])

Mills, C (2000) 'Black Trash' in L Westra and B Lawson (eds), *Faces of Environmental Racism* (2nd edn), Rowman and Littlefield, Lanham, MD

NFU (2000) 'The farm crisis: EU subsidies, and agribusiness market power', policy discussion paper, National Farmers' Union, Ottawa

NRC (1997) *Environmentally Significant Consumption: Research Directions*, National Academy Press, Washington, DC

NRTEE (1998) 'Canada offers sustainable cities solutions for the world' (discussion paper for a workshop), National Round Table on the Environment and the Economy, Ottawa

Ophuls, W and Boyan, S, Jr. (1992) *Ecology and the Politics of Scarcity Revisited: The Unraveling of the American Dream*, WH Freeman and Company, New York

Ormerod, P (1997) *The Death of Economics*, John Wiley and Sons, New York (originally Faber and Faber, London, 1994)

Postman, N (1992) *Technopoly*, Knopf, New York

Radford, T (2000) 'Greenhouse build-up worst for 20m years', *Guardian Unlimited*/*The Guardian* website, 22 August, http://guardian unlimited.co.uk/uk_news/story/0,3604,355070,00.html

Rees, W (1992) 'Ecological footprints and appropriated carrying capacity: What urban economics leaves out', *Environment and Urbanization*, vol 4 (2), pp121–30

Rees, W (1996) 'Revisiting carrying capacity: Area-based indicators of sustainability', *Population and Environment,* vol 17 (3), pp195–215

Rees, W (1999) 'The built environment and the ecosphere: A global perspective', *Building Research and Information*, vol 27 (4/5), pp206–220

Rees, W (2000) 'Global carrying capacity for the human population and for life support systems' in T Munn (ed) *Encyclopedia of Global Environmental Change*, John Wiley and Sons, London

Rees, W (2001) 'Ecological footprint, concept of', *Encyclopedia of Biodiversity, Vol 2*, Academic Press, Amsterdam

Rees, W (2002) 'Socially just eco-integrity: Getting clear on the concept' in P Miller and L Westra (eds) *Just Ecological Integrity: The Ethics of Maintaining Planetary Life,* Rowman and Littlefield, Lanham, MD (in press)

Robins, N and Kumar, R (1999) 'Producing, providing, trading: Manufacturing industry and sustainable cities', *Environment and Urbanization*, vol 11 (2)

Roodman, D M (2001) *Still Waiting for the Jubilee: Pragmatic Solutions for the Third World Debt Crisis* (Worldwatch Paper 155), The Worldwatch Institute, Washington, DC

Rowbotham, M (1998) *The Grip of Death*, Jon Carpenter, London

Sachs, A (1995) *Eco-Justice: Linking Human Rights and the Environment*, Worldwatch Paper 127, Worldwatch Institute, Washington, DC

Shiva, V (1989) *Staying Alive*, ZED Books, London

Smith, G and Naím, M (2000) *Altered States: Globalization, Sovereignty, and Governance*, International Development Research Centre, Ottawa

Spears, J (1989) 'Government regulation and the environment: The role of the civil code' in *Current Problems in Real Estate* (the Meredith Memorial Lectures) Les Éditions Yvon Blais, Cowansville

Stern, P (1997) 'Toward a working definition of consumption for environmental research and policy', in NRC *Environmentally Significant Consumption: Research Directions*, National Academy Press, Washington, DC

The Ecologist (1992) 'Development as enclosure: The establishment of a global economy', vol 22, no 4, pp131–56

Tuxill, J (1998) *Losing Strands in the Web of Life: Vertebrate Declines and the Conservation of Biological Diversity*, Worldwatch Paper 141, The Worldwatch Institute, Washington, DC

UN (1945) *The Charter of the United Nations*, San Francisco, http://www.un.org/aboutun/charter

UN (1995) *World Urbanization Prospects: The 1994 Revision*, The United Nations, New York

UNCED (1992) *Report of the UN Conference on Environment and Development, Annex 1: Rio Declaraion*, The United Nations, Washington, DC

UNCHS (2001) *Cities in a Globalizing World: Global Report on Human Settlements 2001*, Earthscan Publications, London (for United Nations Centre for Human Settlements)

UNDP (1998) *Urban Transition in Developing Countries*, United Nations Development Program, The United Nations, New York

UNDP (2001) *Human Development Report 2001*, Oxford University Press, New York and Oxford (for United Nations Development Program)

UNEP (2000) 'Global Perspectives', in Geo-2000 Overview, UNDP website, 1 October, 2000, http://www.unep.org/unep/eia/geo2000/ov-e/0002.htm

VanderZwaag, D (2001) personal communication (e-mail letter to Rees), 25 January

Vitousek, P, Mooney, H, Lubchenco, J and Melillo, J (1997) 'Human domination of earth's ecosystems', *Science,* vol 277, pp494-99

Wackernagel, M and Rees, W (1996) *Our Ecological Footprint: Reducing Human Impact on the Earth*, New Society Publishers, Gabriola Island, BC and New Haven, CT

Wackernagel, M, Onisto, L, Bello, P, Linares, A, Falfán, I, Garcia, J, Guerrero, A and Guerrero, M (1999) 'National natural capital accounting with the ecological footprint concept', *Ecological Economics,* vol 29, pp375–90

Wenz, P (1995) 'Just garbage' in L Westra and P Wenz (eds) *Faces of Environmental Racism*, Rowman and Littlefield, Lanham MD

Wernette, D and Nieves, L (1992) 'Breathing polluted air', *EPA Journal,* vol 18 (1), pp16–17

WRI/UNDP/UNEP/World Bank (2001) *World Resources 2000–2001 People and Ecosystems: The Fraying Web of Life*, WRI, Washington, DC

WWF (2000) *Living Planet Report 2000*, Worldwide Fund For Nature International, Gland, Switzerland

Chapter 6

Race, Politics and Pollution: Environmental Justice in the Mississippi River Chemical Corridor

Beverly Wright

INTRODUCTION

Environmental racism in the US is spawned from a history of human slavery and is a by-product of racial segregation and discrimination legitimated in the southern US with the enactment of Jim Crow laws that made all forms of segregation and discrimination legal. These practices were not precluded in customs and practices in other areas of the US even without the strength of the law for compliance.

The disenfranchisement of an entire race of people was the law in southern states and was practised throughout the land and included discrimination in housing, education, public transportation, recreational facilities and restaurants. Environmental racism is merely one of the vestiges of the overall pattern and practices of racism and discrimination in the US.

The Mississippi River Chemical Corridor in Louisiana, better known as Cancer Alley, challenges nearly every environmental policy and regulation in this nation. For these reasons, an examination of communities and their struggles in the corridor exemplifies many of the environmental conflicts facing numerous communities in this country and abroad. This paper will examine the impact of environmental racism and pollution on people living in the corridor.

LOUISIANA: THE DEVELOPMENT OF AN OIL ECONOMY

Louisiana is a state that is rich in natural resources and almost to its detriment this has served as the most influential factor in its development. The French, over 300 years ago, recognized Louisiana's economic potential, which was further enhanced by its location at the mouth of the Mississippi River. The Mississippi River created the greatest opportunities for economic development by serving as a means of connecting the state not only with the nation, but also internationally. Moreover, 4 of the 11 largest ports in America are located in Louisiana and each year 4500 sea-going vessels and 100,000 barges travel the state's waterways. The Mississippi River is the heart and soul of the Louisiana economy linking it and much of the US with over 191 countries around the world (Thanos, 2000, p15).

Louisiana state policies have their beginnings with the French, but have also continued in the direction of attracting international trade and service businesses along the river. With the Mississippi River serving as the main attractor, domestic exporters recognized the benefits of locating along the Mississippi. The river served as an easy means of transport for heavy industries.

With the collapse of the sugar plantation system after World War Two, Louisiana became a prime location for the petrochemical industry. Access to the oil fields in the Gulf of Mexico has catapulted Louisiana's position to the second largest producer of refined oil in the US. Louisiana can boast 11 per cent of the petroleum reserves in the US. Louisiana's petroleum industries are extremely important to the state and the nation. Moreover, Louisiana's 19 refineries produce approximately 17 billion gallons of gasoline annually (LDED, 2000).

Petrochemical industries in the state of Louisiana, for decades, have made significant economic contributions to the state. The fact that their fate is intrinsically tied to the state's economic and political climate is not lost on management. Corporations in turn have a vested interest in state economic policies and have been absolutely shameless in the use of their power to ensure that their interests are adequately represented in the state legislature.

Corporations in Louisiana are extremely well organized. The Louisiana Association of Business and Industry (LABI) is a coalition of business leaders who wield tremendous power in the state. It is arguably the most powerful lobby in the state. This power is the result of the implementation of a skillfully chartered plan to mentor and support political candidates who support their interest. They embrace what could be described as a 'cradle-to-grave' approach to political lobbying. Behaving much like a political organization, they actively 'recruit and train potential candidates throughout the state' (Thanos, 2000, p19).

LABI is a recognized force to be reckoned with in Louisiana politics. They do not limit their influence to the grooming and training of potential candidates but expend much energy in their efforts to establish connections with rising politicians. Their efforts are further strengthened through their alliance with other industry-specific political coalitions such as the powerful Louisiana Chemical Association in their attempts to influence and promote a pro-industry economic development model to support their expansion (Thanos, 2000, p19).

LABI has been very successful in presenting believable rationales for the continuation of Louisiana's economic development policy and regulations to support its continuation. Specifically, LABI proposes that Louisiana be in direct competition with other states for capital and jobs. Moreover, they also frequently compare Louisiana's development strategies and incentive programmes to those in the developing world and other states in the southern US.

LABI purports that 'Louisiana is not an island unto itself; legislative policies should not be based only on what happens within our borders since our businesses must compete nationally and internationally for jobs and capital' (LDED, 2000). The pro-industry theory similar to the more popularly known 'trickle down' theory suggests that large manufacturers stimulate development of small businesses that feed off, service and supply mother companies. As large businesses prosper it is expected that they will generate more jobs, in turn increasing tax revenues for state and local government.

LABI's argument begs the question: why is Louisiana such a poor state? If the model purported by LABI was sufficient or, in fact, yielded the benefits that would be expected from such a large and important industry, why have the citizens of Louisiana not benefited from the industry's prosperity? The answer is both simple and complex. The simple answer is that the model does not work for Louisiana citizens as a whole. In 1991, a local newspaper headline read 'Louisiana and Other Oil States are in a position to grow faster than any other part of the economy in coming years' (*The Times-Picayune*, 16 April 1991). But as we embark upon a new millennium it is obvious that this new prosperity is a dream deferred. Louisiana has failed to grow at a projected rate and has failed miserably in efforts to diversify the economy. The state government has chosen to remain committed to old economic development plans that look to the petroleum industry as a means of growing the economy. By subsidizing both the extraction and refinery process of the petrochemical industry, Louisiana has condemned its own future by inhibiting the growth of other sectors of the economy (Thanos, 2000).

Although several factors combine to determine Louisiana's failing economy, the Industrial Property Tax Exemption Program (IPTEP) is by far the greatest contributor to the ill-fated Louisiana economy. The IPTEP is the result of a marriage of ideas on economic development between corporations and politicians and is also the largest of all tax incentives offered by the Louisiana government. The Program provides for the payment relief of local parish property taxes on buildings, machinery, and equipment to manufacturing corporations for up to ten years. Following the exemption period, corporations pay on property taxes at a depreciated rate. From the Program's inception in 1936 to 1988, 11,000 IPTEP exemptions were granted. Between 1988 and 1998, US$2.5 billion in exemptions have been granted (Nauth and the Louisiana Coalition for Tax Justice, 1990).

The IPTEP's legal status is based upon provisions made in the Louisiana state constitution that allows tax exemptions to be granted to industry if it is in the best interest of the state. The question is whether or not these exemptions have been granted in the state's best interest. A 1996 study conducted by the Louisiana Coalition for Tax Justice basically concludes that the IPTEP exemptions and other incentives the Louisiana government offers to big businesses promotes an economic development paradigm that actually stifles economic growth, expansion and diversity. In defence of the IPTEP, even in the face of dismal economic statistics, the Louisiana Department of Economic Development still maintains that the IPTEP encourages new businesses to come to Louisiana and emphasizes the fact that the programme was 'specifically

designed to attract manufacturing jobs' (LABI, 1998, p8). A serious omission in
the mandate however, that has been identified by critics of the Program, is the
absence of a local hire clause that would ensure some return in human form on
the state's investment. Moreover, 94 per cent of all contracts in the 1980s went
to existing businesses and 75 per cent of all exemption contracts produced no
new jobs. While one could easily surmise that the state's investment is not
yielding a sufficient return, their environmental devastation is even more
disturbing. Louisiana's industries produce four times the amount of pollution
per manufacturing job annually than does the average state (Templet, 1997).

The IPTEP could best be described as a corporate welfare programme paid
for by the poor residents of Louisiana. Since the exemption relieves businesses
of local property taxes, operating funds to maintain local roads, parks, libraries
and schools are not received by local governments. 'We're the only state in the
US that lets education subsidize businesses,' explained Cleo Fields, a state
senator from Baton Rouge, Louisiana (Nauth, 1990; Thanos, 2000, p21). There
have been numerous attempts to introduce bills to the legislature to repeal this
act, including legislation that would give parishes (counties) a local option to
exclude education taxes from the IPTEP. To date, all of these efforts have failed
and most bills cannot even get out of committee. Most of the credit for the
defeat of these efforts can be given to LABI and its Political Action Committee
(PAC). This powerful lobby has been extremely successful in their campaign
against what they describe as 'anti-business special interest groups' (LABI, 2000,
Thanos, 2000).

While negative political headliners are not a new phenomenon for Louisiana,
the unjust and possibly unconstitutional nature of Louisiana's tax incentive
programmes has gained national attention. *Time* magazine published a three-
part expose on what they described as American corporate welfare. In that issue
Louisiana took top billing by being named the largest corporate welfare state in
the country (Karmatz and Labi, 1998).

THE MISSISSIPPI RIVER CHEMICAL CORRIDOR

The economy of Louisiana slowly began to change in the early 1900s from an
agricultural and fishing economy based on its cypress swamps, waterways and
fertile soil because of successful oil exploration, which led to the construction
of a refinery in Baton Rouge. The Mississippi River was a pull factor for
petrochemical companies due to its capacity for access to barges and disposal of
chemical waste. In the 1940s, the state's population could be seen shifting in the
direction of jobs created by this new oil-based economy. By 1956, 87,200
Louisianans were directly employed by the petrochemical industry.

The industrial inducements programme implemented by the then Governor
John McKeithen in the 1960s attracted petrochemical companies to the state by
offering very generous tax exemptions. By the 1970s, the Louisiana Industrial
Corridor along the Mississippi River was lined with 136 petrochemical plants
and seven oil refineries. This approximates nearly one plant or refinery for every
half mile of the river. The air, ground and water along this corridor were so full

of carcinogens and mutagens that it has been described as a 'massive human experiment'.

In 1982 the state's petrochemical industry employed 165,000 people and industrial taxes accounted for one out of every three tax dollars collected by the state. In 1991, the then Governor 'Buddy' Roemer cancelled US$30 million dollars of tax exemptions given to petrochemical companies. In 1994 petrochemical industries employed 97,600 people or 5 per cent of the state's total, and paid US$530 million dollars in state taxes. This 85-mile industrial corridor that produces one fifth of the US petrochemicals, transformed one of the poorest, slowest growing sections of Louisiana into communities of brick houses and shopping centres. This growth has not come without a price. The narrow corridor absorbs more toxic substances annually than do most entire states.

In 1980 the Mississippi River chemical corridor was emitting as much as 700 million pounds of toxic chemicals into the air, water and soil. The total releases and transfers reported by Louisiana Toxic Release Inventory (TRI) facilities for 1995 were 185,102,963 pounds. Air releases represented 84,671,835 pounds; water discharges represented 28,269,936 pounds; land discharge represented 4,660,001 pounds; and deep well injection represented 54,494,453 pounds. The portion of the total representing the sub-group of chemicals reported for the first time was 5,177,318 pounds of the air releases; 6,546,753 pounds of the water discharges; 66,705 pounds of on-site land releases; and 7,550,659 pounds of the deep well injection releases.

In 1995, 12 parishes in the Mississippi River Industrial Corridor and the industrialized areas in Calcasieu and St Mary Parishes contributed 82 per cent of the total releases and transfers occurring in Louisiana. There are 169 TRI reporting facilities located in the 14 parishes. Those 169 facilities comprise 55.59 per cent of the facilities in the state, and account for 153,295,438 pounds of the 185,102,963 pounds of chemicals released and transferred in the state in 1995.

In Louisiana, 132 facilities, or 78.11 per cent of all facilities, are located in eight of the top ten ranked parishes for total releases. They are Ascension, Jefferson, St James, St Charles, Calcasieu, East Baton Rouge, St Mary and Iberville. These eight parishes (counties) account for 146 million of the 153 million pounds, or 95.35 per cent of releases and transfers in the industrial corridor and industrial area parishes, and 78.96 per cent of all state total releases and transfers (Louisiana Department of Environmental Quality Toxic Release Inventory, 1995).

In 1995 ten parishes combined made up 85.81 per cent of the state total TRI releases and transfers. The remaining parishes combined contributed less than 15 per cent of the total remaining TRI releases and transfers reported by facilities in the state. It is clear from these data that some areas in Louisiana have greater toxic emissions than others. While the state of Louisiana has had great difficulty in acquiring a high ranking in almost all quality of life indicators, regretfully Louisiana ranks first nationally in per capita toxic releases to the environment. Moreover, the state ranks second in the nation in total chemical releases and wastes injected into the ground and third in air releases, (EPA, 1997 Toxics Release Inventory State Fact Sheets).

The pollution burdens for the citizens of Louisiana are immense but in no case are these burdens greater than for persons who inhabit these lands where the plants are now located. It should be noted that most communities existed long before the plants made the corridor their home.

A Geographic Information Systems (GIS) analysis of toxic release inventory reporting facilities was conducted to determine the relationship between race and facility siting within nine parishes (counties) along the Mississippi River Industrial Corridor. Specifically, what is the distance of minority communities (specifically, African-Americans) from 'toxic' facilities and how does it compare to that of whites? The analysis focused on nine parishes along the Mississippi River Chemical Corridor. They were: (1) Ascension, (2) Jefferson, (3) St James, (4) St Charles, (5) East Baton Rouge, (6) Iberville, (7) St John, (8) West Baton Rouge and (9) Orleans.

Geographic Information Systems mapping for all nine parishes shows clusters of air-polluting facilities largely located in areas with high concentrations of African-Americans. A clearly discernable pattern of discrimination was found in the siting of polluting facilities in close proximity to predominantly African-American communities in the industrial corridor. Specifically, the study finds that approximately 80 per cent of the total African-American community live within 3 miles or 5 kilometres of a polluting facility. (See Table 6.1).

What has been the impact of those siting patterns on corridor communities? In an effort to more closely examine the effects of living so close to industry, this paper will focus on Norco, Louisiana, a 'fence line' community in the corridor.

Table 6.1 *Percentage of minorities relative to the total population*

Parish	Pop size *n*	Percent of minority population relative to total distance (kilometres) from polluting facility			Pop size *n*	Percent of non-minority population relative to total distance kilometres) from polluting facility			Minorities vs non-minorities (p-values) distance (km)		
		5km	3 km	1 km		1 km	3 km	5 km	1 km	3 km	5 km
Ascension	13,649	5.29	34.36	83.98	32,058	5.02	26.05	56.01	0.759	0.000	0.000
East Baton Rouge	128,287	10.01	51.51	82.06	207,691	4.38	25.21	53.77	0.000	0.000	0.000
Iberville	12,162	8.97	36.45	85.21	13,152	6.84	43.07	75.43	0.000	0.000	0.000
Jefferson	97,088	12.24	58.11	87.05	349,144	9.44	37.58	67.10	0.012	0.000	0.000
Orleans	281,558	5.31	21.18	55.65	126,073	2.82	18.85	38.97	0.000	0.096	0.000
St Charles	10,413	15.79	77.11	98.48	28,672	9.35	67.27	96.08	0.000	0.000	0.000
St James	9271	7.40	61.17	70.60	8126	23.11	64.10	79.23	0.000	0.000	0.000
St John	14,956	25.28	72.06	91.76	24,858	15.74	49.72	86.27	0.000	0.000	0.000
West Baton Rouge	4827	5.30	74.95	86.37	6959	0.68	36.40	68.69	0.000	0.000	0.000

TOXIC NEIGHBOURS

The residents of the Diamond Community in Norco, Louisiana are sandwiched between the Shell Chemical Plant and the Shell/Motiva Refinery. Loud noises, noxious odours and deadly chemicals are the constant neighbours of the citizens of Norco. The view from their front windows is of the pipes, storage tanks, and production towers of a petrochemical plant. Flares erupt noisily and unpredictably, sometimes roaring through the night. Delivery trucks and vehicles servicing the chemical plant move deadly chemicals through their communities. Unexplained booming noises shake their homes in the night. Strange smells waft into their homes producing headaches and breathing difficulties. (Concerned Citizens of Norco, 1999)

Norco, Louisiana, is located in St Charles Parish, approximately 40 miles northwest of New Orleans. The town name of Norco is derived from the New Orleans Refinery Company. The Diamond Community is more than 100 years old. As a result of Shell's industrial expansion, the size of the Diamond community has been reduced and now only extends across four remaining streets.

The residents of the Diamond community live in the midst of a Shell/Motiva manufacturing complex that reported over 2 million pounds of toxic emissions to air on the 1997 Toxic Release Inventory. This represents over 50 per cent of the toxic air releases in the entire parish. Among the petroleum industry, the Shell refinery is the second largest emitter of toxic chemicals to air in Louisiana. The Shell refinery also releases more recognized carcinogens to air than any other refinery (EDF, 1997). Residents living near the Shell facilities are exposed to pollution in three ways: permitted emissions, that come out of the stacks; fugitive emissions that come from leaky pipes and valves; and accidental releases that have off-site effects on residents.

In 1998 the citizens of Norco, were exposed, in addition to the hazardous chemicals reported on the Toxic Release Inventory, to other dangerous chemicals emitted by the Shell-Norco facilities, for example: 35 million pounds of nitrogen dioxide (eye, skin and lung irritant); 3.8 million pounds of carbon monoxide (neurological and respiratory toxins); 6.8 million pounds of sulphur dioxide (respiratory irritant; toxic at high levels); and 1.8 million pounds of total suspended particulates (LDEQ, 1998).

A total of 2,562,210 pounds of toxic chemical were emitted into the air from all facilities in Norco in 1997. The Shell-Norco facilities reported 827,352 pounds of toxic air emissions from their stacks. According to the 1997 Toxic Release Inventory, industries operating in Norco released between two (Shell/Motiva Refinery) and seven (Union Carbide-Norco Polypropylene Plant) times more pollution through leaky valves, flanges and pipes than through smoke stack emissions. The massive scale of these emissions is alarming. In the case of the Shell/Motiva Refinery, the fugitive emissions were estimated at 1.14 million pounds in 1997, compared to 570,000 pounds from stack emissions that same year.

Thus, twice as much pollution is released through fugitive emissions at the Shell/Motiva Refinery than is released though the stacks. The Union Carbide

Polypropylene and Catalyst plant reported nearly 150,000 pounds of fugitive emissions, compared to only 22,766 pounds of emissions from stacks, suggesting a dangerous pattern of accidental releases at Shell.

For many of the residents who live near the plants, accidental releases are of great concern. In 1973 an eruption of a flammable pipeline at the Shell chemical plant killed an elderly woman and teenage boy, who lived less than 20ft from the plant. Relatives and neighbours witnessed the flames igniting the house and the young man who was outside cutting grass. Also unforgettable to many Norco residents is the 1988 explosion at the Shell Oil Refinery that killed seven workers and injured neighbouring residents.

Although few accidents at the Shell plants in Norco have resulted in deaths, the frequency of gas leaks, chemical spills and fires are cause for alarm. In 1998 and the first six months of 1999, Shell reported 66 accidents that resulted in releases in excess of its permitted levels at its plants in Norco.

The residents of Norco are clearly at great risk of exposure to toxic chemicals. What then will residents do in response to a plant accident? What provisions have been made by the parish government or the state to insure a logical plan of escape for citizens? Norco citizens are told that in the event of an accident, they should 'shelter-in-place'.

The procedure requires them to run to the nearest building and seal off all outside air. Given the structural conditions of homes in the poor neighbourhoods of Norco, shelter-in-place falls far short of preventing toxic exposure. Both the EPA and LDEQ have approved the use of 'shelter-in-place'. However, this approval is not based on an evaluation of its effectiveness in protecting residents form toxins released during facility accidents.

Research actually shows the shelter-in-place method to be a most ineffective way to prevent toxic exposure (Blewett and Anch 1999). A test building was exposed to a chemical release for three hours. Measurements of air pollutants were taken outside the building and inside as well to determine how much gas was infiltrating. After three hours, the levels of chemicals inside the building were equal to that outside. In addition, it took 36–48 hours for the air inside the house to 'off-gas' the pollution. Moreover, the study suggests that even if 'shelter-in-place' procedures were meticulously followed by residents they would still risk exposure to toxic air pollution long after the incident ended.

THE CASE FOR RELOCATION

In 1973 a Shell pipeline erupted and blasted a home of an elderly African-American woman. She was asleep inside and died from burns received in the fire. A teenage boy mowing the grass outside her house was engulfed in flames. He died three days later in a hospital. Despite these deaths, Shell's pipeline remains in place and is clearly visible above ground where it sprawls almost the entire length of Washington Street, a boundary of the Shell chemical plant nearest residents.

In 1988 a catalytic cracker used at the Shell Oil refinery exploded killing seven plant workers, injuring 48 people, and damaging property for several miles.

Residents had to immediately evacuate the area (*The Times-Picayune*, 4 May 1988, ppA-1, B-1).

According to reports by Shell to the Louisiana Department of Environmental Quality, an average of 3.5–3.75 accidents per month occurred at Shell facilities in Norco from 1998–1999 (Concerned Citizens of Norco et al, 1999, p5).

Shell's dismal accident record has resulted in their being ranked among the worst oil refineries in the nation for the highest releases of toxic chemicals into the surrounding community (EDF, 1999).

The Environmental Protection Agency has also determined that Shell ranks highest among facilities that have frequent accidents. Records show that more than 60 per cent of all accidents occurring at Shell and other industrial facilities are caused by 'equipment failure' and 'operator error'. Moreover, these problems could have been avoided with investments in maintenance, equipment upgrades and safety training. Industrial accidents by Shell and other industries have caused a reported total of 99 deaths and 1800 injuries (*The Times-Picayune*, 8 November 1999, EPA Region 6, 1987–1997).

Throughout Shell's operations, residents have complained of noxious odours and ill-health effects, but Shell denies any responsibility. It is routine for Shell to report an accidental leak and assert that there are 'no off-site impacts' or that the leak is 'non-toxic' (Harden, 2000). However, there is now compelling evidence to support the complaints of residents.

In December 1988 the residents of Norco were once again unnerved by a white gas entering their homes, causing a burning sensation in their eyes and throat and feelings of nausea. This incident resulted in some community residents being treated in hospital emergency rooms and school children being evacuated to areas farther away from the Shell plant. Shell's response to this incident was to immediately notify the local news media to report, 'there were no chemicals released in the community' (*The Times-Picayune*, 9 December 1998, pB–1; Shell's Notice to Our Neighbours, 12 August 1998: Report to the National Environmental Justice Advisory Council).

Shortly after the announcement by Shell, assuring residents that the situation was 'stabilized', members of the California-based Communities for a Better Environment were in town and assisted residents by taking air samples in the neighbourhood. An EPA approved laboratory analysed these samples. The results showed high levels of a toxic chemical, methyl-ethyl ketone (MEK), which is known to cause irritation of eyes and nausea and other health effects, (*L'Observateur*, 9 January 1999, p1-A, Communities for a Better Environment, 12 August 1998). In response to these findings, Shell conceded that a tank containing MEK was over-pressurized, but continued to assert their position of 'no chemical releases from the tank' (*The Times-Picayune*, 9 December 1998, pP-1).

This incident resulted in the scheduling of a meeting by the residents with EPA and the Louisiana Department of Environmental Quality and Shell officials. At issue was Shell's continued denial of MEK releases into the community even in the face of scientific evidence to the contrary. At this meeting the residents of Norco restated their demand for relocation based on this new evidence of health risks caused by the release of dangerous chemicals into the environment where they live (*The Times-Picayune*, 15 January 1999, pB–1).

Norco residents have continued to take air samples in their neighbourhood, and since the MEK leak have consistently detected cancer-causing chemicals that are above the state's health standards. The samples have shown, for example, that even on a clear day there are 20 toxic chemicals in the air that residents breathe daily (*The Times-Picayune*, 21 June 1999, pB–1, Concerned Citizens of Norco et al, 1999, pp6–7).

These findings prompted an EPA investigation of the Shell plant. EPA inspectors found that the Shell/Motiva oil refinery has 'massive' problems meeting environmental regulations and that the plant's senior management was evasive about disclosing the troubles at the Norco plant (*The Times-Picayune*, 6 June 2000, pA–1). Further evidence of Shell's negligence was revealed when a whistleblower at the Shell/Motiva oil refinery reported that the company routinely violates environmental regulations, which has triggered a major investigation by the Environmental Protection Agency (*The Times-Picayune*, 21 June 1999, pA-1).

THE BUCKET BRIGADE: A COMMUNITY RESPONSE

The citizens of Norco in December 1998 launched an offensive against their polluting neighbours and have scored points. The Bucket Brigade is a simple but powerful tool that enables ordinary citizens to take EPA approved air samples. The materials used to take the Bucket Brigade's air samples are a teller bag, a small vacuum pump, and an ordinary plastic bucket with a few attached valves and a vacuum seal. The citizens have taken eight samples since its inception. The samples have recorded a dangerous array of chemicals present in the air that people in Norco breathe every day. These chemicals include toluene, benzene, carbon disulfide, styrene, methyl tert-butyl ether and methyl-ethyl ketone (MEK). A determined group of Norco residents have used the Bucket Brigade to bring attention to air pollution in their community.

The power of the Bucket lies in comparing its results with other public information. The 1997 Toxic Release Inventory Report submitted to the EPA by facilities in Norco proved to be a useful tool for indicating likely sources of chemicals detected in bucket air samples. The Shell/Norco facilities reported releasing a majority of the chemicals detected in the bucket samples. This implicates the Shell-Norco facilities as the source of these chemicals in the community. The air samples have provided scientific evidence that confirm the community's worst fears: even on 'good' days, the air in Norco is a toxic soup of hazardous chemicals (Concerned Citizens of Norco et al, 1999).

In addition to TRI reports, EPA monitoring data and accidental release reports can be compared in a similar way to the results of the bucket samples. Each bit of available information is fitted together to determine the sources of the pollution. Shell's facilities have shown to be a significant source of pollution in the air in Norco.

The date 19 June 1999 is one that most Norco residents will never forget. The Bucket Brigade usually takes samples on days when there is a particularly bad smell or a suspicious gas. However, on 19 June 1999, a day on which everything at Shell appeared to be normal, nine harmful chemicals were detected in the sample, including toluene, acetone, MEK, and carbon disulfide. The citizens of Norco now know that it is likely that on a daily basis, and even on a good day of no smells, it is likely that they are being exposed to chemicals.

The Bucket Brigade is an extremely powerful tool for residents of Norco to document their exposure the Shell's toxic emission. It gives the citizens power by allowing them to identify air pollutants when they suspect a problem and demand responsible action from the Louisiana Department of Environmental Quality, the Environmental Protection Agency and local institutions.

CORPORATE WELFARE: SHELL NORCO, IS THIS ILLEGAL?

Royal/Dutch Shell Corporation is the tenth largest in the world. Their operations can be found in over 120 nations and provide approximately 12 per cent of the world's oil (Nauth, 1990, Thanos, 2000). In Louisiana Shell can boast an income of over 26 billion dollars, an income larger than the economies of most countries of the world. Shell also rank fourth in the state in receipt of tax exemptions and in 1980 alone, received 67 tax breaks, totalling US$96,280,906 (Nauth, 1990, Thanos, 2000). Shell has very skillfully used the Industrial Property Tax Exemption Program to their advantage. Following the May 1988 explosion, the Shell-Norco plant received penalty fines from only the Occupational Safety and Health Administration.

They received a mere fine of US$3,630 for a 'deficient pipe inspection, insufficient monitoring and testing of the fluids that flowed through the pipeline, and deficient engineering design of the system' (Nauth, 1990). Additionally, it is important to note that seven workers were killed and 48 others were injured in this explosion. The Shell plant was also significantly damaged due largely to their own negligence in following occupational health and safety guidelines. This fact did not stop Shell from filing for aid receiving a new IPTEP exemption from the state of Louisiana: a US$450 million tax exemption was granted. Additionally, Shell applied for a US$2500 tax credit for each new employee hired after the explosion. Ironically, Shell received a US$2500 tax credit for each employee hired to replace the seven killed. This ought to be illegal.

THE GLOBAL CONNECTION

During one of the many 'toxic' tours of Cancer Alley (ie a stretch of land along the Mississippi River located between New Orleans and Baton Rouge, Louisiana) that are conducted by the Deep South Center for Environmental Justice, a young Nigerian man expressed his dismay in the following manner: '… I cannot believe that this is happening in the US. I know that the oil companies exploit my people and degrade and devastate the environment, but I had no idea that this was being done in the US'. I replied, 'It is done over and over again and to the same people'. Most people directly affected by the pollution from the petrochemical industry in Louisiana are African-Americans. The young Nigerian could not believe the similarities of the environment degradation and that the people affected were in both Africa and the US.

Environmental degradation and exploitation are transported globally by transnational corporations. Just as has been shown in the US, the people and communities most affected are the poor and disenfranchised minorities of a nation. The situation is curiously redundant, whether in the US or abroad. It is the poor and minority groups that bear the biggest burden of pollution. Dirty extraction industries (ie oil, timber and minerals) that devastate the ecosystem and destroy lives are located in communities that are powerless and represent the path of least resistance for corporations.

Several factors have been identified that interact and support environmental injustice around the world. According to Dr Deborah Robinson, Executive Director of International Possibilities Unlimited, these factors are as follows: (1) oil; (2) minority group status; (3) poverty; (4) multinational corporations; (5) human rights' violations; and (6) environmental devastation. To what extent do these factors exist to promote environmental injustice in both Africa and the US? To this end, what will follow is a brief review of the conditions and quality of life that exist for two communities: the Ogoni tribe in Ogoniland, Nigeria, and the old Diamond Plantation residents of Norco, Louisiana.

Nigeria's oil economy

The country of Nigeria has a population of over 100 million and represents the largest population in all of Africa. Nigeria's economy has changed from a mostly agricultural and mineral producing economy to one that is based on a single export commodity. Today, 80 per cent of Nigeria's total revenues and 90 per cent of its foreign exchange comes from petroleum. With oil now occupying a superior seat of importance to the Nigerian government and its economy, just how that money is spent bears directly on the quality of life for different ethnic groups of this society. Who will benefit and who will bear the burden of the pollution from this massive extraction industry goes to the heart of any discussion of environmental justice.

There are three major ethnic groups in Nigeria (Hausa/Fulani, Yoruba and Ibo), but their ancestral homes or areas that they can claim as their homeland are not oil producing lands. The oil rich lands of Nigeria are located in the Niger Delta and are home to Nigerian minority groups. Consequently, a great disparity exists in revenue allocations from the oil-rich producing lands of minority groups. The majority argues that the wealth of the country should be 'distributed for the greater good of the country, regardless of where or whose lands it is derived from'. In contrast, minority groups whose lands represent the oil rich revenue producing lands of the Niger Delta argue 'they should have more control over the resources from their land, particularly since they have to bear the brunt of the pollution the petroleum industry generates' (see Robinson, 1998).

Ogoniland: The oil-rich minority

The Ogoni people represent a small minority in Nigeria of approximately 500,000. They occupy 404 square miles of the Niger Delta and are located in the south-eastern region of Nigeria (see Agbola and Alabi, Chapter 13 in this book). The Ogoniland ecosystem is considered one of the most endangered habitats in the world due mainly to the operation of the petroleum industry in the region. Beginning with the discovery of oil in 1956 and oil production beginning in 1958, an estimated '30 billion dollars of petroleum has been extracted from Ogoniland' (Robinson, 1998). The prosperity of the oil industry however, has not trickled down to the Ogoni people, but the devastation has done so. The living conditions and overall quality of life for the Ogoni people are very low. The Ogoni people do not have 'pipe-borne water, telephones, electricity or proper healthcare facilities'. However, they have borne the pollution burden of this dirty extraction industry, including health consequences for years. Gas flaring, oil spills, seepage from drilling waste and waste from petrochemical companies are commonplace and the impacts have not been assessed. Deborah Robinson, 1998 reports that this, however, is what we do know:

- It is estimated that only 20 per cent of the population in the Niger Delta have access to clean drinking water.
- Of the total number of spills recorded from Shell, a company that operates in more than 100 countries, 40 per cent were in Nigeria.
- In January 1998, 40,000 barrels of crude oil spilled from an offshore pipeline operated by Mobil Oil, affecting over 120 communities, with immediate impact on plant and animal life. There are speculations that the government of Nigeria may ask Mobil to pay US$9.2 million as compensation to the communities affected.
- Gas flares are often situated less than 300 metres from settlements and the local population has to live with the noise, the constantly flickering light, and the soot. This soot settles on the skin and gets into the mucous membranes and respiratory tracts, as well as into fields and rivers.
- Cases of asthma, bronchitis, pneumonia, skin diseases, gastroenteritis, as well as emphysema, are much more common in Ogoniland and other oil-producing areas compared to Nigeria as a whole.
- Gas flaring in the Niger Delta produces 12 million tons of methane gas per annum. Methane gas is regarded as the main cause of the greenhouse effect and is reportedly much more dangerous in relation to global warming than carbon dioxide (CO_2).
- Ninety-five wellheads, five flow-stations, one oil refinery, a petrochemical plant and the largest fertilizer company in the whole of West Africa are all located in Ogoniland.
- Oil spills and dumping oil into waterways have been extensive in the Niger Delta, resulting in the destruction of aquatic life and mangrove forests, as well as the pollution of drinking water. A recent study found total petroleum hydrocarbons in a stream in Ogoniland, tested at 18 parts per million (ppm), which is 360 times higher than levels allowed in Europe.

- Oil spills in Ogoniland have continued even after Shell Oil withdrew its staff in 1993. Between 1993 and 1994 there were 24 spills.

It is very important to note that the Ogoni people have suffered the lost of their farming economy supported by the rich soils of the Niger Delta. Their economy, culture and quality of life have been devastated by the petroleum industry. Race/ethnicity, political power and pollution have all combined to exploit and oppress the Ogoni people. The end result is extreme poverty. In response to the years of oppression and resultant poverty, on 4 January 1993, Ken Saro-Wiwa and the Movement for the Survival of the Ogoni People (MOSOP) launched a protest of 300,000 against the Nigerian government and the Shell Petroleum Development Corporation (SPDC). They openly charged the Nigerian government with being responsible for their economic marginalization and resultant lack of development, and human rights violations. Additionally, they charged the Shell Petroleum Development Corporation as being responsible for the environmental degradation of their land and the devastation of their livelihood. The protest by the Ogoni people was sustained and eventually gained international attention and support.

The Ogoni people had taken a bold step and with an international audience, confronted the status quo. There was much at stake for the Nigerian military regime. The leaders of the protest (MOSOP) were captured by the regime, jailed and tried by a special military tribunal that flagrantly violated international standards and norms of due process. Sadly, on 10 November 1995, the Ogoni Nine, as they were called, including Ken Saro-Wiwa, a poet and Goldman Award winner, were hanged in full view of the world. This act of murder speaks to the arrogance of this regime and the political and economic importance that is attached to the petroleum industry.

The assault on the Ogoni people that was launched in 1993 continues today and has had a devastating impact on their future. The military regime terrorized the people and their assaults have resulted in over 2000 deaths. Thirty thousand more have been displaced and over 1000 have fled to neighbouring countries and beyond. Although a new government is now in control of Nigeria, only time will tell if this will have any effects on the lives of the Ogoni.

ENVIRONMENTAL JUSTICE AND SUSTAINABLE ENERGY

The link between environmental justice and the production of fossil fuels is instructive when one understands the relationship between dirty extraction industries and environmental degradation and pollution. The exploitation of land and the people who inhabit it are inevitable. This section reviews the intersection of environmental racism and fossil fuel production. It attempts to show the importance of the development of sustainable renewable energy sources for the achievement of environmental justice. It strongly suggests that the struggle against environmental racism is a building block in the fight against the continuation of fossil fuel production or unsustainable, non-renewable energy production.

It may be hard to imagine, but at one time the US had about as much oil as Saudi Arabia and freely used this abundant fossil resource to produce our modern society. These reserves allowed the rapid development of farming, manufacturing, housing, transportation and military technologies.

In 1997 the world energy industry produced 26.4 billion barrels of petroleum, 81.7 trillion cubic feet of natural gas and 5.2 billion tons of coal. The combustion of these carbon-based fuels released 6.2 billion tons of carbon into the atmosphere (Natural Resource Defense Council, 1997). Astonishingly, only 122 companies are responsible for 80 per cent of carbons from these fossil fuels. Moreover, 22 per cent of the carbon pollution comes from fuels produced by only 20 private companies. Even more bothersome is the discovery that just five private global oil corporations – Exxon Mobil, BP Amoco, Shell, Chevron, and Texaco – produce oil that contributes approximately 10 per cent of the world's carbon emissions (Brotsky et al, 1999). The annual global burn rate presently exceeds 1 million years of fossil accumulation (Sustainable Energy Independence, 2001). Energy prices are lower than replacement costs and each year over one million years of fossils are burned. This has been described as 'energy bankruptcy' (Sustainable Energy Independence, 2001).

The US imports about 75 per cent of the nuclear fuel, 65 per cent of the oil and 16 per cent of the natural gas it uses. One could project that with the expanding demands for oil and other fossil fuels, the world demand for oil will exceed production capabilities.

This process is not sustainable. Although we are not predicting that the world will run out of oil, we are predicting that the difficulty in retrieving it will increase as the demands for fossil fuels increase and exceed production.

The demand for fossil fuels is steadily increasing and will continue to increase as other nations attempt to modernize. OPEC has 75 per cent of the remaining oil reserves and represents only 10 per cent of the world's population. China now uses about 18 per cent of the US demand for energy but plans to soon equal the US in productivity and energy consumption. India uses somewhat less energy than does China, but also plans to increase productivity and energy consumption. This new demand by Asia for using energy 'like America' is already causing friction and will continue as world production of fossil fuels fail to meet supply demands. This system is not sustainable (Sustainable Energy Independence, 2001).

The production of fossil fuels is the greatest contributor to global warming and subsequent climate change. These companies that are the refiners and marketers of oil and gas are extremely powerful. They are accused of using their political power to prevent technological transformation and maintain the status quo and have also been accused of buying public and scientific opinion (Brotsky et al, 1999).

Presently, over US$20 billion dollars in subsidies is provided each year by the US government for fossil and nuclear fuels. Not to be overlooked are the military programmes provided to assure the delivery of oil from foreign countries to the US. The costs to assure the delivery of foreign oil are said to exceed US$100 billion each year. Additionally, US$10 billion is spent for surveillance, intelligence collection and related 'efforts' to thwart anti-

Americanism, including terrorism aimed at disruption of offshore oil commerce (Sustainable Energy Independence, 2001).

The latest proposal advocated to reduce the US dependency on foreign oil is the proposal by Congress for drilling in the Alaska National Wildlife Refuge (ANWR) and the addition of US$30 billion in new federal incentives for fossil and nuclear energy projects. This proposal will increase the already record breaking profits of fossil energy companies, but it will in no way affect America's dependency on foreign oil or sustainable energy independence. Moreover, it further illustrates the problem of the demand for fossil fuels exceeding production. If the ANWR was completely mined, it would only supply the US with approximately 16 months of oil at our present consumption rate. It would also take 10 to 15 years for delivery and it is expected that the demand would have increased over that period of time. It seems obvious that our continued dependency on unsustainable supplies of oil, natural gas, coal and uranium can only lead to world conflict, economic inflation and hardship (Sustainable Energy Independence, 2001).

OIL AND GAS EXPLORATION IMPACTS

The US makes up only 5 per cent of the world's population, but consumes nearly 25 per cent of the global energy production. US citizens enjoy a better quality of life that most other citizens of the world. This is due in part to abundant endowments of oil, natural gas and coal. This has allowed the US to grow more crops, manufacture more goods, travel more often and maintain a military might that is the greatest in the world (Sustainable Energy Independence, 2001). These benefits, however, have come at great cost to minorities and the poor in the US and to other people of colour around the world.

This chapter investigates the effects of oil and gas production on the lives of the Ogoni people in the Niger Delta and the Norco community in Louisiana. The devastating results, however, have been the same for both communities. The exploration efforts have resulted in environmental pollution and degradation of the soil, air and water. Even more disturbing has been the tremendous health effects the communities have suffered because of environmental pollution.

SHELL'S OPERATIONS

The Shell Petroleum Development Company manages Shell Oil operations in Nigeria in a joint venture agreement with the Nigerian National Petroleum Company. This operation is responsible for nearly 14 per cent of Shell's production, second only to the US (Fact Sheet on the Ogoni Struggle, 2001). The impact of oil production on the Nigerian economy cannot be overstated. Approximately US$30 billion in revenue has been generated from oil production in the Niger Delta. The Nigerian government has become almost completely dependent on oil sales with oil accounting for a whopping 80 per cent of government revenue (Fact Sheet on the Ogoni Struggle, 2001).

The Ogoni people have historically maintained an agricultural and fishing society, living peacefully with their neighbours, trading food grown on the rich soil of the Niger Delta for other goods. This soil, which maintained the economic and cultural lifestyle of the Ogoni people for hundreds of years, is being destroyed. The Ogoni people have been forced to live with over 100 oil wells on their land. The environmental effects have been severe, resulting in almost 3000 separate oil spills between 1976 and 1991 averaging 700 barrels each spill. Responses by Shell to oil spills in the Niger Delta have been reported by the Ogoni people as being most often slow and sometimes very damaging. They report that a major spill in 1970 was torched, causing irreparable damages to the land and is even now leaking oil into surrounding water supplies (Fact Sheet on the Ogoni Struggle, 2001). A more recent experience is reported from a farmer who is waiting for a pipeline to be fixed six weeks after it began leaking (Fact Sheet on the Ogoni Struggle, 2001).

Oil exploration in the Niger Delta has resulted in numerous oil spills that have rendered the land and the crops useless. The complete degradation of the land and the destruction of villages and displacement of people for access roads and plant expansions have destroyed the lives of many people (Shell in Nigeria, 2001).

The environmental devastation of oil exploration has also resulted in water pollution in the Niger Delta. The importance of clean water for human survival is clearly understood by the Ogoni people. They are also subsistence fishers. This increases the importance of water for survival. The Ogoni water supply has been compromised by oil exploration. The contamination of lakes and streams has destroyed the natural habitat for fish, resulting in the loss of not only the historic occupation of fishing for the Ogoni people, but also the loss of a rich source of protein for subsistence. Their health and their lifestyle have been severely compromised by oil exploration.

One of the most visible affects of oil exploration in the Niger Delta has been the loss of mangrove trees in the swamps. The mangrove forests of the Niger Delta are one of the most important wetlands ecosystems of the world (Shell in Nigeria, 2001). The mangrove served as the major source for firewood and a habitat for seafood. It was an abundant source of seafood such as oysters, crabs and mussels. This important food source, however, has been unable to survive the toxicity of the water from the oil spills. The fishermen from these areas, who also caught crabs from the mud banks now destroyed by crude oil, have been unable to survive. They have lost the source of their occupation and are now impoverished (Shell in Nigeria, 2001).

The Ogoni people of the Niger Delta bear the greatest similarity to the Norco community of Louisiana in the devastating effects of air pollution from oil refineries. The Ogoni people live in the shadow of huge gas flares that burn 24 hours a day. Some flares have been burning for more than 30 years. The villagers, as do the Norco community, live with the constant noise of the flares. The air pollution produced by the flares results in acid rain and the surrounding community suffers from respiratory problems similar to those described by the Norco community. Shell pipelines on Ogoni land are above ground and pass through villages and over what was once agricultural land. Shell pipelines in Norco are fence line and border the community.

Environmental racism can be shown to be a factor in both the Ogoni and Norco cases. Disparate treatment as it relates to responses by Shell to the community or environmental protection by the government can be seen in both Ogoniland and Norco. In both cases, the government supports the oil exploits of Shell. This point was clearly made in a case in the UK where a pipeline required 17 different environmental assessment surveys before construction. Although Shell operates in 110 countries, 40 per cent of its spills worldwide have occurred in the Niger Delta. To the contrary, the Ogoni have never seen a single environmental impact assessment. The fights between the Norco community and the Louisiana Department of Environmental Quality are legendary. As discussed earlier, the Norco community has had to take extraordinary steps to prove to government agencies responsible for environmental protection that Shell has violated their permits and have exposed the community to toxic air pollution. It has only been after extraordinary efforts like the 'Bucket Brigade' that communities have got the attention of the government (Sustainable Energy Independence, 2001).

What is clear from this analysis is that discernible differences are made between people of different racial and ethnic backgrounds in regards to who will bear the burden of this non-sustainable processing of fossil fuels. It is also clear that persons of the same racial or ethnic groups are less likely to bear the brunt of the burden of the processing of fossil fuels and that sometimes extraordinary efforts are made to assure the environmental protection of communities that represent the racial or ethnic groups of the industry. It is therefore easy to deduce that racism plays an important role in the perpetuation of environmental injustice. In fact, racism legitimizes the exploitation of people and the degradation of their land by proclaiming them less worthy than others. Human differences become a marker for the people and their land. Environmental racism is one factor that perpetuates the continuation of the non-sustainable fossil fuel industry.

It is also clear that environmental racism is global and that only a concerted effort that is transnational in scope can impact the reduction of pain suffered by humans and all other species through this processing.

One possible answer to this dilemma is a worldwide campaign for sustainable energy independence to avoid economic difficulties, environmental exploitation and degradation and international conflicts (Sustainable Energy Independence, 2001). The US has already fought one war over oil in Kuwait. This could become the war of the future – 11 September 2001 has already put us on alert for such occurrences. This world could be held hostage by terrorist attacks on oil reserves. In the words of one writer, 'we taught the world how to base politics and economics on energy that's distributed at far less than the cost of replacement' (Sustainable Energy Independence, 2001). We should therefore become leaders in the transition to sustainable energy technology and policies.

CONCLUSION

As discussed earlier, Louisiana holds many similarities to Nigeria with its shift in its economy to an almost total dependency on the petrochemical industry. Likewise, the poverty and environmental devastation that seems synonymous with large dirty extraction industries like the petroleum industry bears an even greater similarity. The people affected by the oil industry in both Louisiana and Nigeria represent minority groups within those countries. For the same reasons, a lack of political clout and power, in the Nigerian incident the oil is stolen from the people and processed on their land. In the Louisiana incident, the plants necessary for processing the oil (refineries) are placed in the communities of the poor and disenfranchised minority. In both cases, transnational corporations are largely responsible for the environmental devastation, destruction of a culture and livelihood of the people whose lands they inhabit.

In both cases it can be argued that Shell and others have violated the human rights of the people whose lives they have destroyed. Ken Saro-Wiwa lost his life in his attempt to make these human rights' violations known to the world. While his attempts were thwarted, the citizens of Louisiana, including Norco, are taking their case to the United Nations Commission on Human Rights. Their basic argument is that the environmental racism that occurs in Louisiana is a violation of the International Convention of the Elimination of All Forms of Racial Discrimination, to which the US is a signatory. While most Americans tend to think locally, more and more, in every facet of life it will be necessary to think globally.

Most Americans have already made these connections when we speak, for example, on subjects related to the education of our children. How do we measure up with the rest of the world, particularly in science and mathematics? Most Americans would reply that we are falling behind and that is not a good thing. As we observe the development of transnational corporations and the increasing political and economic weight of the World Bank, thinking globally is the way of the future. In the area of the environment, there is a direct relationship between the 'increasing globalization of the economy and the living spaces for many of the world's people' (Robinson, 1998).

Increasingly, these types of environmental degradation or environmental racism are being seen as human rights violations. As we as a society begin to think globally, making the connection between what happens locally as in Norco and what happens globally as in Nigeria, the society will be better capable of responding to the government and their policies that support the exploitation of transnational corporations.

Race, politics and pollution all combine to place minorities and the poor at greater risk from exposure to toxic chemicals around the world. Moreover, it is more likely that they, in contrast to their white counterparts, will live in a place that has been devastated by large dirty extraction industries and will live in closer proximity to these polluting facilities. This pattern is being repeated around the world and is supported by larger governmental policies.

As citizens' groups make the connection and as they begin to think globally in terms of the environment, governmental policies will be challenged as they are presently and charged with human rights violations. The difference is that these charges will cease to be viewed as an anomaly, but will be viewed as the norm for exposing environmental degradation and racism as the world's most serious threat to human life and peace. This is a battle for the lives of our communities, our children, our livelihood and our culture. It is a battle that we must win!

REFERENCES

Bell, R (1998) 'Norco panics as Shell chemical tank rumbles', *The Times-Picayune*, 9 December, pB-1

Blers, J M (2000) 'Blown off', *The Times-Picayune*, 6 June, pA-1

Blers, J M (1999) 'EPA puts LA & Texas chemical plants on notice: Facilities ordered to reduce accidents', *The Times-Picayune*, 8 November

Blers, J M (2000) 'Motiva broke air, water rules EPA officials say', *The Times-Picayune*, 6 June, pA-1

Blewett, W K and Anch, V J (1999) *Experiments in Sheltering-in-Place: How Filtering Affects Protection against Sarin and Mustard Vapor*, ECBC-TR-034, Edgewood Chemical and Biological Center, Aberdeen Proving Ground, Maryland

Brotsky, C, Bruno, K, & Karliner, J (1999) *Greenhouse Gangsters vs. Climate Justice*, Transnational Resource & Action Center, San Francisco

Concerned Citizens of Norco, Sierra Club-Delta Chapter, Xavier University Deep South Center for Environmental Justice, Earth Justice Legal Defense Fund – New Orleans Office (1999) *Shell-Norco Toxic Neighbor: The Case for Relocation*

Communities for a Better Environment (1998) *Preliminary Results of Bucket Air Samples Taken in Norco, LA*, 8 December

Edgewood Chemical and Biological Center June (1999) *Experiments in Sheltering-in-Place*

Environmental Defense Fund (EDF) (1997), EPA TRI data: *EDF Scorecard*, http://www.environmentaldefensefund/scorecard.org

EDF (1999), *Pollution Prevention Performance Rankings Among Oil Refinery Facilities*, EDF, New York

Environmental Protection Agency (EPA) (1997) *Toxics Release Inventory*, State Fact Sheets, Louisiana, Government Printing Office, Washington, DC

Fact Sheet on the Ogoni Struggle November (2001) http://www.ratical.org/corportations/OgoniFact.html

Gray, L (1999) 'Test Claims Shell Released Toxins', *L'Observateur*, 9 January, pA-1

Harden, M (2000) '*Beyond Shell Oil Company's PR: Dangerous Ground in Norco, LA*', Earthjustice, New Orleans, Louisiana

Karmatz, L and Labi, A (1998) 'Special Report: Corporate Welfare', *Time Magazine*, vol 152, p9

Louisiana Association of Business and Industry (LABI) (1998) '*An Open Letter to Louisiana Business People*', http://www.labi.org, 27 April

Louisiana Department of Economic Development (LDED) (2000) '*National Marketing*', www.lded.state.la.us, November

Louisiana Department of Environmental Quality (1995) *Toxics Release Inventory*

Louisiana Department of Environmental Quality (LDEQ) (1998) *Criteria Air Pollutant Data*

Natural Resources Defense Council (1999) '*King pins of Carbon: How Fossil Fuel Producers Contribute to Global Warming*', available: www.nrdc.org/ndrcpro/carbon/kocinx.html July (Accessed: 2001)

Nauth, Z and the Louisiana Coalition for Tax Justice (1990) 'The Great Louisiana Tax Giveaway', Baton Rouge, Louisiana Coalition Inc

Report to the National Environmental Justice Advisory Council (1998) *Shell's Notice to Our Neighbors*, 8 December, NEJAC, Washington, DC

Robinson, D (1998) 'Environmental devastation in Nigeria: Why we should care', unpublished monograph, *International Possibilities Unlimited*, pp41–42, Washington, DC

Robinson, D (1999) 'Environmental devastation at home and abroad: The importance of understanding the link', unpublished monograph, *International Possibilities Unlimited*, Washington, DC

Shell in Nigeria (2001) http://www.maangstavat.fi/oileng/chartity.htm

Sustainable Energy Independence (2001) http://www.clean-air.org

Swerczek, M (1999) 'Plant's neighbors bag their air', *The Times-Picayune*, River Parishes Bureau, 21 June, pB-1

Templet, P (1999) 'Grazing the commons: An empirical analysis of externalities, subsidies and sustainability', *Ecological Economics*, vol 12, pp141–59

Templet, P (1997) *The Full Economic Costs of Louisiana's Oil/Gas and Petrochemical Industries*, People First: Developing Sustainable Communities, http://www.leanweb.org/pub

Thanos, N D (2000) *Economic Development, Corporate Accountability and the Environment: Comparative Case Studies from Costa Rica and Louisiana*, unpublished thesis, Tulane University, New Orleans, LA, http://www.tulane.edu/~eaffairs/thanos.html (cited May 9, 2000)

Thibodeaux, R (1988) 'Refinery blast is traced to ruptured pipe', *The Times-Picayune*, 4 May, pB-1

Thibodeaux, R & Frazier, L (1988) 'Explosion hit town like a hurricane and six still missing in NORCO blast', *The Times-Picayune*, 7 May, pA-1

US Environmental Protection Agency (1999) *Selected Information on Hazardous Releases, 1987–1997*, EPA, Washington, DC

Chapter 7

Identity, Place and Communities of Resistance

Devon G Peña

INTRODUCTION

This essay is a contribution to the critical study of discourse and identity politics in the environmental justice movement. My focus is on a particular set of environmental justice struggles that linked the restoration of traditional land and water rights with ecosystem management and social justice. I am especially concerned with outlining certain political and ideological processes that are associated with the formation of multiple, shifting identities and 'hybrid' project strategies. I am also concerned with exploring place-based identities and their role in the formation of the ideological discourses of environmental justice.

The case study involves the Sangre de Cristo land grant in the Rio Culebra watershed located in south central Colorado's San Luis Valley (see Figure 7.1). This study allows us to examine how the concepts of sustainability, ecosystem management and social justice were framed in the context of discourses of place, identity and legal rights to contested space (ie property rights to land or water). How do differently positioned social actors seek to attain legitimacy in articulating their claims to contested space? Which forms of spatial logic do they invoke and seek to legitimize? What knowledge bases underlie such claims? What are the local, regional, national and global conditions that influence the process of mobilization of resistance? How are different organizational forms shaped by the discursive regimes of place and identity in relation to the legal and political definition of land, water, and other rights? What are the constraints and contradictions facing different actors? To address these complex issues, this chapter presents a site ethnography focused on the dynamic interactions of identity and place-making. But this is done in the context of a theoretical discussion of framing discourses in environmental justice movements (see Taylor, 2000, Benford and Snow, 2000).

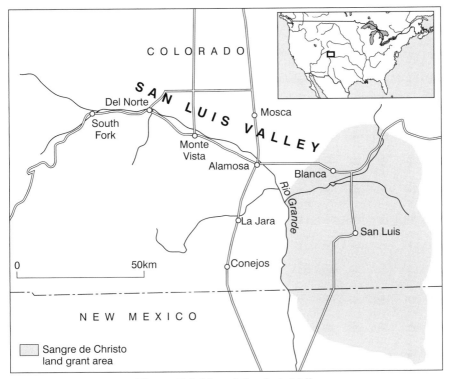

Figure 7.1 *Map of San Luis Valley*

Of particular relevance to this study is the theory of identity, networks and social movements first outlined by Castells (1997). Castells juxtaposes two types of spatial logics, that of the 'space of flows' and that of the 'space of places'. The space of flows 'organizes the simultaneity of social practices at a distance, by means of telecommunications and information systems', while the space of places 'privileges social interaction and institutional organization on the basis of physical contiguity' (1997, p123). In the so-called network society, the 'most dominant processes, concentrating power, wealth, and information, are organized in the space of flows'. But

> [m]ost human experience, and meaning, are still locally based. The disjunction between the two spatial logics is a fundamental mechanism of domination in our societies, because it shifts the core economic, symbolic, and political processes away from the realm where social meaning can be constructed and political control can be exercised. Thus, the emphasis of ecologists on locality, and on the control by people of their living spaces, is a challenge to the basic lever of the new power system. (Castells, 1997, p124)

The conflict between the space of flows and the space of places underlies the processes that have come to be called globalization. But local cultures and communities can also produce their own space of flows on a global level. The transnational suburbs of the Zapotec Indians that gave rise to the term

'Oaxacalifornia' are a good example of these reticular subaltern organizational forms (Kearney, 1996, Davis, 2000, Peña, 2001a, b, c).[1] My study is a further affirmation of this concept of 'place-based identity politics'.

Castells discusses three identity formations in what he conceives as the discursive politics of domination and resistance: (1) legitimizing identity, (2) resistance identity, and (3) project identity. Legitimizing identities are produced by the dominant institutions of society to extend and rationalize domination by particular social actors. Members of dominated social groups who seek to revalue their marginalized positions and stigmatized livelihood practices generate resistance identities. Project identities are created when social actors involved in resistance articulate new identities to redefine their positions in society and, by doing so, transform the overall structure of power through strategic projects for social change (Castells, 1997, p8).

The interplay among these three identity formations produces a set of 'framing discourses' that are used by differently positioned social actors to articulate contested claims to socio-spatial power (Taylor, 2000). An example is the framing of the legal and political economic authority to direct the organization of the space of places. The discursive politics of place in this way has significant legal, political and socio-economic ramifications. In this work, I am interested in exploring how local communities construct and navigate these contradictory identity formations to coalesce strategies of resistance to globalization and modernization.

The setting: Historic acequia communities and globalization

Over five generations preceding enclosure, acequia farmers in the Culebra basin valued the upland forests of the Sangre de Cristo land grant as the source of water for their snow-fed irrigation system. They undertook no effort at extraction of timber or metals. The montane and sub-alpine forests and meadows provided local space for the exercise of usufruct rights to graze, hunt, fish, gather wood and wildcraft on common lands. La Sierra was 'inhabited wilderness', a wild but familiar place intricately woven into the everyday lived experience and material culture of the local community (Peña and Martínez, 1998, 2000, Peña, forthcoming, 2001b). The Sangre de Cristo land grant was the last of the great Rio Arriba ejidos to be legally enclosed or partitioned. In 1960 descendants of President Zachary Taylor furtively bought the last unfenced portion of 77,000 acres. The intervening 40 years witnessed a veritable 'range war' and legal conflict, the infamous Rael vs Taylor land rights case, waged over the common lands the locals call 'La Sierra'. By the 1990s the struggle had evolved into one of the most closely studied environmental justice movements in a land-based Chicana/o community (Peña and Gallegos, 1993, 1997, Peña and Mondragon Valdez, 1998, Peña, 1998). Some observers consider the Rael vs Taylor land rights case to be the longest remaining shadow cast by the 1848 Treaty (Weinberg, 2000). The more recent campaign against the logging destruction of La Sierra was characterized by the New York Times as the 'hottest environmental dispute' in the Rocky Mountain region (Brooke, 1997).

La Sierra looms large above the ribbon-like strings of acequia farming communities that inhabit the high altitude, arid land environment of the San

Luis Valley along the banks of the Culebra River and its tributaries. Local multi-generational farmers have developed an irrigation system, the acequia, which is based on the spring snowmelt cycle, gravity-flows, earthen ditches, cooperative maintenance and operation of the ditches, and communal self-governance. The acequia farms of the Culebra watershed are renowned for their sustainable agriculture and unique, biodiversity-friendly, overlapping patchwork landscape mosaics (Peña, 1998, 1999, 2001a, Rivera, 1999).

Costilla County may be one of the last strongholds of the historic multi-generational acequia communities but it is also one of the most ruthlessly sub-divided areas in Colorado and is the only county in the state without any public lands. There are approximately 60,000 sub-division plots in the county (ranging in size from tens to thousands of acres). The enclosure of the land grant commons by the Taylor Ranch is not the only force underlying privatization and development of the local landscapes. There are about 38,000 different owners of these sub-division properties. Among these are individuals and corporations from places as far-flung as Germany, England, France and Spain in Western Europe; Japan, Malaysia and Taiwan in the Pacific Rim; and Mexico as well as other points in Latin America. Owners reside in places like Los Angeles, Seattle, New York, Dallas, Houston, Chicago, Kansas City, Denver and San Francisco.[2] These qualities make this case study particularly intriguing for the study of local communities of resistance in the context of political economic globalization.

IDENTITY, PLACE AND RESISTANCE

Like local land-based cultures in other parts of the world, the acequia communities of the Rio Arriba recently experienced a qualitative transformation in the environment and their relationship to it. The late 19th–early 20th century arrival of railroad mass markets, enclosure of common lands, proletarianization of rural communities and industrialization of forestry, grazing, mining, agriculture and recreation, greatly accelerated and expanded the process of environmental degradation and posed a direct threat to the continued survival of the acequia communities (Peña and Martínez, 1998). These forces led to the emergence of local and regional social movements for environmental and economic justice that emphasized recovery and protection of the watershed commons. An examination of Castells' model of identity formations and socio-spatial logic in the context of a case study of environmental justice struggles over land and water rights and ecosystem management in an enclosed common land is presented below (see Table 7.1).[3]

Legitimizing identities

La Sierra as private property
The most prominent legitimizing identity positions were staked out by the Taylor family of New Bern, North Carolina, acting as absentee owners of the enclosed commons. The Taylor Ranch managers espoused a spatial logic that defined La Sierra as private property. This position was articulated through

Table 7.1 *Identity formations and socio-spatial logic in the ecological politics of the Culebra watershed*

Identity formations	Socio-spatial logic	Practical knowledge (mètis or strategies)	Sources of legitimizing narratives project
Legitimizing identities:			
private land owner/timber manager	La Sierra as private property: 'Taylor Ranch timber and watershed management plan'	commercial timber extraction; legitimation of private timber plan (scientific and fiscal forestry)	private property law (as enjoined with civil rights); 'tragedy of the commons'; racialized environmental history
state, public lands manager	La Sierra as potential public land (state park or wildlife management area): 'The Crown Jewel of Colorado State Parks'	state parks, open space and wildlife management (timber and wildlife management)	state laws; public welfare values; scientific forestry; history of public land management
Resistance identities:			
acequia irrigators/farmers	La Sierra as water source: 'Sin agua, no hay vida', 'La semilla es es regalo', 'La tierra es familia'	maintaining and operating acequias; conserving and using heirloom land race crops; planting, cultivating and harvesting (agroecology)	law of thirst, customary law of the acequia (watershed commonwealth); inherited local practices; narratives of place
curanderas(os)	La Sierra as habitat for medicinal and edible plants; 'La sierra da remedios'	wildcrafting of remedios (ethnobotany)	oral tradition; ejido rules of commons use (customary usufruct); narratives of place
hermana(o)s	La Sierra as spirit of place: 'La tierra es sagrada', 'Arbol de la vida'	fasting, praying, alabados (ethnoreligious ethics)	Jesus Nazareno; 11th Commandment; narratives of place
Project identities:			
land grant heirs	La Sierra as land grant commons: 'La sierra no se vende', 'Perdimos la libertad, nos encercaron'	hunting, fishing, grazing and wildcrafting rights; after enclosure: trespassing; litigating to restore historic use rights of land grant heirs to commons (cultural nationalism and human rights)	ejido rules of common use; land grant law (negotiations with Anglo law); Treaty of Guadalupe-Hidalgo: international treaty rights
'Brown & green' activists	La Sierra as homeland watershed: 'Sin agua no hay vida y sin tierra no hay paz'	researching/litigating to protect watershed; promoting local/state acquisition to restore common property regime; organizing acequia association and land trust ('applied' ethnoecology)	law of hybrid acequia association; *mutualistas*; environmental justice in local land use planning (convergence of place-based stewardship ethics, local knowledge and conservation biology)
white eco-activists	La Sierra as wilderness, habitat: 'Stop the logging', 'Salva la Sierra' solidarity with environmental justice principles	organizing non-violent protests and other political and legal actions, support and affinity groups, collaboration	ecosystem science and conservation biology in local land use planning; deep ecology and with acequias on scientific research (eco-system science and conservation biology)

scientific and legal discourses to rationalize and legitimize enclosure and activities like commercial timber extraction. A 'timber and watershed management plan' was drafted to define the practical knowledge base and formulate a project strategy. In April 1995 the executor of the Taylor Ranch Estate, Zachary Taylor, sent a letter to the local community outlining the plan. He wrote of a concern for 'water quality and quantity, watershed protection and enhancement [as] a major component of … harvesting efforts' (Taylor, 1995, p1). The willy-nilly extractive industrialization of La Sierra was now a form of benign ecosystem management. More, La Sierra's forests were 'sick' and had to be nursed back to health through selective logging. The logging plan was called tree 'harvesting', as if to render the scale of ecological disturbance more benign. Taylor worked to create the appearance that logging was like farming, an idea that would presumably gain the trust of the acequia irrigators.

But Taylor could not mask the massiveness of the cuts nor the damage from extensive road and skid-trail construction behind the simulacra of habitat improvement and watershed protection. Was the logging benign 'thinning' or another more destructive form of timber extraction? Aerial and ground surveys showed that the operation actually involved a massive 'over-storey' removal of close to 200 million board feet on 34,000 acres of land with 'merchantable' timber rather than 'selective thinning'. This was confirmed as the largest timber cut on private land in Colorado history. Taylor had to defend this plan against scientific experts including local farmers, who were associated with the Culebra Coalition, and who publicly denounced the Taylor Ranch logging operations as a 'world-class case of massive deforestation' (Robert Curry in Wet Mountain Productions, 1998). The impact of the logging operations on the hydrology of the watershed, and on stream hydrographs, became a major point of contention. Taylor quickly sought to de-legitimize this criticism (Taylor, 1995, p1).

This identity invoked six discursive constructs in its quest for legitimacy. First was the idea of La Sierra as private property to be managed by the Taylor family through the combination of the 'new ecology' (top-down ecosystem management) with free market economics (Wolf, 1995). This was related to a second tenet that posited private property as an inviolate civil right. The third construct privileged top-down scientific and fiscal forestry as the more sustainable and fairest model for management of La Sierra. A fourth idea, a derivative of the preceding three, was that local efforts to impose land use regulations on timber and other development activities constituted a violation of the landowner's civil rights to use his private land without undue interference by local government. A fifth and more explicitly ethnocentric idea impugned the presumed destructive ecological behaviour of a lawless and violent local culture. A sixth construct presented the anti-logging campaign and land rights lawsuit as the work of a few disgruntled outsiders who were misleading a handful of noisy locals.

Taylor sought public support, in part, by invoking the work of apologetic environmental historians and pundits who were willing to contribute vituperative attacks against a local 'Hispanic' version of the tragedy of the commons. It was an interloping timber baron, and not the multigenerational farmers, who would save La Sierra from environmental degradation. The public was, in effect, asked to believe that lawless and violent ecological thugs had victimized a law-abiding and innocent land developer and timber baron (Wolf, 1995, Hess and Wolf, 1999, for a critique, see Peña and Martínez, 1998). In the end, Taylor Ranch managers ultimately relied on the legal system to legitimize this identity position. This discursive strategy was important to local ecological politics because it asserted that whoever owns private property rights largely controls the rules encoding processes of spatial organization, resource extraction and environmental change (Goldman, 1998).

La Sierra as public land

The state of Colorado, a prospective buyer of La Sierra, articulated another legitimizing identity. In 1993, Governor Roy Romer signed an executive order establishing the Sangre de Cristo Land Grant Commission (LGC). The state wanted to resolve the vexing land rights case to move forward with plans to showcase La Sierra's 14,000 foot-high Culebra Peak as the centrepiece of a new state park and 'Crown Jewel' of Colorado State Parks. The LGC was convened to resolve the issue by means of a major real estate transaction involving public and private funding. The fact that Costilla County was the only county in Colorado that lacked public lands strengthened the state's position that this was a reasonable and justifiable plan that addressed critical issues of concern to the public welfare.[4]

The LGC brought different local, state and national stakeholders together in an effort to develop a strategy to acquire the property and a plan for a bold experiment in participatory management that incorporated the historic common use rights of the land grant heirs.[5] Despite good intentions, the LGC became embroiled in a number of difficult disputes, not the least of which was a heated debate over the ability of the local community to act in an ecologically responsible manner. The tragedy of the commons argument was invoked throughout the three years of the Commission's work. Some of the governmental agency representatives on the Commission were opposed to the local community's insistence on a major role in managing the land. They recounted the tired story of overgrazing by Hispanic livestock herders and pointed to the degraded condition of the 633-acre remnants of the village common lands known as 'San Luis Vega'. The Vega was overgrazed, but this was more a result of the loss of access to the 77,000-acre upland commons than a consequence of poor stewardship of resources by local acequia farmers (Peña and Mondragon Valdez, 1998, Peña and Martinez, 1998). But the subtleties and ambiguities of a conflicted multicultural environmental history were lost on most of the agency representatives serving on the LGC.

Despite these divisions, the Commission managed to arrive at a rather quick consensus. The State and La Sierra Foundation (LSF), representing local interests, would purchase the land, establish a 500-acre state park and treat the rest of the expansive mountain tract as a wildlife management area. The management plans involved local participation and the restoration of some semblance of historic use rights. Negotiations with Taylor continued for more than four years. In April 1998 the state of Colorado and the LSF made a final offer to Taylor for the sum of US$18 million on a 53,000-acre portion. As expected by many local observers, Taylor rejected the offer and this marked the end of the state's direct involvement with the land rights struggle.

Resistance identities

La Sierra as homeland common
Resistance identities in this struggle revolve around the place-based memories and oppositional narratives of the multigenerational acequia farming families. 'Sin agua, no hay vida' is the local aphorism – 'Without water, there is no life'. Another *dicho* (aphorism) is 'La tierra es familia' or 'The land is family'. These are expressions of a land ethic and deeply felt 'memory maps' that farmers and other local residents have developed over generations. Whatever their principal place-based identities – as irrigators, *curanderas*, or *hermanos* – local people have constructed their positions of resistance through the recollection of multigenerational narratives of place. Thus, locals are commonly heard saying: 'We don't want to become another Taos' or 'We don't want McDonald's or Taco Bell on Main Street'. These identity formations have been used by locals to mobilize the widespread opposition to development projects including the BMG gold mine and proposals for an industrial hog farm, state prison, solid waste dump and assembly line factories (Peña, forthcoming).

Resistance identities extend deep into local subaltern history. In the 1890s absentee land speculators had arranged for the official survey of Costilla County to impose the rigid square grids of the Township-Range-Section system. Speculators moved against local shepherds and filed a series of trespassing lawsuits (Peña, forthcoming). The speculators retreated when the bubble burst on the land market in southern Colorado, and locals persisted in customary use of the common lands. The right to hunt for subsistence, via usufruct, remained intact until the 1960s when it was criminalized as poaching by Taylor Ranch and state and federal wildlife agents. By 1989 and the infamous 'Rambo-styled' midnight raid on San Luis-area 'poachers', the local Chicanas/os had been type-cast as ignorant and lawless. How could one expect a culture deemed to be wasteful with water to be any more proficient at wildlife management? Chicanas/os, it seemed, did not have a conservation ethic. They would sooner poach wildlife than control their overgrazing livestock (Wolf 1995, p272).

The resistance identities of the acequia farmers derive from lifelong experiences with racial discrimination and racist ideology. We have the persistence of a tendency by the state and capitalist stakeholders to treat Chicanas/os as primitive, inefficient and maladapted farmers who are doomed to extinction. Despite these racist projections, the farmers' ethnoecological knowledge of the land, water, plants and animals remains an enduring source of moral authority and ecological legitimacy that buttresses local claims to the traditional 'watershed commonwealth'. Moreover, under the Doctrine of Prior Appropriation, the acequias enjoy the oldest adjudicated water rights decrees in the state of Colorado, a source of both legal power and fortitude. In a continuing battle over these senior surface water rights, both state and capitalist interests, who covet acequia water, view these rights as a 'waste of resources' (Hicks and Peña, 2001).

Figure 7.2 *Acequia farmer's political cartoon*

Project identities and organizational hybridity

Over the course of the 1990s resistance identities were explicitly linked to a strategic project seeking to transform the local social order. Acequia farmers directed narratives of resistance towards a public critique of the scientific discourse used to legitimize the Taylor Ranch timber and watershed management plan. They situated themselves as 'experts' and used a wide range of venues and genres to express their critical views. One particularly interesting form of expression was the use of editorials and political cartoons, published in the local *La Sierra* newspaper. The acequia farmers recognized the timber and watershed management plan as an example of science in the service of power and profits. In one of the many political cartoons circulated during the 1995–1999 anti-logging campaign, Joe Gallegos satirized the claims made by the Taylor Ranch that the Stone Forest Industries timber operations were designed to increase water supplies and create wildlife habitat (see Figure 7.2). His favourite protagonist, Raul, is shown being reassured by a scientific forester that logging creates 'good elk habitat'. Raul replies by bringing attention to a logging truck that had just killed one of the local cows on a road leading to the Taylor Ranch. The scientist slips on the question: 'What's one cow for less water and more elk habitat?'

The public discourse initiated by the acequia farmers sought to de-legitimize the scientific and fiscal forestry of Taylor Ranch management. But the acequia farmers' narratives of resistance extended beyond the issue of historic use rights to the common lands (as in the LRC position) and instead focused more on strategies for local control of land use planning, acequia management of the watershed and engagement with a broader range of environmental justice issues.

This included the formation of a coalition with radical environmentalists to resist logging of La Sierra's watershed and implement a land use code and watershed protection zone based on principles of conservation biology and ecosystem management (Curry et al, 1996). These narratives linked watershed integrity, cultural continuity and social justice, and they were critical in the transition from resistance to project identities. By 1998 the acequias had formally organized themselves into the Colorado Acequia Association (CAA) to contest not just the enclosure of the commons by the Taylor family but to address a whole range of threats to the integrity of the watershed as the material basis of local agricultural livelihoods. This involved a shift from a purely cultural nationalist to an autochthonous bioregional strategy. But the member ditches of the acequia association were not alone among local civic institutions to generate project identities. The emergence and development of project identities followed five principal 'trajectories' involving (1) cultural nationalist, (2) environmental justice, (3) radical environmentalist, and (4, 5) two 'hybrid' organizational strategies with a definite bioregionalist framework (see Table 7.2).

1 Cultural nationalism: La Sierra as lost land grant

Cultural nationalist identity politics figured prominently in the oldest and most deeply grounded forms of local resistance; this was expressed through the identity positions of the Land Rights Council (LRC). The LRC grew out of the *Asociación por Derechos Civiles* (ADC, Association for Civil Rights), an organization established by Apolinar Rael in the early 1960s (see Peña and Mondragon Valdez, 1998). The ADC was part of a region-wide proliferation of land grant movement organizations active in northern New Mexico and southern Colorado during the 1960s and 1970s. The ADC eventually evolved into the LRC with the filing of the original Rael vs Taylor civil lawsuit in 1979.

The discursive politics of this identity derive from a long tradition of 'cultural nationalist' political thought and activism in New Mexico and Colorado. Like other groups in the land grant movement, the LRC connected the loss of land grants to the decline of cultural integrity. And it posited the loss of the Spanish and Mexican land grants within the context of a critique of a well-documented and sustained pattern of governmental violation of basic human and treaty rights. The problem of the enclosure of the Sangre de Cristo land grant was framed in terms of 'human rights violations':

> *While people in the US moralize about human rights violations in other parts of the world, we often avoid dealing honestly with our own violations. The human rights issue is complex. In a nutshell, this discussion involves the ethics of individual owner's [sic] who [sic] monopolize resources which are integral to the underpinnings of culture.* (Mondragon Valdez, 1994, p1, emphasis added)

La Sierra is constructed here as a set of natural resources that are the 'underpinnings of culture'. This is an important aspect of the identity formations that evolved over the course of the 1990s, resulting in the hybrid

Table 7.2 *Project identities and strategies in the ecological politics of the Rio Culebra (1979–)*

Project identity	Membership base	Representative organization	Project strategies
Cultural nationalism	land grant heirs acequia farmers	Land Rights Council (1979–)*	historic Rael vs Taylor case: litigation to restore historic use rights to former common lands; legal research by pro bono lawyers; historical research by land grant heirs and anthropologists
Environmental justice	local leaders acequia farmers research scholars	La Sierra Foundation (1993–) Rio Grande Bioregions Project (1988–)	a local/state partnership to purchase the Taylor Ranch; participative co-management and restoration of historic use rights; formal collaborative ecosystem and agroecosystem research: acequia farmers, natural scientists and social scientists
Radical environmentalism	white eco-activists	Ancient Forest Rescue (1995–) Colorado Greenpeace (1995–) Culebra Earth First! (1995–)	direct action to stop logging and protect wildlife habitat; solidarity with local efforts to restore use rights; ecosystem and conservation biology research by eco-activists (eg document local endangered species)
Hybrid 1	all of the above	Salva Tu Sierra /Culebra Coalition (1995–99)	direct, locally based action to stop logging; political support of Rael vs Taylor; scientific and legal research on La Sierra's biodiversity; fund-raising; land use planning advocacy in public hearings
Hybrid 2	all of the above	Colorado Acequia Association (1998–)	scientific and legal research to empower acequias to participate more effectively in the management of the Culebra watershed; land and water trust to protect historic acequia farmlands, sensitive watershed areas, open space, and habitat; collaborative research among loosely affiliated network of lawyers, scientists, social scientists and local acequia farmers

* The year in parenthesis is the year that the organization first became involved in local land rights and land and water use politics.

form of the Colorado Acequia Association. It is an identity implicitly based on ethical recognition of the watershed as the material or biophysical basis of local 'right livelihoods', and therefore as a 'place inviolate'. However, until October 1996 the LRC clearly emphasized the human rights. The organization did not articulate a clear public discourse on the environmental justice dimensions of the Taylor Ranch struggle.

The LRC was opposed to the purchase of the Taylor Ranch by the LSF/state partnership. The socio-spatial logic of this framework was evident in the LRC's aphorism, 'la Sierra no se vende, la Sierra no se compra' (the Mountain is not sold, the Mountain cannot be bought). This essentialist construct imagines a landscape that is somehow 'decommodified'. It imagines a place for use rights *even* in the context of the limiting and contradictory structure of private property rights, which have inscribed their socio-spatial power on a colonized and enclosed common land. Thus, Taylor's monopoly over La Sierra's resources was reduced to a problem of equity, due process and human rights, under-girded by a critique of the history of unethical and racist violations of international treaties by the federal and state governments and courts (García and Howland, 1995).

With the advent of industrial logging of La Sierra in 1995 the LRC was confronted with the sheer force and magnitude of the environmental justice issues. Prior to the logging the LRC and LSF had been deeply divided over the fate of the mountain and the best strategy to deal with the injustices of enclosure. Key members of the LRC were vehemently and loudly opposed to the LSF plan to purchase the land. They insisted that the Rael vs Taylor lawsuit should remain the principal focus of the community's efforts to restore the historic use rights to the commons. Several LRC activists voiced opposition to any 'deal with Taylor'. They viewed the LSF and state of Colorado plan to buy the land as unethical because it was seen as pre-empting the dues process rights of the Rael vs Taylor plaintiffs in their ongoing litigation. At a critical point in 1995 this opposition led to the defeat of an effort by La Sierra Foundation to establish a 'Regional Service Authority' (RSA). LSF worked to place a ballot question before the voters in the general election that would establish La Sierra Regional Service Authority (LSRSA), a rural development district with the authority to collect a mill levy over ten years to raise an estimated US$8–10 million for the local contribution to the purchase of the Taylor Ranch. One of the LRC activists filed a formal complaint with the Colorado Secretary of State alleging that the Costilla County Conservancy District (CCCD)[6] had misspent public funds by contributing US$250 for the publication of a voter education brochure.

By 1996 the LRC and the LSF had moved toward a symbolic reconciliation. In October the two organizations issued an important statement of solidarity at a time when the LRC had just agreed to rejoin the Governor's Commission to participate in the negotiations for the purchase of the land (La Sierra Foundation, 1993). The solidarity statement between the LRC and LSF was a sign that the two organizations had finally overcome the principal divisions that had undermined the LSRSA campaign in 1995. It also opened the door to a new set of negotiations with Taylor and the state of Colorado. The political and organizational climate at that point allowed the LRC to participate in the Land Grant Commission while moving the Rael vs Taylor lawsuit forward. In the meantime, the Culebra Coalition anti-logging campaign maintained political pressure on the state and Taylor to resolve the conflict.

2 La Sierra as watershed commonwealth

The emergence of La Sierra Foundation, in 1993, signalled an important shift in local strategies of resistance. From the start, LSF worked with an explicit environmental justice framework. The LSF was founded with one mission in mind – empowering the local community to acquire ownership of the Taylor Ranch through a direct purchase. The LSF organizers had a different take on the adequacy of the LRC strategy, with its focus on litigation to restore historic use rights. First, a legal victory in Rael vs Taylor was viewed as a highly unlikely prospect, given the historical lack of success with such litigation in New Mexico and Colorado. Moreover, the LSF believed that the restoration of historic use rights would not eliminate environmental threats to the watershed because such a legalistic approach would fail to resolve the limits imposed (on co-management) by the status of the land as private property. Even if use rights were restored, without a change in ownership, Taylor would continue logging and other commercial and extractive activities detrimental to the watershed. Only local ownership would resolve this vexing complication that resulted from the 'bundled' nature of private property rights. The LSF envisioned restoration of communal ownership through a local land trust organization. Local ownership of the land would give the acequia communities direct control over the fate of watershed protection.

The LSF's efforts to acquire the so-called 'Mountain Tract' were initiated in June 1993 with a planning meeting of a 'community land trust working group' brought together by the CCCD. The goal of the meeting was to outline a strategy to establish the LSF. Shortly thereafter, articles of incorporation and a 501(c)3 (US non-profit organization) application were prepared. In December 1993 the LSF launched a national capital campaign. But Governor Romer had intervened in August 1993 with the signing of an executive order establishing the Sangre de Cristo Land Grant Commission. The LSF was no longer alone in pursuit of ownership of the Taylor Ranch. From that point forward, the local community, including the LRC and LSF, participated in negotiations to purchase the land. As noted earlier, the Commission made a final offer of US$18 million in April 1998, which Taylor rejected. Within the year, Taylor had sold the land in two transactions to Lou Pai for a reported US$20 million.

La Sierra Foundation had planned for precisely this eventuality. LSF staff were concerned that Taylor would refuse *any* offer that involved substantive local participation. For that reason LSF staff had been quietly working on another option – the possibility of an independent condemnation of the Taylor Ranch property by the county, on the basis of precedents established through public nuisance and waste litigation. To initiate a taking (in which private property is condemned and acquired by a governmental entity), the county would have to pay the 'fair' market value for the Taylor Ranch. An independent appraisal prepared in 1993 established US$18 million as the fair market value of the 77,000 acres. An additional US$10–20 million would be needed to buy out the remaining timber contracts. The 1995 campaign to establish the LSRSA was envisioned as a critical component of a successful takings strategy since the funds from the mill levy could become available to compensate for condemnation by the county. Taylor rejected the Commission's offer in 1998, and the county and LSF were left without the resources promised by the LSRSA. In place, the LSRSA would have been collecting the mill levy for two years; it represented a multi-million dollar revenue stream for the local community, an exceptional financial asset that could have been extended beyond the original ten-year period. This strategy also represented an early effort by the community to shift from an 'equity-based approach' (LRC) to a 'community asset-building' strategy (LSF) in the search for a solution to the loss of the common lands. It represented an extension of environmental justice principles through a shift in emphasis from the critique of environmental racism to the elaboration of a 'just sustainability'.

Much of the work of La Sierra Foundation during the most intense direct-action period of the anti-logging campaign (1996–1998) expanded beyond acquiring local ownership of the Taylor Ranch. Of necessity, the foundation also worked with the acequias to address water quantity and quality issues. Since the Taylor (and Forbes) Ranches comprised the headwaters of the acequia watershed, the environmental protection of these areas became a primary concern of the acequia farmers and the LSF. The source of legitimation narratives was the concept of the acequia as a 'watershed commonwealth' (Peña, 1998). This opened the door to the local political mobilization that led to the official work of the Costilla County Land Use Planning Commission (established in 1997). The Costilla County Commissioners adopted a revolutionary county land use code that included a critically important watershed protection zone (Curry et al, 1996). Related to this effort was a set of research and public education projects to document and legitimize the traditional environment knowledge (TEK) of the acequia farming communities (Peña, 1998, 2001a, b). It was this search for ecological legitimacy and moral (qua legal) authority that led the LSF and the acequias to participate in the lengthy county land use planning commission hearings that took place between 1997 and 1998.

The campaigns for land use regulations and watershed protection programmes grew out of the original Taylor Ranch land grant struggle. But the LSF focused on the acequia farmers and their ancient customary law of the watershed commonwealth and not the land grant heirs; the focus was on the links between the montane headwaters and the equally endangered farms in the valley's bottomlands. Influenced by the context of the anti-logging campaign and the coalition with radical environmentalists, this effort eventually led to the establishment, in 1998, of the Colorado Acequia Association (CAA), which represented a 'hybrid' organizational form.

3 Radical environmentalism: La Sierra as wilderness and habitat

The arrival of the first white eco-activists in San Luis during the spring and summer of 1995 definitely represented a new chapter in the history of local environmental justice struggles. Members of Ancient Forest Rescue and

Boulder Earth First! were among the first to arrive, but they were soon followed by activists from Greenpeace, New Mexico Direct Action and other groups. The Wildlands Project briefly interceded when Michael Soulé, the famed conservation biologist, came to the defence of the anti-logging campaign waged by the Culebra Coalition (see Wet Mountain Productions, 1998).

The eco-activists initially defined La Sierra as wilderness and wildlife habitat. They emphasized the threats posed by the Taylor Ranch logging operations to the habitat for endangered species including the Rio Grande cutthroat trout, Mexican spotted owl, and Southwestern willow flycatcher. They worked with the LSF to conduct research on La Sierra's biodiversity and to prepare legal actions under the Endangered Species Act. During 1995–1996, the eco-activists focused their efforts on development of a direct action campaign to protest and disrupt the logging operations and educate the public about the Taylor Ranch.

Eco-activists worked directly with the acequia farmers instead of just with the land grant or environmental justice activists. The acequia farmers themselves made it possible for the eco-activists to settle and work in the area. The environmentalists worked and lived on the acequia farms. In turn, the acequia farmers became a support and 'affinity' group for the anti-logging protestors. One group of elderly Chicano farmers was present at every direct action that took place between June 1995 and July 1999 (Peña, forthcoming).

The white eco-activists educated themselves about the history of the land grant struggle and the efforts of the LRC to restore the use rights of the heirs. Environmentalists learned about the LSF plan to buy the Taylor Ranch, and they also learned about the sustainable and regenerative qualities of the acequia farming system. The eco-activists accompanied locals on cattle drives, helped with ditch maintenance and clean-up and provided labour for all manner of farm work in return for meals and a place to sleep. By the end of 1996 the eco-activists had taken a position in support of the land rights case but had not yet endorsed the proposed LSF/state buy-out of the ranch.

The organizing relationships that brought the radical environmentalists and acequia farmers together during the anti-logging protests and 'lock-downs' led to the formation of the Culebra Coalition in 1996. This represented the first 'hybrid' organizational form in this struggle. The radical environmentalists embraced the acequia farming community as a bona fide indigenous community with valid rights to the common lands and a legacy of ecologically sound stewardship. They could pursue the formation of an anti-logging and environmental justice coalition with the acequia farmers without 'sacrificing' their environmental ethics (Wet Mountain Productions, 1998).

4 Hybrid organizational forms

The Culebra Coalition had an intriguing precursor, the Salva tu Sierra Coalition. The Salva tu Sierra Coalition was organized in June 1995 to support the first direct actions against the logging operations.[7] The Culebra Coalition succeeded Salva tu Sierra in the winter of 1996. The principal objective of the Culebra Coalition was to stop the logging through direct and legal actions. Eventually, this expanded to include organizing around the development, adoption and enforcement of the Costilla County Land Use Code. The Coalition played a critical role in linking the county commissioners with legal and technical experts in support of the land use regulations.

But by the summer of 1998 it appeared that the logging operations were not going to be profoundly altered by the endless protests, blockades and other direct actions. It was out of this recognition that the Culebra Coalition turned to the land use regulations as a route to controlling the more pernicious impacts of the logging operations.

The disbanding of the Governor's Land Grant Commission and collapse of the LSF effort to purchase the Taylor Ranch, the limited political and tactical gains resulting from the direct action anti-logging campaign, and the continuing legal quagmire of Rael vs Taylor all led to the emergence of another hybrid organization, the Colorado Acequia Association (CAA).

The CAA, established in 1998, in many ways signalled a return to the roots of the land grant movement. The acequia farmers recognized the legitimacy of the efforts to restore the historic use rights to the enclosed common lands. Recognizing the limits of litigation, the CAA embraced the view that the community should organize against the threats posed by logging, mining and sub-dividing. The acequia organization focused on finding alternatives to protect the ecosystem integrity of the watershed and long-range viability of the acequia farms as the basis for an asset-based social movement for environmental justice. The CAA defined its mission as to organize and conduct scientific and legal research to empower the acequias to manage and protect the Culebra watershed, regardless of the ownership of La Sierra. The CAA is developing a land and water trust to protect acequia water rights, historic family farmlands, sensitive watershed areas, open space and wildlife habitat. These activities are carried out through the collaborative research and advocacy of a loosely affiliated network of lawyers, natural and social scientists, sustainable agriculture advocates and local acequia farmers.

The CAA recognizes that threats to the acequia system may start but do not end with the logging of the Taylor Ranch. The increasing pressures on farmlands posed by sub-division development are a critical problem. This represents a significant shift in focus for local environmental justice struggles – from the upland commons back down to the bottomlands that are the heart of the acequia farming landscape mosaic. In this manner the CAA is a hybrid organizational form because it draws its ethical vision and legitimizing narratives from the customary law of the acequia and traditional mutual aid practices. But it also draws new ideas, research methods and organizational inspiration from conservation biology, ecosystem management, environmental anthropology and agroecology. The association incorporates lessons developed over more than four decades of land grant and water rights activism and research. The hybrid nature of the association is also apparent in the office staff. The first two directors of the CAA were highly respected white eco-activists who settled in San Luis during the 1995–1999 anti-logging campaign. Yet the CAA draws its board only from among the ranks of the grass-roots acequia farmers. The CAA counts 73 acequias as its membership base and constituency. But it also works with radical environmentalists and research scholars from around the country on all its programmes.

FROM COMMUNITIES OF RESISTANCE TO SUSTAINABLE INHABITATION

The emergence of the CAA was the culmination of a long historical process of local organizing and circulation of struggle that began in the 1960s when Apolinar Rael first founded the Asociación por Derechos Civiles. Such local organizations emerged from the persistence of the customary law of ejidos (common lands) and acequias. After enclosure of the common lands the acequias were the sole local institution directly empowered to practice watershed-based self-governance. The invention of La Sierra as the Taylor Ranch ended the commons regime. Privatization criminalized the full range of historic use rights including the subsistence hunt for elk. Local people responded with subaltern forest crimes of 'trespass' to continue the traditional hunt, wood-gathering foray or search for *remedios*. This subaltern resistance was etched on the living landscape of La Sierra through the 'memory maps' the locals have constructed of their world (Peña, 2001b).

In 1995 the land grant movement tradition of resistance to enclosure was suddenly facing the direct and massive challenge of rapid deforestation of La Sierra. The industrial scars of logging became another inscription of power on the landscape: that of the legible, fiscal forest, a landscape made safe for profit. Clearly, some tried to avoid this path by recognizing that the land rights case would not be settled in time to stop logging destruction of La Sierra's watershed. The Culebra Coalition tried direct action to block the logging trucks and gates. They tried using the Endangered Species Act (ESA) but couldn't find the species, since these were largely extirpated from their historic range in the Culebra. The limits of direct action notwithstanding, one consequence of these protest activities was to focus more public and media attention on La Sierra. This often happened in ways that involved ecologists and journalists romanticizing the land and the farmers.

The most serious contradictions and divisions were 'internal' to the acequia and land grant community; and these were deeply rooted. One of the divisions involved LRC opposition to the LSF plan to buy the Taylor Ranch. The episode over the Regional Service Authority is an example of the dangers posed by essentialist ideologies; the construction of dogmatic and inflexible identities like 'cultural nationalism' can hamper the formation of networks and strategies of resistance. In the LRC the notion that the historic use rights could only be negotiated for a select group of multi-generational land grant heirs is an example of such essentialism. In contrast, the LSF envisioned a definition of use rights that at a minimum included all acequia farmers, all of whom were also land grant heirs and many plaintiffs in Rael vs Taylor.

Was multi-generational status imposed as an essentialist precondition for membership in the class of individuals with claims to rights of usufruct? This issue was never resolved in the discussions that took place between the LRC, LSF and the Governor's Commission. In the end this was meant to thwart the efforts of the LSF to place the RSA before the voters. The LRC blocked this strategic flexibility and thwarted a planned response by the community in the event that Taylor rejected the purchase offer and won the land rights lawsuit.

These local organizations formed a network that developed out of interactions between environmental justice and radical environmental movements and the land grant and acequia associations. But these latter are institutionally embedded in the local civil society and inherited from late antiquity. The acequia is really a community of 'practice' derived from customary laws establishing local self-governance. But the acequias have had to confront more than just enclosure. They have had to confront the aftermath of ecological degradation and displacement from their traditional role in watershed management. These communities are also struggling with urban and modernized identities that pull people away, ironically at the very moment that others (researchers) document the extraordinary ecological and economic value of the acequia farms. We raise the possibility of compensating the acequia farmers for their traditional environmental knowledge and ecosystem services (Peña, 2001a). Such demands spring from an appreciation of the agroecological and hydrological value of the acequias; but the Sociedad Española Ornitología (SEO) has yet to embrace this as a legitimate claim. Yet, the most lucid examples of successful ecosystem management are the numerous local watershed communities like the acequias. These are undergoing a transition: from organizing struggles against environmental racism to campaigns for autonomous, sustainable and equitable communities rooted in local natural, social, cultural and financial assets. One path to sustainable inhabitation of watersheds lies in the direction of the work advanced by the CAA that recognizes how land-based people are the keystone communities of a culture-nature ecotone; these communities are sustainable because they remain committed to inhabitation instead of development of places. This ethic of inhabitation supports wildness and biodiversity as well as human welfare. The identity politics of the historic acequia communities thus represent an example of the role of place as an anchor for identity and the articulation of claims to local and communal ownership of the watershed's biophysical resources (land and water rights). The establishment of a land and water trust is a critical component in the emerging community asset-building strategy that focuses on revaluing the natural and cultural assets of the historic acequia farms.

Finally, the acequias can be considered an important example of existing grass-roots alternatives to top-down environmental managerialism (Redclift, 1987). The acequias are an example of autochthonous democratic models for local self-management of ecosystems. They represent an important example of social movements that combine the principles of self-determination (autonomy) with environmental justice (sustainability). But the micro-mobilization of identity in such communities is not just based on shared beliefs and norms. It also seems to be supported by the inherited customary laws for local self-management of the watershed. The acequia institution represents a unique community of practice in which local political assets (the power granted to acequias under an otherwise ambivalent Colorado water law) may convey additional moral and legal authority to the local culture (Hicks and Peña, 2001). The stability of the acequias as a local civic institution may strengthen the prospects for multigenerational reproduction of ecosystem knowledge and management values. This suggests that a new social movement can emerge from the immediate connection that local communities in resistance make between material livelihood (economy) and the natural conditions of existence (ecology), and this may be a defining quality of place-based identities.

NOTES

1   The emergence of 'Oaxacalifornia', with its church-based local governance assemblies in Los Angeles among the Zapotecs, is an important example of reticular organizational forms. Mike Davis (2000) discusses these new transnational communities and shows how the Zapotecs 'have transplanted traditional village governments *en bloc* to specific inner city Catholic parishes' to create a system of local self-governance and communal economic power:

> *The Zapotecs outmaneuvre the slumlords by buying apartment buildings – which the church dutifully blesses – listing multiple names on the titles and paying for them jointly. Their councils come up with parochial school tuition and send their kids to college as a rate that defies the poverty and illiteracy of their parents. The small percentage of Zapotec youths in gangs are often exiled to a year in Oaxaca* (*Los Angeles Times*, 25 March 1998) as cited by Davis (2000, p85)

2   This profile is based on documents in the author's vertical file collection for La Sierra Foundation of San Luis.
3   Castells does not develop an interpretation of the environmental justice movement in the chapters dealing with ecological movements.
4   But there were plenty of pundits, outdoor recreation, hunting and fishing groups and other interests that were opposed to a public land acquisition of the Taylor Ranch. Some decried it as a case of 'special rights' for a disgruntled and unworthy ethnic minority; others questioned the ecological value of the land; and others depicted the sale as having little value to hunters and other user groups with claims to the public domain. Some feared that Coloradoans would not enjoy the acquisition, since the proposed management plan would favour local users. Another source of fear derived from the reputation of the San Luis area as a lawless and violent community. A few locals, including members of the Land Rights Council (LRC), which was involved in the heated and protracted Rael vs Taylor lawsuit, were at first also opposed to the purchase of the land, an issue explored later in this essay (for further discussion, see Peña, forthcoming).
5   The proposed management plan is in the Commission's final report (Sangre de Cristo Land Grant Commission, 1993). Also see the alternatives outlined in La Sierra Foundation (1993) and Peña and Green (1993).
6   The CCCD is a sub-county level local governmental unit organized under Colorado statutes as part of the state's institutional framework to promote flood control and conserve water resources. The CCCD was first established in 1974 by local farmers to protect the acequias from a proposed coal slurry line that would have operated with water from the unconfined aquifer underneath southern Costilla County farmlands. The CCCD played a critical political role in supporting the development of La Sierra Foundation, the Costilla County Land Use Planning Commission, and the Colorado Acequia Association (CAA). See Peña (forthcoming).
7   The original members of this coalition included: La Sierra Foundation, San Luis Peoples Ditch, Ancient Forest Rescue, Greenpeace-Boulder, New Mexico Direct Action, Culebra Earth First!, Forest Guardians, The Wildlands Project, Forest

Conservation Council, Sangre de Cristo Parish, Costilla County Economic Development Council, San Luis Town Council, Vega (Village Commons) Board, Costilla County Conservancy District, Committee for Environmental Soundness and Acequia Advisory Board (a group organized under the auspices of the CCCD and LSF).

REFERENCES

Benford, R D and Snow, D A (2000) 'Framing processes and social movements: An overview and assessment', *American Review of Sociology*, vol 26, pp611–39

Briggs, C and Van Ness, J (1987) *Land, Water, and Culture: New Perspectives on Hispanic Land Grants,* University of New Mexico Press, Albuquerque

Brooke, J (1997) 'In a Colorado valley: Hispanic farmers battle a timber baron', *New York Times*, 24 March

Castells, M (1997) *The Power of Identity*, Blackwell, London

Curry, R, Soulé, M, Peña, D and McGowan, M (1996) *Critical Analysis of Montana Best Management Practices and Sustainable Alternatives*, Technical consultants' report presented in October to the Costilla County Land Use Planning Commission, prepared for the Costilla County Conservancy District by La Sierra Foundation staff, available from the author

Davis, M (2000) *Magical Urbanism: Latinos Reinvent the US Big City,* Verso Books, London

Esteva, G and Prakash, M (1999) *Grassroots Postmodernism: Remaking the Soil of Cultures,* Zed Books, London

García, R and Howland, T (1995) 'Determining the legitimacy of Spanish land grants in Colorado: Conflicting values, legal pluralism, and demystification of the Sangre de Cristo/Rael case', *Chicano-Latino Law Review* 16, p39–68

Goldman, M (1998) *Privatizing Nature: Political Struggles for the Global Commons*, Rutgers University Press, New Brunswick

Green, R (1996) 'Land Rights Council and La Sierra Foundation unite', *La Sierra: Costilla County Edition,* vol 2 (28), 11 October

Hess, K and Wolf, T (1999) 'Treasure of La Sierra: Colorado's embattled Taylor Ranch is the West writ small. Here's how capitalism may conserve it', *Reason Oneline*, October, available online at: http://reason.com/9910/fe.kh.treasure.shtml

Hicks, G and Peña, D (2001) 'Acequia customary law, community, and labor in the age of privatization and prior appropriation: Toward a critical anthropology of water law in the intermountain West'. Unpublished research report prepared for a grant from the Labor Studies Program, University of Washington, Seattle

Kearney, M (1996) *Reconceptualizing the Peasantry: Anthropology in Global Perspective,* Westview Press, Boulder

La Sierra Foundation (1993) *Revenue Potential and Ethical Issues in the Management of the Culebra Mountain Tract as a Common Property Resource,* La Sierra Foundation, San Luis

Mondragon Valdez, M (1994) 'The story of La Sierra', *Valley Voice,* vol 4 (1), pp1–8 (Winter)

Peña, D (1998) *Chicano Culture, Ecology, Politics: Subversive Kin,* University of Arizona Press, Tuscon

Peña, D (1999) 'Cultural landscapes and biodiversity: The ethnoecology of an Upper Rio Grande watershed commons' in V Nazarea (ed) *Ethnoecology: Situated Knowledge/Located Lives,* University of Arizona Press, Tuscon

Peña, D (2001a) 'Rewarding investment in natural capital: The acequia commonwealth in the Upper Rio Grande' in J Boyce and B Shelly (eds) *Natural Assets: Democratizing Environmental Ownership,* Russell Sage Foundation, New York

Peña, D (2001b) '*Nos encercaron*: Identity, place, and community in ecological politics' in R Stein (ed) *Environmental Justice: Politics, Poetics, Pedagogy,* University of Arizona Press, Tuscon

Peña, D (2001c) 'Globalization, transnationalism and transborder studies: A short view from an "isolated rural village in the hinterlands"', paper presented at the Interdisciplinary Workshop on Transborder Peoples and the Interactions Between Latino Studies and Latin American Studies, Indiana University, Bloomington, IN (1–2 June)

Peña, D (forthcoming) *Gaia en Aztlán: Endangered Landscapes and Disappearing people in the Politics of Place,* University of Arizona Press, Tuscon

Peña, D and Gallegos, J (1993) 'Nature and Chicanos in southern Colorado' in R Bullard (ed) *Confronting Environmental Racism: Voices From the Grassroots,* South End Press, Boston

Peña, D and Green, R K (1993) 'Revenue potential and ethical issues in the management of the Culebra Mountain Tract as a common property resources', unpublished position paper prepared for La Sierra Foundation and the Costilla County Conservancy District, submitted to Governor Roy Romer's Sangre de Cristo Land Grant Commission

Peña, D and Gallegos, J (1997) 'Local knowledge and collaborative environmental action research' in P Nyden et al (eds) *Building Community: Social Science in Action,* Pine Forge Press, Thousand Oaks, CA

Peña, D and Martínez, O (1998) 'The capitalist tool, the lawless, and the violent: A critique of recent Southwestern environmental history' in D Peña (ed) *Chicano Culture, Ecology, Politics: Subversive Kin,* University of Arizona Press, Tuscon

Peña, D and Martínez, O (2000) *Upper Rio Grande Hispano Farms: A Cultural and Environmental History of Land Ethics in Transition, 1598–1998,* final report, National Endowment for the Humanities, Grant #RO-22707-94, Rio Grande Bioregions Project, San Luis, CO and Seattle, WA, available from the author

Peña, D and Mondragon Valdez, M (1998) 'The brown and the green revisited: Chicanos and environmental politics in the Upper Rio Grande' in D Faber (ed) *The Struggle for Ecological Democracy: Environmental Justice Movements in the United States,* Guilford Press, New York

Pulido, L (1998) 'Ecological legitimacy and cultural essentialism: Hispano grazing in northern New Mexico' in D Peña (ed) *Chicano Culture, Ecology, Politics: Subversive Kin,* University of Arizona Press, Tuscon

Pulido, L (1996) *Environmentalism and Economic Justice: Two Chicano Struggles From the Southwest,* University of Arizona Press, Tuscon

Redclift, M (1987) *Sustainable Development: Exploring the Contradictions,* Routledge, London

Rivera, J (1999) *Acequia culture: land, water, and community in the Southwest,* University of New Mexico Press, Albuquerque

Rosenbaum, R (1987) *Mexican Resistance in the Southwest: The Sacred Right to Self-preservation,* University of Texas Press, Austin

Sangre de Cristo Land Grant Commission (1993) *Sangre de Cristo Land Grant Commission Report,* Governor's Office, Denver, CO

Taylor, D (2000) 'The rise of the environmental justice paradigm: Injustice framing and the social construction of environmental discourses', *American Behavioral Scientist,* vol 43, pp508–80

Taylor, Z (1995) *Letter to the San Luis community regarding timber and watershed management plans* (April), in 'Taylor Ranch' vertical file collections, Rio Grande Bioregions Project, Department of Anthropology, University of Washington, Seattle

Weinberg, B (2000) *Homage to Chiapas: The New Indigenous Struggle in Mexico,* Verso Books, London

Wet Mountain Productions (1998) *Sin agua no hay vida*. Video documentary, available from: Ancient Forest Rescue, PO Box 762, San Luis, CO

Wolf, T (1995) *Colorado's Sangre de Cristo Mountains*, University Press of Colorado, Niwot, CO

Chapter 8

Environmental Justice in State Policy Decisions

Veronica Eady

INTRODUCTION: THE TIME IS NOW

The son of a slave, Paul Robeson was a martyr and crusader who suffered at the hands of what he called 'the Establishment', to ensure human dignity for all Americans. Many of us remember him as the celebrated singer and actor.[1] Fewer know that he was a brilliant lawyer, scholar (speaking 20 languages) and professional football player. Fewer yet realize that he was silenced, had his passport revoked by the American government and was systematically and effectively deprived of earning a livelihood in the arts because of his unyielding advocacy for the rights of humankind.

'The time is now,' Paul Robeson said in his 1958 biography *Here I Stand*, calling to action all those seeking social and political justice. He preached that Americans need not wait decades or generations before 'civil wrongs can become civil rights' (Robeson, 1958, p75).

Since Robeson's prime, the US has seen a Civil Rights Movement, a Civil Rights Act and the sacrifices of many to achieve equality for themselves and their progeny. Have we succeeded in achieving this elusive sense of justice, fairness and equality? Many people of colour would answer in the negative. Many poor people would answer in the negative as the march for justice continues in the areas of housing, employment, occupational health and safety, voting rights, the environment and many, many other areas.

This chapter speaks to environmental justice, borrowing Robeson's immediacy concept to demonstrate that states play a critical role in preserving environmental rights and moving society towards environmental justice. This chapter is a snapshot of environmental justice policy-making in one state, Massachusetts, and may hold value and truth for other states. The time is now indeed. States should not mark time awaiting a favourable political climate or for further technological or scientific advancements. State environmental policy must drive the political climate and the advancement of science. Environmental justice, like any civil right, is a moral imperative. As with all moral imperatives, the time for action is now and always.

THE EPA MODEL

EPA's environmental justice strategy

The US Environmental Protection Agency (EPA) has served over the last 30 years in the unenviable role of model for state environmental regulatory programmes. With some variations, most state regulatory programmes have followed EPA's format for enforcement, risk assessment, pollution prevention and more, sometimes repeating and sometimes learning from the inevitable missteps made by the federal EPA's regulatory trailblazing. In a sense, the success or failure of environmental protection in this country has rested on the shoulders of EPA. Similarities with EPA's model are so close that most states are, in fact, implementing programmes delegated by EPA or implementing their own programmes that are at least as stringent and often the verbatim regurgitation of EPA programmes. As the nation's environmental regulatory prototype, it is no wonder that EPA was first to forge new ground in carving out a regulatory role for environmental justice.

EPA embraced the notion of environmental justice after the *National Law Journal* published its 1992 indictment of EPA, charging that EPA's enforcement strategy was discriminatory (Lavelle and Coyle, 1992). That same year, EPA's Office of Environmental Justice (OEJ) (known then as the Office of Environmental Equity) opened its doors for business. OEJ made great strides, even prior to former President Clinton's Executive Order on Environmental Justice 12898, to institutionalize environmental justice policy in EPA programmes.

In April of 1995, EPA issued its final Environmental Justice Strategy (Office of Environmental Justice, 1995). The strategy called in essence for enhanced public participation,[2] enhanced health and environmental data research, enhanced access to public information, enhanced environmental protection in Indian country and enhanced enforcement and compliance reviews. During the years that followed, states have slowly fallen in line behind EPA in crafting their own environmental justice or environmental equity policies and strategies. Coincidentally, the two states that were first to adopt formal environmental justice (equity) policies were both in New England – Connecticut and New Hampshire.[3]

The National Environmental Justice Advisory Council

In September 1993, EPA chartered its formal federal advisory committee, the National Environmental Justice Advisory Council (NEJAC), a multi-stakeholder advisory body that provides independent advice, consultation and recommendations to the EPA Administrator on matters related to environmental justice.[4] The NEJAC model has been fairly successful at achieving broad stakeholder buy-in and gathering public input on a variety of environmental justice matters. A number of states, including Massachusetts, have imported the model to guide state environmental policy-makers through the process of crafting their own environmental justice policies and programmes.

One might argue that state versions of the NEJAC model have been more successful than NEJAC itself because they have been empanelled for the discernable goal of drafting environmental justice policy. While NEJAC has provided an ongoing public forum for community activists and a process for making concrete policy recommendations, one flaw in NEJAC and all federal advisory committees is that EPA is not obligated to take its advice.

States, on the other hand, may ostensibly be more inclined to implement the advice of advisory bodies that have come together for a clear and specific purpose – to develop an environmental justice policy.

The Massachusetts model

During the early part of 2000, Robert Durand, the Secretary for Environmental Affairs in the cabinets of Governors Argeo Paul Cellucci and Jane Swift, created an advisory body called the Massachusetts Environmental Justice Advisory Committee (MEJAC) to assist him in the development of a broad-range environmental justice policy that would steer the environmental justice agenda for all of the Commonwealth's environmental agencies for the first time with cohesion and formality. He also assembled a separate working group of state officials from each of the state's environmental agencies. The two bodies together shaped Massachusetts' environmental justice policy.

Modelled after EPA's NEJAC, MEJAC coordinated all public outreach, held public meetings, conducted neighbourhood tours and coordinated presentations by activists from across the state in order to expose state policy-makers to the diverse world of environmental justice issues and communities at risk in Massachusetts. The MEJAC guided the development of the state's philosophical environmental justice policy and made recommendations for implementation. In the second phase of policy-making, the state working group was charged with developing an implementation strategy based on that philosophical policy.

Massachusetts' environmental justice policy-making found support in a number of pre-existing provisions of the state constitution.[5] These were anti-sprawl legislation, known as the Community Preservation Act,[6] the Massachusetts Brownfields Act (Chapter 206 of the Acts of 1998) and Executive Orders under both former Governors William F Weld and Argeo Paul Cellucci calling for sustainability (Executive Order No. 385, Planning for Growth) and affordable housing (Executive Order 418, Assisting Communities in Addressing the Housing Shortage), respectively. Not surprisingly, Massachusetts regulators did not have to look hard to find the tools and legal authority to support an environmental justice programme. Policy-makers also met enthusiastic if sober[7] public support for the MEJAC process.

With Massachusetts placing all of its environmental agencies under the umbrella of a single cabinet office, the Executive Office of Environmental Affairs (EOEA), the Commonwealth was in a position to make a broad policy statement on environmental rights. Not only was the secretariat positioned to make a statement protecting all individuals from a disproportionate share of the environmental burden with respect to new siting decisions, it was able to ensure an equal share of environmental benefits, including access to open space, parks, state-run ice skating rinks and swimming pools, and funding.

Also quite significantly, Massachusetts took the opportunity to promote sustainable technologies and businesses focusing on clean production in the most environmentally distressed communities. Through its Office of Technical Assistance, EOEA pledged to pilot sustainability programmes and to provide technical assistance in the area of toxic use reduction to existing businesses such as autobody shops in the communities bearing this largest share of the environmental burden. EOEA recognized that existing businesses, particularly smaller sources, that were unregulated due to their size or because they pre-existed the more modern stringent environmental regulations (through 'grandfathering'), were often more of a nuisance than new facilities meeting tighter pollution standards.

Massachusetts' environmental justice programme was designed to address existing sources of pollution with workshops, the introduction of newer technologies and through programmes under Massachusetts' Toxic Use Reduction Act (see Conclusion for details of The Toxics Use Reduction Institute (TURI) at the University of Massachusetts, Lowell, and The Lowell Center for Sustainable Production at the University of Massachusetts, Lowell).

Further, EOEA introduced a programme where it partnered strategically with the insurance industry to provide lower environmental insurance rates for businesses that use fewer toxics. EOEA's environmental justice programme in many ways broke new ground across programmes to begin to reduce the existing environmental burden.

Three weeks after Massachusetts released its environmental justice policy in a public comment draft, Daniel Faber of Northeastern University and Eric Kreig of Buffalo State University unveiled their long anticipated report, 'Unequal Protection: Ecological Injustices in the Commonwealth of Massachusetts'. Using enforcement and other public data from Massachusetts' own regulatory agencies, Professors Faber and Kreig demonstrated that Massachusetts' communities of colour and low-income communities were, in fact, bearing a considerably larger environmental burden than higher income and predominantly white communities. While the report surprised few in the environmental justice and regulatory worlds, it sent shockwaves through neighbourhoods across Massachusetts where residents sensed an unfair pollution burden but did not know what to call it. All of a sudden, 'environmental justice' was a widespread battle cry, not just among the state's communities of colour and low-income neighbourhoods, but all across the landscape.

The Faber/Kreig report created a media splash that produced two key results. First, residents began to look closely at the draft environmental justice policy. At the policy's initial release, media paid little or no attention to the precedent setting policy. The Faber/Kreig report drew careful scrutiny to the draft policy not only by residents, not only by the media, but also by other states and by industry.

People overwhelmingly became interested in the draft policy and engaged in the public comment period. The second key development induced by the Faber/Kreig report was that activists demanded legislation that would in effect make any environmental justice policy adopted by the EOEA applicable to state agencies across the board, not just environmental agencies.[8]

IDENTIFYING COMMUNITIES AT RISK

The challenge to define community

New England is largely perceived to be a homogeneous, European-American region of the country. Moreover, Boston's reputation for racial strife dating back to the court-ordered busing conflicts of the 1970s has labelled the region as less than welcoming to people of colour and non-European immigrants. Perceptions aside, Massachusetts and New England have growing minority populations, much like the rest of the US. Additionally, Massachusetts and New England have significant Native American populations. Much of this Native American population goes uncounted because the large majority of tribes in New England have not received federal recognition, or tribal members are living away from their tribal land.

Like so many other parts of the nation, Massachusetts has its share of neighbourhoods that are populated predominantly by people of colour. The Roxbury and North Dorchester neighbourhoods of Boston, for example, are nearly all African-American, Caribbean and Latino in race and ethnicity. In these same neighbourhoods, the air pollution is visible as over-crowded diesel buses choke narrow streets and diesel trucks course through the neighbourhoods to reach waste transfer facilities, food production and distribution facilities, auto salvage yards and other noxious land uses. The environmental injustice in these neighbourhoods is as palpable and profound as that in East St Louis, Illinois, for example, another notoriously industrialized and blighted northern city.

The challenge for the state policy-maker is not in identifying the North Dorchesters or the East St Louis's. The challenge is in identifying communities with smaller at-risk populations where environmental injustice hides more easily. This begs a series of questions. First, what percentage of a community must be 'of colour' in order to comprise a community of colour or minority community? The Executive Order on Environmental Justice, 12898, signed by former President Clinton suggests that a minority population should be found 'where ... the minority population of the affected area exceeds 50 per cent'.[9]

Second, what borders define a community? Under the paradigm guiding many environmental protection programmes, municipal boundaries define 'community'. In that case, many communities in the nation, including the City of Boston, do not meet the Executive Order 12898 threshold and neighbourhoods like Roxbury and North Dorchester would not be reached by environmental justice protections.

The groundbreaking report *Toxic Wastes and Race* (United Church of Christ, 1987) examined the distribution of environmental threats using postal zip codes.

While zip codes seem a more viable definition of community than a municipal boundary for the purpose of evaluating environmental justice or injustice, zip codes have their own flaws. Postal zip codes are based on the volume of mail delivered. So in smaller towns, the entire town may have a single zip code compared to larger cities that have numerous zip codes.

Census tracts, blocks and block groups, on the other hand, are grouped for reasons that have sociological factors. Assuming that using census units is the most useful grouping tool to define community boundaries, we are yet faced with the challenge of determining what percentage of census grouping should comprise a community of colour or a low-income community. So far, no state or federal agency has unearthed the to key to the percentage figure. Some might argue that even using the smaller census grouping as opposed to municipal boundaries, Executive Order 12898's 50 per cent threshold is still too high.

For example, Freetown is located in the south-east region of Massachusetts, bordering New Bedford and Fall River. Freetown is home to one of the Commonwealth's oldest Cape Verdean communities, located on a two-mile stretch of road. This 102-year old Cape Verdean community numbers about 200 to 300 people. In the year of 2000 alone, the neighbourhood was slated to host an asphalt plant, a concrete operation and most recently an 800,000 square foot warehouse owned by a large retail department store.

Using Executive Order 12898's 50 per cent approach and using the census block grouping, this Cape Verdean community would not make up 50 per cent of the census block of which it is a part. Even using a percentage as low as 15 per cent, this neighbourhood is still invisible in its census block. It is communities like this where environmental injustice or racism can most easily hide. The challenge for the regulator is to devise a system that will afford these hidden 'pocket' communities appropriate protections within a framework that is objective. Distilled to its utmost simplicity, the difficulty lies in translating an issue fraught with instinct, passion and visceral reflex into a robust system of analysis that would be compatible with permitting processes and other environmental assessment processes.

Discerning disproportionate impacts

Even more confounding than how one defines the limits of communities is how one develops a standard methodology for defining the term 'disproportionate'. When making a reference to an impact as 'disproportionate', the implication is that the impact is being compared to something. The Faber/Kreig report concluded, for instance, that 'communities where people of colour make up 25 per cent or more of the total populations average nine times more hazardous waste sites per square mile than communities where less than 5 per cent of the population are people of colour' (Faber and Kreig, 2001, piii). By contrast, EPA (2000, p88) 'compares the affected population to an appropriate comparison population to determine whether disparity exists that may violate EPA's Title VI regulations'.

Clearly, that communities of colour host nine times the number of hazardous waste sites than predominantly white communities is Massachusetts is a disproportionate impact. In this case, the comparison population is the entire state, which seems to make good sense. Yet, EPA (2000, p88) reasonably suggests that comparison populations may be different in every case, stating 'since there is no one formula for analysis to be applied, [the Office of Civil Rights] intends to use appropriate comparisons to assess disparate impact depending on the facts and circumstances of the complaint'.

Recently, Logan International Airport in Boston prepared an environmental impact report (EIR) analysing the environmental impacts of a proposed additional runway. At the request of EOEA's Massachusetts Environmental Policy Act (MEPA) Office, the EIR included an environmental justice impact assessment. The EIR identified the affected population as the neighbourhood where the airport is located, East Boston, and an adjacent city, Chelsea.[10] As the comparison community, the EIR used the entirety of Suffolk County where the airport is located.

One could easily argue that Suffolk County was not the appropriate comparison population. Certainly, a larger population than Suffolk County accepts the benefits of Logan International Airport. But there is no more logic in using the entire state of Massachusetts as a comparison population than there is in using Suffolk County. And it would be quite a chore to determine where the boundaries of the benefits of Logan International Airport might be drawn. A distinction between the limits of benefits and burdens, of course, assumes that an exercise of weighing benefits and burdens is the key to determining disparate or disproportionate impact.

Yet another perplexing question is how much worse an impact must be on a community in order for it to be a disproportionate impact. In Roxbury, asthma rates soar to six times the state average, while in the Merrimack Valley region the asthma rate is two and a half times the state average. Is one and a half orders of magnitude bad enough to assign the label 'disproportionate' to an impact? Professors Faber and Kreig created a point system for rating existing sources of pollution in communities to weigh the cumulative environmental burden. Under their system, a Superfund site would score the highest number of points as compared to a gas-fired power plant that would rate lower. Some critics of the Faber/Kreig report called the system arbitrary. Environmental justice science is inexact at best, and Faber and Kreig may likely have given the most definition to disproportionate impact that we will see for the time being.

EPA is right in its suggestion that there is no ironclad formula that is appropriate for all cases. While reports like the Faber/Kreig report make compelling observations about the disproportionate distribution of pollution and waste that must be corrected, there is a clear subjectivity to environmental justice that is not easily overcome by policy or any science that can predict scientific or legal outcomes with certainty. And as long as environmental justice analysis remains overly subjective, there are no guarantees that state policies can sufficiently protect communities like Freetown or even Roxbury from environmental racism or discrimination.

Pollution without borders

Massachusetts is divided into 351 cities and towns. Rivers and watersheds and airsheds web the state, crossing in and out of cities, towns and Indian country with no respect for political sub-divisions. Carried by water and air currents, pollution naturally snakes in and out of cities, towns and tribal lands demonstrating the irrelevance of artificial boundaries where resource protection is concerned.

Federal programmes have, to some extent, been successful in protecting resources through regional or ecosystem approaches. States have also realized benefits from working together with other states to protect shared resources.[11] On a local level, however, cross-boundary environmental protection techniques begin to complicate issues. Neighbouring city, town and tribal interests may be so divergent (or perceived to be divergent) that collaboration becomes impossible. Holyoke, a city with profound economic distress, borders wealthier Amherst that enjoys a thriving economy, supported by academic institutions such as the University of Massachusetts flagship campus. Disparate economies can be reflected in environmental and planning staff, programme resources, and incentives or lack of incentives for developing certain industries. Indeed, Holyoke is largely industrial, in clear contrast with Amherst's education-based economy.

Often, state programmes can address quite effectively regional environmental issues within state borders. Massachusetts' Watershed Initiative, for example, has provided state environmental agencies with a framework within which to protect resources that cross local boundaries by watershed. Nevertheless, public participation in permitting and siting decisions is often thwarted by political boundaries. Take, for instance, the town of Dracut, Massachusetts, which approved the siting of a new power plant. The location of the power plant was actually near the town border, in an industrial neighbourhood.[12] Project proponents prepared an EIR as required under the Massachusetts Environmental Policy Act (MEPA). Residents from a host of nearby communities commented during the MEPA process. Despite strenuous public criticism of the project, EOEA in the end found the EIR adequate for the purposes of MEPA.[13] Among other things, residents from neighbouring municipalities argued that air emission impacts from the power plant would be more significant in neighbouring municipalities, especially in Lawrence, which has the largest Latino population[14] in the state and is the 24th poorest city in the nation. Town boundaries impeded the ability of neighbouring residents who were placed at highest risk by the siting decision from influencing local municipal processes. Activists argue quite persuasively that this impediment has led to clustering of facilities in their region.

Environmental justice may easily call for a restructuring of regulatory systems to provide a vehicle for all stakeholders who may be impacted by a facility to participate meaningfully in siting decisions. Realistically, elected officials at all levels may be loath to allow radical change to the current regulatory format. In the end, an enormous challenge lies at the feet of state environmental justice policy-makers to fine-tune existing processes like MEPA and to invent new processes that will provide a path for more meaningful public participation and serve as a conduit for incremental regulatory change.

Local zoning conflicts

Zoning did not exist in Freetown, Massachusetts, until 1995. Nearly 40 years before zoning arrived, Freetown's Board of Selectmen allowed the first industry, a trucking business, to locate in the centre of Freetown's Cape Verdean Village. Over the ensuing years, the Town Fathers again and again permitted additional industrial facilities in the Cape Verdean Village. These industrial uses included the siting of a state salt barn to store salt used on roads during the winter. The salt barn resulted in the contamination of the ground water, which was the drinking water source for local residents. This led to officials connecting the village to an alternate water supply, piped in from neighbouring New Bedford. However, rather than being able to tap water sources on their property through their own ground water wells, residents would now have to pay a water bill. Complicating matters, improper piping from the street water line into homes has made water pressure in the village even today far from adequate. Anecdotal accounts reveal that it takes 15 seconds to fill an eight-ounce glass full of water from a household tap. There have also been reports that fire hydrants in the village are 'dead', meaning they lack water pressure necessary to fight a fire. When zoning finally arrived in Freetown, the town reserved as residential two-tenths of a mile swath on the two-mile road running through the spine of the historic Cape Verdean community. That tiny residential reservation is sandwiched between industrial zoning and general (mixed-use) zoning. Moreover, several pre-existing homes fall within both the industrial and the general zones. In essence, the residential reservation is meaningless.

Residents have waged fight after fight to halt incompatible land uses proposed for their neighbourhood. But even if they are successful on both accounts, zoning has guaranteed that residents will be engaged in a constant battle to defend industrial facilities, pollution and habitat loss from encroaching on their neighbourhood and history. Discriminatory or otherwise flawed zoning is where state environmental justice policy-makers may find themselves in checkmate. State constitutions and legislation enabling cities and towns to zone, rather than environmental laws, are the culprits putting Freetown residents at risk. Even a successful challenge under Title VI of the federal Civil Rights Act of 1964 could not relocate the 19 businesses sited in the Cape Verdean community over the last few decades. Without funding to support technical assistance and finance lawsuits, Freetown's historic Cape Verdean community, like so many communities of colour, may be forever subject to environmental battles. A lesson from Freetown may be that environmental justice policy-makers have an important role in building and supporting the environmental justice movement. That is, environmental justice policy-makers must direct funding (both public and private), access to decision-makers and other resources to communities in order to reach a proper result in instances where the policy-maker may lack a legal handle to reach the same result.[15]

TITLE VI AS A SWORD

Title VI of the federal Civil Rights Act of 1964 'prohibits discrimination based on race, colour or national origin under any programme or activity of a Federal financial assistance recipient' (EPA, 2000, p14). States are a primary category of federal financial assistance recipient to whom this prohibition applies. States all over the nation are the subject of Title VI complaints currently pending before the EPA that claim discrimination in a panoply of state decisions, including facility siting and rule-making.[16] Massachusetts is one of those states.

Title VI may or may not be a useful tool for the environmental justice movement. That debate aside, Title VI begs the question how many (or how few) regulators consider whether their decisions will create a discriminatory impact? Most likely, it is a rare regulator, indeed, who gives Title VI serious thought *before* the discriminatory decision is made.

The year 2001 marked unprecedented legal decisions involving Title VI rights in environmental decision-making. *South Camden Citizens in Action v New Jersey Department of Environmental Protection*, 145 F.Supp.2d 446 (D. NJ 2001), involved the siting of a cement processing facility in a predominantly people of colour neighbourhood of South Camden, New Jersey. In its decision, the federal district court for the District of New Jersey went through a lengthy discussion of the existing health conditions of the South Camden neighbourhood as well as a discussion of the plethora of other noxious land uses in the community. The district court held, in sum, that the New Jersey Department of Environmental Protection (NJ DEP) had failed to perform an adequate Title VI analysis to determine whether permitting the new facility would create a disparate impact on the host community. Moreover, the district court granted the plaintiffs' request for a preliminary injunction, blocking the fully constructed facility from opening its doors for operation. In granting the preliminary injunction, the court recognized that the facility's operation could cause imminent and irreparable harm.

Four days after this landmark decision, the US Supreme Court held in an unrelated case, Alexander v Sandoval, __US__, 121 S.Ct. 1511 (April 24, 2001), involving a driver's licence exam in Alabama that there was no 'private right of action' under the Title VI regulations that allowed plaintiffs to show that a *discriminatory effect* had resulted from a decision (in this case a decision not to offer a driver's licence test in Spanish) rather than showing unveiled *intent* to discriminate. In other words, no longer would a showing of *discriminatory effect* be enough to win a Title VI action in front of a federal judge. The Supreme Court closed the doors of federal courts for all litigants attempting to bring a Title VI action without making the virtually impossible showing of intentional discrimination.

The Supreme Court decision in *Sandoval* was a devastating blow to the environmental justice movement. Nevertheless, the South Camden case remains significant in that the District Court recognized the obligation of the NJ DEP to evaluate questions regarding race, existing health conditions of the host community and cumulative risks in the neighbourhood.[17] The South Camden case serves as a warning to state regulators that they must ask the right questions at the right time – before a permit is issued.

The end of regulation in the abstract

As in the South Camden case, many environmental regulators have become accustomed to implementing their programmes without really knowing the communities threatened by undesirable land uses. This is a posture dangerous to the public that public environmental agencies are charged to serve. Without experiencing the neighbourhoods that regulatory decisions impact, without seeing the people who live in those neighbourhoods, the bureaucratic role, at best, is an abstract set of numbers, test methods, technologies and in some cases political influence.

Every regulator has a vital responsibility to know about the jurisdiction (s)he is serving. That jurisdiction may be a county, it may be a city, or it may be an entire state. Regardless of size, when drafting a permit, making a siting decision, reviewing an EIR – every regulator should know as much as possible about the demographics of the relevant community.

As with Freetown, census data or geographic information system (GIS) data may not reflect the true demographics of the community. In towns like Freetown, where the minority or low-income population is small, it is much easier to hide even blatant environmental discrimination. Regulators, therefore, should be compelled to read signs that a community may be disproportionately impacted by environmental decision-makers. For instance, clustering of sources is perhaps the most common indicator that a regulator should visit a neighbourhood proposed for a siting and speak with neighbouring residents. Again, using Freetown as an example, three locations within less than one mile on a single road proposed for new industrial facilities within a matter of months would give a regulator a clear clue that the decision-maker should look beyond the paperwork to ensure that a Title VI violation is not about to occur.

The district court in South Camden was on target in considering the existing health conditions in South Camden. Public health data can be a strong warning sign that a community is bearing a disproportionate environmental burden or that a Title VI violation may be about to happen. Like most states, Massachusetts makes public health data available to the general public as well as state regulators. It would be a simple task to check asthma rates in Freetown or elevated blood lead levels in Lawrence, or cancer rates in the town of Aquinnah, which is home to the Wampanoag Tribe of Gay Head (Aquinnah). Being aware of public health data should be an integral part of the environmental decision-making process.

The right questions at the right time

The South Camden case, above, has affirmed EPA's recent suggestion that states should consider proactive approaches to environmental protection using Title VI. In other words, states should ask at the front-end rather than in response to a Title VI complaint whether or not a decision conforms to the requirements of Title VI (EPA, 2000, p35). EPA's confirmation of the notion that decision-makers should be asking more questions about communities that may be impacted by environmental decisions helps lay the framework for what regulators then do with what they learn.

Common sense would dictate that a state decision-maker conduct an analysis once (s)he learns that a decision may impact a class protected by Title VI, to determine whether an affirmative decision would violate Title VI. Since we know that the Cape Verdean Village in Freetown is a community of colour and we know that there is a clustering of industrial facilities in a residential neighbourhood, we should analyse in detail whether permitting of an asphalt plant will cause a disparate impact. Is the Cape Verdean Village the site of more industrial facilities and pollution than any other part of Freetown?[18] If the answer to the former is yes, the asphalt plant should not receive a permit.

The filing of a Title VI complaint does not enjoin a project. Thus, if the asphalt plant were permitted, the Cape Verdean Village may suffer irreparable harm even if the residents file a timely Title VI complaint. EPA is swamped with a backlog of Title VI complaints. Furthermore, the EPA investigatory process is headed by the EPA headquarters Office of Civil Rights in Washington, DC, where evaluators may not be familiar with the subtle intricacies of a Massachusetts community. It could easily be years before EPA renders a decision. During that time, the asphalt plant would likely be in full operation making initial hopes of using Title VI to block the plant abjectly moot.[19]

Accordingly, Title VI holds the most hope for communities as the basis for pre-decision analysis by states. States have the added benefit of conserving precious staff resources required to mount a Title VI defence. Even if a state finds that its action will not violate Title VI, and it is subsequently hit with a Title VI complaint, the state will already have a detailed foundation for its defence. As matter of public policy, environmental decision-makers adopting a proactive Title VI approach will be making better-informed, better reasoned and more responsible decisions.

CONCLUSION: WE MUST BEGIN ANEW

State environmental justice policy continues to be new and largely experimental territory. Some states have chosen to leave the mechanical underpinnings of disproportionate impact unanswered, delving directly into implementation. Implementation, in the end, is critical. A state with an award-winning definition of environmental justice and system for analysing disproportionate impact has failed all of its residents without a solid implementation programme. Nevertheless, these questions must be answered and Massachusetts has attempted, at least, to provide some answers along with implementation.

Regulators must be flexible in environmental justice policy-making, willing to improve the current system while risking the chance of being imperfect. They must be willing to continually review environmental justice policies and programmes and make adjustments to correct imperfections that may not be apparent until the policy or programme is in place. Through a series of incremental steps, state regulatory agencies can evolve into more forward-thinking, social conscious and socially responsible bodies that truly work for the public good.

The Reverend Dr Martin Luther King, Jr once said, 'world peace through non-violent means is neither absurd nor unattainable. All other methods have failed. Thus, we must begin anew'. State policy-makers must look at their challenges squarely and be prepared and willing to try innovative approaches, regardless of the political climate and regardless of bureaucratic recalcitrance and even regardless of a judicial mandate, to help communities achieve the ultimate goal of the environmental stewardship, environmental justice.

NOTES

1   Paul Robeson is best remembered for his moving rendition of the Negro spiritual 'Old Man River' that he performed in the film and stage versions of the musical *Showboat*.
2   On 16 August 2000, President Clinton signed Executive Order 13,166 for Improving Access to Services for Persons of Limited English Proficiency. The Order binds state agencies through Title VI of the Civil Rights Act of 1964 to enhance public participation in languages other than English where appropriate.
3   Connecticut was the first state in the nation to adopt an 'environmental equity' policy, pre-dating EPA's EJ Strategy by nearly a year.
4   The NEJAC's current charter has been extended until September 2001, and it may be renewed.
5   Article 97 of the Massachusetts constitution provides that: 'The people shall have the right to clean air and water, freedom from excessive and unnecessary noise, and the natural, scenic, historic, and esthetic qualities of their environment; and the protection of the people in their right to the conservation, development and utilization of the agricultural, mineral, forest, water, air and other natural resources is hereby declared to be a public purpose'.
6   Secretary Bob Durand served in the state legislature for many years before being appointed to the cabinet of Governor Argeo Paul Cellucci. As a senator, he first introduced The Community Preservation Act 15 years prior to Governor Cellucci signing it into law in September 2000.
7   Environmental justice activists reserved final judgment on the exercise until a finished policy was in place.
8   As a matter of history, two years before the completion of the Faber/Kreig report, Massachusetts state Senator Dianne Wilkerson introduced a bill that would have created 'areas of critical environmental justice concern' (Senate Bill 1060 (1999)) that would receive heightened protections during permitting and environmental assessment processes. The Wilkerson bill died in July 2000, never having come to a vote. Senator Wilkerson re-introduced her bill (Senate Bill 1145 2001) in January 2001 sharing the hope with Faber, Kreig and the activist community that the Faber/Kreig report would raise the profile of her environmental justice bill and give it the thrust of momentum it had lacked in the previous session. As proposed, the Wilkerson bill would ground EOEA's environmental justice policy framework firmly in enforceable legislation.
9   Text of Executive Order 12898 'Federal Actions to Address Environmental Justice in Minority Populations and Low-Income Populations', Annotated with Proposed Guidance on Terms in the Executive Order, Section 1–1 Implementation.
10   One could argue very persuasively that the EIR erred in limiting the affected population to those communities.
11   For example, the six New England governors have joined with Eastern Canadian Premiers to devise a regional plan for the elimination of anthropogenic mercury. 'Northeast States and Eastern Canadian Provinces Mercury Study: A Framework for Action', February 1998.

12 Dracut is located in the Merrimack Valley region, which has a long history of clustering of sources.

13 Many environmental justice scholars and activists have held generally that the National Environmental Policy Act (NEPA) and state 'mini-NEPA' statutes (such as MEPA) are not appropriate for addressing environmental justice issues. In sum, they suggest that the NEPA-type process is tailored to find that the proposed project is the preferred alternative. As such, the NEPA process is rarely a useful or timely process for project opponents.

14 By percentage, Lawrence has a larger Latino population than Boston. Latino students comprise nearly 90 per cent of Lawrence's public school population.

15 Residents of East Freetown may be well served by a suit under the Fair Housing Act (Title VIII of the federal Civil Rights Act of 1968) claiming that the zoning in their community has interfered with habitability of their homes.

16 Since 1993, EPA has decided only one of the more than 60 Title VI complaints filed with the agency on the merits. That case, Select Steel, upheld the state permitting decision.

17 The South Camden decision also held that states could not rely on EPA's decision in Select Steel that compliance with air quality standards was sufficient to support a finding of no disparate impact.

18 Denying the asphalt plant an air permit would not obviate the pre-existing industrial burden in the village. But a denial based on Title VI concerns may slow the onslaught of proposed facilities in the area.

19 Again, the US Supreme Court held in Alexander vs Sandoval that an individual has no private right of action under the discriminatory effect regulations.

REFERENCES

EPA (2000) *Draft Revised Guidance for Investigating Title VI Administrative Complaints Challenging Permits.* EPA, Washington, DC

Faber, D and Krieg, E (2001) *Unequal Exposure to Ecological Hazards: A Report on Environmental Injustices in the Commonwealth of Massachusetts*, Northeastern University, Boston

Lavelle, M and Coyle, M (1992) 'Unequal protection', *National Law Journal*, Special Issue, 21 September

Commonwealth of Massachusetts, *Massachusetts Environmental Policy Act,* Mass. G.L. c. 30, sec. 61-62H, Commonwealth of Massachusetts, Boston

Northeast States for Coordinated Air Use/Northeast Waste Management Officials Associations/New England Interstate Water Pollution Control Commission/The Ecological Monitoring and Assessment Network (1998) *The Northeast States and Eastern Canadian Provinces Mercury Study*, Portland, Maine

Office of Environmental Justice (1995) *Environmental Justice Strategy: Executive Order 12,898,* Pub. No. EPA-200-R-95-002, EPA, Washington, DC

Robeson, P (1958) *Here I Stand*, Beacon Press, Boston, MA

United Church of Christ Commission for Racial Justice (1987) *Toxic Wastes and Race in the United States*, United Church of Christ Commission for Racial Justice, New York

Part 4

Selected Regional Perspectives on Sustainability and Environmental Justice

In a volume such as this, it is essential to look at the nexus between environmental justice and sustainability in a regional context. There are two primary reasons for this. First, some environmental justice and sustainability problems and solutions may be regionally, culturally or worldview-specific and as such may preclude generalization or wider applicability outside that region, culture or worldview. An example might be challenges facing the Maori in Aotearoa New Zealand who see environmental justice and sustainability as inseparable. Second, the opposite is also true, that there are some environmental justice and sustainability problems and solutions through which we can generalize. An example of this might be Friends of the Earth Scotland's focus on the primacy of socio-economic, rather than racial inequities, as the driver of its national campaign for environmental justice and sustainability. Part 4 therefore presents some selected regional, national and sub-national perspectives on sustainability and environmental justice.

In Chapter 9 Roberts uses her experience as a South African practitioner to argue that the notion of equity is an overarching theme in both sustainable development and environmental justice debates. She suggests however, that the current interpretation of the concept is flawed and indeed iniquitous given its largely anthropocentric focus. The need for practitioners to deepen their understanding of equity to embrace a more holistic concern for the non-human component of the global ecosystem is emphasized as a prerequisite for achieving global sustainability and justice. The evolution of environmental management in a newly democratizing South Africa – a country challenged by a legacy of extreme unsustainability and injustice – is examined at both the national and local level (using Durban's Local Agenda 21 programme as a case study) to determine the extent to which this more ambitious interpretation of equity is finding root in African soil.

Martinez-Alier investigates some historical and contemporary mining conflicts in Chapter 10 and the international environmental liability of mining corporations. Comparisons are made between conflicts in the US and South Africa which fall under the aegis of the international environmental justice movement. Such conflicts are fought in many languages, and the economic valuation of damages is only one of such languages. He asks a series of questions. Who has the power to impose particular languages of valuation? Who rules over the ways and means of simplifying complexity, deciding that some points of view are out of order? Who has power to determine which is the bottom line in an environmental discussion? He concludes that there is a clash in standards of valuation when the languages of environmental justice, or indigenous territorial rights or environmental security, are deployed against monetary valuation of environmental risks and burdens.

The role that gender plays in environmental justice struggles is addressed by Wickramasinghe in Chapter 11. The wide spectrum of issues of development concern that have been discussed at various levels suffer either from lack of integrity or inadequate attention given to women's rights in local environments, particularly for those who manage these local environments for their livelihood security. In this regard, South Asia as a geographical region provides a wide range of scenarios on controversial issues of development concern. Justice for women's claims to local environments as presented in this chapter is associated with two conditions; the first is related to women's citizenship rights; the second to their rights to manage the sustainability of local environments and to have security over their livelihood sources.

In Chapter 12 Rixecker and Tipene-Matua argue that current problems implementing sustainability point to fragmentation based upon (Western) modernity's propensity towards atomism and reductionism in contending with the environmental problematique. Rather than acknowledging the inherent connections between ecological and social health, modern practitioners focus upon 'ecological bottom lines' and continue to maintain the arrogance that humankind has the ability to manage planet Earth. Other peoples and cultures, such as Maori in Aotearoa New Zealand, do not see planet Earth, or Papatuanku, as divisible into separate spheres or dimensions. Thus, ecological and social justice are inseparable; one is inherently entangled with the other. In this chapter, the authors focus on the complexity of post-colonial challenges in Aotearoa New Zealand through an examination of contemporary responses to genetic engineering and bio-prospecting evident in the Aotearoa cultural, socio-political and biotechnological landscape.

Agbola and Alabi present a case study of environmental injustice in the Niger Delta region of Nigeria. They describe the unfair management of petroleum resources by the Nigerian state, which has resulted in a general consensus that the federal government has violated the basis of equity, fair play and social justice on which modern federalism is based. The neglect of the Niger Delta region's environment by the state has brought about a discourse of environmental injustice. Thus, petroleum extraction with its attendant environmental problems has made the people of the Niger Delta feel that the Nigerian state has repeatedly failed to protect their lives and property from environmental pollution while pollution costs are being unfairly imposed on them. Chapter 13 showcases the link between human rights abuse and environmental degradation as exemplified through the process of selective victimization.

In Chapter 14 Costi asks whether environmental protection, economic growth and social justice are compatible in Central and Eastern Europe. He describes how the concept of environmental justice in Central and Eastern Europe has not been studied exhaustively. The chapter examines whether Central and Eastern European countries can reconcile the need to achieve rapid growth based on post-Cold War economic liberalization, with their commitment to protect the national, regional and global environment in order to provide a democratic and sound environment for human development. The analysis provides a survey of the evolution of the environmental legislative framework in the region and of the emergence in recent years of a public awareness as to the impact of humankind on the surrounding environment. A review of the costs and benefits behind sustainable development follows, addressing the social, economic and political implications of the intervention of the state. Finally, Costi assesses the present and future role of the state, citizens and industry in working towards, and promoting, environmental justice.

Scotland makes an interesting context from which to analyse environmental justice, argue Dunion and Scandrett in Chapter 15. It is a nation which has been both victim and perpetrator of imperialism. Within the UK for 300 years it has recently achieved a significant devolution of power to an elected parliament. It reflects many of the economic and environmental attributes of being on the periphery of the European economic bloc. High levels of poverty are reflected in both post-industrial urban and depopulated rural societies, at the same time as excessive resource consumption consistent with Northern patterns. Drawing on the experience of black communities across the world, they argue that Friends of the Earth Scotland has positioned itself as 'a campaign for environmental justice'. This combines a demand for social justice with the requirements of environmental space, and is encapsulated in the strap line: 'no less than a decent environment for all; no more than our fair share of the Earth's resources'. The chapter provides case studies of community action, initial outcomes of research and an account of the influence of an environmental justice analysis on Friends of the Earth's approach to the developing policy context in Scotland.

* * *

Chapter 9

Sustainability and Equity: Reflections of a Local Government Practitioner in Southern Africa[1]

Debra Roberts

What we most need to do is to hear within ourselves the sounds of the Earth crying

Thich Nhat Hanh (quoted in Seed et al, 1988, p23)

INTRODUCTION

A chapter addressing issues of sustainable development and environmental justice in southern Africa should write itself. In Africa, images of poverty, exploitation and political conflict are commonplace. From the war-torn landscapes of Rwanda and Angola to the asbestos mines and toxic waste facilities of South Africa – the litany of environmental destruction and suffering is well known. Africa shows how the witch's brew of inequity and unsustainability can cast a spell of death and despair over an entire continent, afflicting some of the world's poorest and most vulnerable people. The recent United Nations Environment Programme's (UNEP) Millennium Report on the Environment – Global Environmental Outlook (GEO) 2000 – concludes that (UNEP, 1999, pxx):

> *The global human ecosystem is threatened by grave imbalances in productivity and the distribution of goods and services. A significant proportion of humanity still lives in dire poverty, and projected trends are for an increased divergence between those who benefit from economic and technological development, and those that do not. This unsustainable progression of extremes of wealth and poverty threatens the stability of the whole human system, and with it the global environment.*

The ideals of equity and sustainability are key because they benchmark our concern for the way in which we access, manage and distribute the limited resources of the globe. In other words the manner and extent to which we control 'access to environmental "goods" and environmental "bads"' (Haughton, 1999, p63). The most widely recognised of all sustainable development definitions, produced by the Brundtland Commission in 1987, defines sustainable development as: 'development that meets the needs of the present without compromising the ability of future generations to meet their own needs' (World Commission on Environment and Development, 1987, p8). This is synergistic with the definition of environmental justice, ie: 'the fair treatment and meaningful involvement of all people regardless of race, colour, national origin or income with respect to the development, implementation and enforcement of environmental laws, regulations and policies' so that 'no group of people, including racial, ethnic or socio-economic groups, should bear a disproportionate share of the negative environmental consequences resulting from industrial, municipal and commercial operations or the execution of federal, state, local and tribal programmes and policies' (Bullard, 1999, p7). A key difference between the two approaches is the weight assigned to issues of *inter*-generational versus *intra*-generational equity. Sustainable development requires that we give consideration to our own developmental needs, as well as those of generations still to come, while environmental justice prioritizes accountability to those currently alive. This is an important distinction as it influences the manner in which we will respond to the challenges that face our species in the years ahead.

GOVERNMENT AS A CHAMPION FOR SUSTAINABILITY AND EQUITY

In proposing solutions to the global environmental crisis, international policies such as Agenda 21[2] and the Habitat Agenda[3] have all placed government at the centre of the sustainability and equity debate. Both of these policies propose that 'environmental governance' can be used as a tool for establishing 'a new partnership between governments and civic society that can foster the eradication of poverty and an equitable distribution of environmental costs and benefits' (UNEP, 1999, p20). They also emphasize the role of local government in addressing global challenges: 'Because so many of the problems and solutions being addressed by Agenda 21 have their roots in local activities, the participation and cooperation of local authorities will be a determining factor in fulfilling its objectives' (Johnston, 1993, p423). This local focus has led to the development of programmes, such as Local Agenda 21, which mandate local governments to work with their stakeholders in the preparation of sustainable development programmes to address globally significant issues of equity, poverty and environmental degradation.[4]

CHANGING THE CONCEPTUAL GOALPOSTS

Given the enormous expectations placed on local government by the international community, it is important to understand that differences in worldviews will significantly influence the way in which these international mandates are interpreted and implemented at the local level. Of particular concern is the selective and anthropocentric (ie human-centred) manner in which the concepts of sustainability and equity are generally understood. For most people, the assumption of human self-importance (that has dominated Western philosophical, social, political and economic traditions for the past several centuries) means that we live in a society in which the non-human world has been devalued and where the only legitimate rights to equity and a sustainable existence are vested in the human species.

The understanding that anthropocentrism is linked to the modern day environmental and development crisis is a fundamental tenet of a branch of environmental philosophy known as 'deep ecology'. According to the deep ecologists, this situation has emerged from 'prescientific views that saw humans as dwelling at the centre of the universe, as made in the image of God, and as occupying a position well above the "beasts" and just a little lower than the angels on the Great Chain of Being. And while the development of modern science … served to sweep these views aside – or at least those aspects that were open to empirical refutation – it did no such thing to the human-centered assumptions that underlay these views' (Fox, 1990, p10). By putting ourselves at the centre of the universe, we became the yardstick by which the value of the non-human world is assessed: 'We argue that the non-human world should be conserved or preserved because of its value to humans (ie its scientific, recreational or aesthetic value) rather than for its own sake or for its value to non-human beings' (Fox, 1990, p11). Everything that is non-human has been turned into an 'alien entity' (Evernden, 1992, p85) destined – and as in all good science fiction movies – is subject to exploitation and/or elimination.

It is this obsession with our own self-importance and the relative unimportance of the non-human world that laid the foundations for the social and economic world order that has produced global environmental degradation and injustice as its by-products. The key question posed by the deep ecologists is: how can we hope to deliver justice or sustainability to members of our own species if we are capable of denying it to the millions of other with whom we share the Earth? It is clear that until we realize that the world was not made for us and act accordingly, the goals of achieving of equity and sustainability for our own species outlined in documents such as Agenda 21 and the Habitat Agenda must remain idealistic and remote.

Although the deep ecological analysis of the global environmental crisis has been labelled by some as anti-humanist, fascist or misanthropic, eco-philosophers such as Naess, Fox, Evernden, Sessions, Devall and Macy contend that we must reinvent our society and the way we see the world if we are to progress beyond the unsustainable and inequitable status quo. They encourage us all to develop a deeper and wider identification with nature and acknowledge that humanity and other entities are all aspects of a single, interwoven reality. It requires us to understand that we are the Earth and part of the reality that created it! The following expressions capture the spirit of this alternative worldview:

> *Deep ecology ... requires openness to the black bear, becoming truly intimate with the black bear, so that honey dribbles down your fur as you catch the bus to work.* (Aitken, 1980 quoted in Roberts, 1990, p54)

> *I am part of the rainforest protecting itself. I am that part of the rainforest recently emerged into human thinking.* (Seed et al, 1988, p36)

The message of deep ecology is that by liberating Nature we liberate ourselves. By respecting and valuing our role within Nature, as part of Nature, we begin to value the diversity of needs and values of other members of our own species. These concerns lie at the heart of achieving local level sustainable and equitable development.

Deep ecology has not, however, gone unchallenged. Other worldviews have offered alternative explanations for inequity and unsustainability that surround us all. Most notable among these are the 'social ecologists', who hold that the human domination of the world is linked to the existence of hierarchical and dominating relationships among human beings, and the 'ecofeminists', who point to androcentrism (ie male-centredness) as the main culprit of the current global dilemma (Mathews, 2001). Deep ecology, however, answers these critiques by emphasizing the need for an egalitarian attitude on the part of *all* humans towards *all* members, entities and forms in the ecosphere as the only solution to all these dilemmas.

Despite their differences, deep ecology, social ecology, eco-feminism and a variety of related philosophical viewpoints have been collectively grouped together under the umbrella term 'radical ecology'. Merchant (1992, p1) explains that: 'radical ecology confronts the illusion that people are free to exploit nature and to move in society at the expense of others, with a new consciousness of our responsibilities to the rest of nature and to other humans. It seeks a new ethic of the nurture of nature and the nurture of people'. This approach has also been critiqued. A notable review by Keulartz (1998, p21) cites the moral monism of radical ecology as problematic, arguing that 'it orients itself one-sidedly towards systems ecology's account of nature, thereby doing scant justice to the multiplicity of views of nature circulating both in science and in society ... The repeated appeals to ecology is not only misleading in view of the selective use made of this science, it also has a stifling effect on all those voices trying to make themselves heard in the social debate about a future sustainable society that base their case on other than ecological considerations'. Keulartz promotes a more pluralist approach which 'would create openings for a fairer and more equitable debate on future nature and environmental policies, with room for a wide variety of positions and perspectives' (Keulartz, 1998, p21).

While sympathetic to Keulartz's call for pluralism in the environmental debate, we must also acknowledge the urgent need to act now against a future where continued 'progress' leads to a devastated Earth and a human race that has lost it humanity. Given the range of philosophical options available, and in the absence of a single agreed moral majority, it is clear that each individual must choose his or her own preferred path. My choice is for an approach that advocates the need for equity and sustainability that addresses and serves the human and non-human world equally. By pretending that issues of environmental justice and sustainability relate only to the rights and needs of *Homo sapiens*, we perpetuate the very social and economic systems that have created human oppression and environmental degradation in the first place.

Can local government practitioners afford not to be deep ecologists, is the question uppermost in my mind.

THE SOUTH AFRICAN EXPERIENCE

While philosophical debates are vital if we are to progress our thinking, we must also be pragmatic and move our dialogue beyond a discussion of 'the ideal state' to an analysis of the progress currently being made towards greater sustainability and equity. Because of its background, South Africa provides an ideal case study for such an analysis.

South Africa is still best recognized internationally for the abandoned system of 'apartheid', the system of government that made a racial distinction between those entitled to enjoy the benefits and 'goods' of the country and those involuntarily subjected to the resulting 'bads'. Apartheid produced one of the most extreme examples of environmental injustice and unsustainable development ever known, ensuring 'that there was a skewed distribution of access to natural resources favouring the white minority. Under apartheid, economic development was not guided by concern for the sustainability of the natural resource base, but went full steam ahead without regard for the effect on human health or the environment. That inheritance can be seen in the degraded homelands, in the toxic mine tailings and waste dump sites, in the pollution in the streams draining the sugar plantations and hanging in the air above industrial communities' (International Mission on Environmental Policy, 1995, p4).

The implementation of 'separate development' meant that the black communities of South Africa were disproportionately impacted by unsustainable development, a fact that has not changed significantly during the first seven years of democratic rule. In the urban areas black communities are still the predominant victims of industrial pollution, while in the rural areas landlessness and lack of tenure continues to force people onto marginal land, resulting in severe land degradation. Black women particularly, as the custodians of family health and welfare, have (and continue) to bear the brunt of these environmental injustices. Despite this, progress is being made.

Democracy has brought with it a growing realization of the importance of equity and sustainability to good governance. As a sign of this the 1996 South African Constitution includes an environmental right which states that: 'Everyone has the right to an environment that is not harmful to their health or well-being' and 'to have the environment protected, for the benefit of present and future generations through reasonable legislative and other measures' (The Government Printer, 1996, p11). This right embodies the notions of equity and sustainability, that is, the need to minimize the environmental 'bads' for present South Africans and ensure that there are enough environmental 'goods' to pass onto future South Africans. Similarly, the country's 1998 National Environmental Management Act (NEMA) stipulates that development must be 'socially, environmentally and economically sustainable' (The Government Printer, 1998, p10) and that 'environmental justice must be pursued so that adverse environmental impacts shall not be distributed in a manner as to unfairly discriminate against any person, particularly vulnerable and disadvantaged people' (The Government Printer, 1998, p12).

Although both the Constitution and NEMA mark important steps forward, they still perpetuate the concept of human self-importance critiqued by the deep ecologists. The management principles outlined in NEMA, for example, state that: 'Environmental management must place people and their needs at the forefront of its concern, and serve their physical, psychological, developmental, cultural and social interests equitably' (The Government Printer, 1998, p10), effectively entrenching the paradigm of human/non-human dualism that has frustrated attempts at achieving sustainability and equity in the past.

Given that national level policy and legal frameworks in South Africa offer no immediate challenge to the prevailing anthropocentric worldview, has greater progress been made in implementing a more egalitarian approach to sustainability and equity at the local level?

The Durban case study

In South Africa the Local Agenda 21 mandate has become the rallying call for many local governments seeking a more sustainable development path. Durban was the first of the major metropolitan areas in South Africa to initiate a Local Agenda 21 programme as a corporate responsibility in 1994.[5] Because of the extensive restructuring of local government in South Africa between 1994 and 2001, Durban's Local Agenda 21 programme has been undertaken in a phased approach. The first phase of the programme (1994–1996) culminated in the production of the first State of the Environment and Development report and was followed by the initiation of a package of strategic projects in Phase 2 (1996–1999) aimed at addressing the city's most urgent environmental and development challenges. Two projects among these are particularly significant from a sustainability and equity perspective:

- the Durban South Basin Strategic Environmental Assessment (SEA), and;
- the Durban Metropolitan Open Space Framework Plan.

Durban South Basin Strategic Environmental Assessment (SEA)
The Durban South Basin is the economic heartland of the Durban Metropolitan Area (DMA) and is the second most important manufacturing region in the country. The Basin contributes over 60 per cent of the DMA's gross geographic product (GGP) and contains 30 per cent of the industrial land in the city. At a national level, the area generates about 10 per cent of all manufacturing jobs and about 10 per cent of GGP.

Apartheid policies have impacted on the Basin area producing a mixed land use of residential areas juxtaposed with heavy industry (eg petrochemical and paper industries) and transport infrastructure (ie international airport and port). Not surprisingly, the proximity of these land uses has impacted negatively on the social, economic and natural environment of the area. These problems have been further exacerbated by poor environmental planning and management on the part of both industry and government. The result is that local communities have paid the price of regional and national wealth generation with the degradation of their local environment. The resulting mobilization of these communities around environmental justice issues now poses a severe constraint to further industrial development and expansion in the Basin.

Given this context, the Durban South Basin SEA was undertaken as a way of addressing community dissatisfaction and providing a mechanism for working constructively with all key stakeholders – communities, business, industry, government – towards (CSIR, 1999, pii):

1   producing a baseline assessment of the Durban South Basin that identified opportunities and constraints for future development in terms of sustainable development requirements;
2   identifying key Strategic Development Criteria to assess current and future development in terms of their overall sustainability;
3   evaluating various types of future development in terms of their sustainability; and
4   developing a policy planning framework for sustainable development to guide management and planning in the Durban South Basin.

In terms of these objectives the five most likely future development scenarios for the Durban South Basin were assessed, ie:

•   The maintenance of the Status Quo situation.
•   The Mixed Use Option (combining further housing and light industrial development).
•   The Petrochemical Option (an expansion of the existing petrochemical development).
•   The Second Port Option (creation of a dig-out port on the current airport site following airport relocation).
•   The Combined Second Port and Petrochemical Option.

Each option was reviewed in terms of its equity and sustainability impacts using the sustainable development principles outlined in Agenda 21, the Council's then Metropolitan Vision and a position paper prepared by an alliance of South Basin communities – the South Durban Community Environmental Alliance (SDCEA). (See Box 9.1.)

As a result of this assessment the following conclusions were drawn by the SEA Project Team (CSIR, 1999, px):

BOX 9.1 SUSTAINABLE DEVELOPMENT PRINCIPLES USED IN ASSESSING THE DURBAN SOUTH BASIN

Agenda 21 Principles interpreted for the Durban South Basin:

- Development must not degrade the biophysical, built, social, economic resources or the system of governance on which it is based.
- Current actions should not cause irreversible damage to natural and other resources, as this may preclude future development options.
- In cases where there is uncertainty of the impacts of an activity on the environment, caution should be exercised in favour of the environment.
- The cost of pollution should be paid by the party causing the pollution.
- The needs of land use, environment and economic planning need to be integrated.
- A development framework and paradigm which promotes resource generation rather than resource degradation must be created.
- Immediate and long term actions need to be identified and planned for, to address urgent needs while still progressing towards longer term sustainable solutions.
- Resources must be utilized more effectively through each sector applying its resources to support other sectors rather than acting in an uncoordinated or competitive manner.

The **Durban Metropolitan Council's Strategic Vision** indicated that all development actions should be:

- supportive of economic development;
- supportive of human development;
- supportive of community development;
- supportive of community participation;
- supportive of affirmative action;
- democratic and transparent;
- cooperative and coordinated;
- equitable;
- fair to all;
- goal-orientated;
- accountable;
- effective and efficient;
- environmentally sustainable;
- financially sustainable;
- affordable to the consumer; and
- address historical imbalances.

Community Development Objectives established by the South Durban Community Environmental Alliance outlined the need for:

- Development that is collectively agreed upon with the people of the Durban South Basin.
- Development that enhances the socio-economic profile of South Africa.
- Development that recognizes the need for work opportunities for the unemployed living in Durban South Basin.
- Development that occurs with the aim of reversing the impacts of apartheid planning of the Durban South Basin.
- Development that ensures that the natural heritage of Durban South Basin is always protected.
- Development that should instil a sense of pride in the residents of communities living in Durban South Basin.

Source: CSIR, 1999, pp38–43

- Although likely to be unpopular with residents of the South Basin area, the Petrochemical Option performs best at all levels. Economically it provides a major economic driver to support the economy of the city and the Basin, and by allowing expansion in a controlled manner provides the capital to finance improved environmental management as well as the opportunity to introduce newer and cleaner technologies and processes. Socially, a significant number of new jobs will be created by the expansion and the option allows the continued existence of a residential/heavy/light industrial mix in the area. Ecologically, the expansion is not land hungry and will allow the development of extensive green corridors for recreation and ecological use.
- In terms of providing the greatest economic benefit to the Durban Metropolitan Area, the combination of petrochemical and port development is the best option. This scenario combines the financial benefits of both options as well as the air quality improvements of the petrochemical development. Unfortunately it also exacerbates the problems associated with the transformation of the Basin, ie the need to displace existing residential areas and open space areas to facilitate industrial expansion.

Based on these findings, it was concluded that the Basin was likely to remain industrial in the foreseeable future and that limiting, reducing or changing the current development profile of the area in the short term would not only undermine the extensive capital and infrastructural investment in the area, but would jeopardize local and national economies, exacerbate unemployment and poverty levels and contribute to a downward spiral of environmental degradation by promoting urban decay. It was acknowledged that while the implementation of some of the development scenarios might lead to the loss of some of the existing residential areas (probably over a period of decades) as the result of industrial expansion, this could be compensated for by:

- the benefits accrued to the metropolitan and national populations in terms of economic growth and poverty alleviation;
- an improvement in environmental management in the area through investment in new and cleaner technologies;
- the possibility of financial compensation for affected parties through appropriate planning tools (for example the re-zoning of affected properties from residential to industrial); and
- government's allocation of specific and dedicated resources to improved management and enforcement in the area.

Given government's limited financial resources, the high level of political support for economic development and the urgent need to improve environmental planning and management in the Basin, the SEA Project Team considered these findings to be realistic, if not ideal. The findings have, however, been strongly challenged by local communities, particularly the possibility of industrial expansion disrupting well-established residential areas. Some components of the local community have vowed to fight any form of relocation, suggesting that such a process would be no different to the forced removals experienced under apartheid. Others have called for a clean-up of the existing industries unaccompanied by expansion and argued that if industrial expansion were to occur (outside of the residential areas) that it should focus on 'clean and green' industries (as opposed to the existing heavy industries). The SEA considered this option, but found that it would not be economically viable and hence would result in increased urban decay and exacerbated environmental degradation. Others community elements have requested that if government is to support industrial expansion then this decision should be made public and suitable time frames, consultation and financial compensation should be agreed to by all stakeholders. Two years on, local government is still to take a final formal decision on the recommendations of the report.

Reflections on the Durban South Basin Strategic Environmental Assessment

In a situation as complex as the Durban South Basin, it is difficult to see how the principles of deep ecology can be constructively used to ensure that future developments enhance sustainability and equity for local, metropolitan and national stakeholders. Arne Naess, the founder of deep ecology, himself conceded that it is very difficult for deep ecological principles to take root in a materialistic world where modes of consumption and production are 'secured by the inertia of dominant ideas of growth, progress and standard of living', and that in the real world 'narrow, utilitarian, short-range arguments' are often necessary 'in order to get things done' (Naess, 1986, quoted in Roberts, 1990, p55). Certainly, the SEA's finding that industrial expansion and economic growth are appropriate vehicles for improving environmental management falls firmly into the category of a narrow, utilitarian, short-range argument! Given this finding and Naess' challenge that experts and people with knowledge 'repeatedly and persistently deepen their argument' (Naess, 1986, quoted in Roberts, 1990, p55) we are forced to interrogate the SEA process more deeply.

Such an interrogation prompts numerous questions. If we identified more closely with the world around us rather than treating it merely as resource, would we support development that involves the exploitation of limited, non-renewable resources, which creates some of the most globally significant environmental problems and endangers the survival of hundreds if not thousands of species around the world though climate change, habitat loss and pollution? The answer is clearly no. By the same token, would we be satisfied with a solution where improved environmental management comes with the price tag of industrial expansion and possible community displacement? Probably not. While questions such as these were debated by the SEA Project Team, the urgent need to improve environmental conditions in the Basin, the lack of an audience among local decision-makers for the 'deeper' debate and the inability (within the project's time and financial limitations) to determine realistic and economically viable development options that addressed broader sustainability and equity concerns, effectively prevented this deeper discussion from influencing the outcome of the project. Does this outcome signify the end of the road for deep ecology in Durban?

No, for while the SEA project was unsuccessful in tacking these issues, it is clear that local communities (particularly as watchdogs of existing processes and institutions) provide a potentially fertile ground for advancing the debate in the future. The project itself never built capacity or encouraged deeper dialogue among stakeholders around future development choices because of resource limitations and the distrust that existed between the different parties. It is hoped, however, that as more constructive partnerships are forged under South Africa's new democracy, that education and outreach programmes initiated by local government might provide a vehicle for the beginnings of a new, deeper level of debate. In recognition of this, the intended restructuring of the environmental management function in Durban will include the creation of an environmental education division. The work of this division will be guided by the city's new Environmental Management Policy which includes 'Education, Training and Awareness' as one of the six key policy areas and specifies the need for 'a culture of learning about the environment to enable the effective participation of all in managing the environment' (Common Ground Consulting, 1998, p27). Thus, while doors may be shut in the short term, longer term education and capacity-building programmes must be used to foster a more egalitarian, forward-looking debate around the way we plan and manage our city.

Durban Metropolitan Open Space System (D'MOSS) framework plan

As well as being an industrial and manufacturing centre, Durban is also a tourist city located within a region of high biodiversity on the East African coast. The need for an open space plan to ensure the conservation of the city's important natural resource base has been a component of the city's strategic planning for over 20 years. In 1997, in response to the demarcation of the expanded Durban Metropolitan Area (DMA), a revised open space framework plan was commissioned by the (then) Durban Metropolitan Council.

This project confirmed the fact that past apartheid policies have created a social division around issues of environmental management, whereby concern for the 'green' environment is seen as the prerogative of an elite, white minority and of little concern to the larger, poorer black majority. In order to bridge this gap and secure support for the new open space plan from all stakeholders, the Project Team had little choice but to exploit the predominant 'shallow' worldview to its own advantage.

Acknowledging that the non-human world is seen primarily as a resource by most people, the Project Team framed the rationale for the new open space plan in terms of the goods and services that the natural ecosystems deliver to the residents of the DMA (Markewicz English, 1999), ie: gas regulation, climate regulation, disturbance regulation, water regulation, water supply, erosion control, soil formation, nutrient cycling, waste treatment, pollination, biological control, refugia, food production, raw materials, genetic resources, recreation and cultural value. The replacement value of these 17 services was calculated using international research in the field of resource economics and estimated to be equivalent to approximately half the value of the operating budget of the then Durban Metropolitan Council, ie US$0.25 billion (R2.24 billion) per annum (US$1 = R9).

This use of resource economics was an obvious and targeted response to a local government system that is strongly influenced by economic and political forces. Because open spaces were now seen to have 'real' (ie monetary) value it was possible to secure increased political and stakeholder support for the plan to protect Durban's open space system. Unfortunately, this success has come in the form of a double-edged sword.

On the one side, Durban's use of 'resource economics' in open space planning has been hailed as innovative and cited as an example of sustainable development best practice. On the other, it has served to reinforce the idea that the non-human world is only of value when it provides goods and services to the human world. Despite the initial short term advantages of protecting open areas under immediate threat from ad hoc and inappropriate development because decision-makers are now more cautious about taking decisions that impact unfavourably on the open space system, in the long term it makes a mockery of the concepts of environmental equity and sustainability espoused by the deep ecologists. It means that in the instances where the value of open areas cannot be clearly identified or where their economic value is not viewed as sufficiently substantial, areas will be treated as valueless. Simply put: 'If they don't pay they cannot stay!' By giving natural areas a dollar value, we have effectively divested them of any other worth.

Reflections of the Durban Metropolitan Open Space System plan

As argued previously, real equity should accord all elements of the world the right to exist as part of a broader reality, regardless of their value to the human species. Similarly, real sustainability should acknowledge the interconnectedness and importance of all parts of the global ecosystem. The result is that the D'MOSS project, unquestionably sincere in its attempt to raise the profile of conservation issues at the local level, has entrenched the most dangerous of world views, that: 'Man becomes the measure of all things, and the world becomes nothing but a collection of things for man to measure' (Evernden, 1985, quoted in Roberts, 1990, p50). The difficulty remains, however, that in the absence of a paradigm shift to a global recognition of the moral considerability of the non-human world, the use of tools such as resource economics is a necessary 'holding operation' without which the short term losses to the ecosphere would be unconscionable.

CONCLUSION

So where do we go to from here? Based on the Durban's Local Agenda 21 experiences it seems that the potential for local government to act as a true champion for sustainability and equity is not realized without difficulty. Operating within the confines of resource-poor, conservative, conflict-laden, pressurized institutional environments, there appears to be a real danger that initiatives such as Local Agenda 21 could (in a worst case scenario) help perpetuate unsustainable and iniquitous development. Even those practitioners with a sympathy for the deeper questions are cowed by a system were 'their jobs are often in danger, or they tend to lose influence and status among those who are in charge of overall policies' (Naess, 1998, p194), if they were to open articulate deeper beliefs.

Do we then despair of ever making progress? No, for as mentioned previously, education and capacity-building programmes offer important opportunities to deepen our debate around the issues of sustainability and equity. We must also acknowledge that even if the human world is the current focus of sustainable development work, improvements in quality of life and environmental protection are being achieved through initiatives such as Local Agenda 21. These interventions are often urgent and vital and cannot wait for a worldwide paradigm shift. Given that over 3000 Local Agenda 21 programmes exist globally, these (and related) processes provide a useful and influential forum in which to initiate longer term debate around alternative worldviews.

Ultimately, however, it will be up to individuals within these programmes to begin to raise and champion the alternative debate, itself a difficult and onerous task in a world that favours conformity. It seems that Fox's (1990) choice of the term 'transpersonal' to describe the paradigm shift required of us – from the current superficial awareness to a deeper understanding of the world and our place in it – is therefore both insightful and instructive, emphasizing the need for *trans*cendence and the *personal*. In these terms it is up to each one of us to make the change in ourselves first and then encourage similar change in those around us through personal interactions and debates. If as Rorty (quoted in Evernden, 1992, p133) suggests 'a talent for speaking differently, rather than arguing well, is the chief instrument of cultural change', then this paper represents part of my contribution to that transformation.

> *If what a tree or bush does is lost on you,*
> *You are surely lost. Stand still. The forest knows*
> *Where you are. You must let it find you.*
> '*Lost*' by David Wagoner, 1976. Source unknown

NOTES

1   The views expressed in this paper are not necessarily those of the eThekweni Council.

2   The global plan of action aimed at achieving sustainable development in the 21st century endorsed at the 1992 United Nations Conference on Environment and Development.

3   The global plan of action aimed at achieving sustainable development and improved living environments in human settlements endorsed at the second United Nations Conference on Human Settlements in 1996.

4   The Local Agenda 21 mandate contained in Chapter 28 of Agenda 21 states that: 'By 1996, most local authorities in each country should have undertaken a consultative process with their populations and achieved a consensus on a local Agenda 21 for the community.'

5   Durban is the largest city and port on the east coast of Africa with a population of some 2.8 million people, covering an area of 2297 km^2.

REFERENCES

Bullard, R D (1999) 'Dismantling environmental racism in the USA', *Local Environment*, vol 4 (1), pp5–9

Common Ground Consulting (1998) *Environmental Management Policy for the Durban Metropolitan Area*, Durban Metropolitan Council, Durban

CSIR (1999) *Durban South Basin Strategic Environmental Assessment: Final Integrated Report*, Durban Metropolitan Council, Durban

Evernden, N (1992) *The Social Creation of Nature*, Johns Hopkins University Press, Baltimore

Fox, W (1990) *Towards a Transpersonal Ecology*, Shambhala, Boston, MA

The Government Printer (1998) 'National Environmental Management Act 107 of 1998', The Government Printer, Pretoria

The Government Printer (1996) 'The Constitution of the Republic of South Africa, Act 108 of 1996', The Government Printer, Pretoria

Haughton, G (1999) ' Environmental justice and the sustainable city' in D Satterthwaite (ed) *The Earthscan Reader in Sustainable Cities*, Earthscan, London

International Mission on Environmental Policy (1995) *Building a New South Africa: Environment, Reconstruction and Development*, International Development Research Centre, Ottawa

Johnson, S P (1993) *The Earth Summit: United Nations Conference on Environment and Development (UNCED)*, Graham and Trotman/Martinus Nijhoff, London

Keulartz, J (1998) *The Struggle for Nature: A Critique of Radical Ecology*, Routledge, London

Markewicz English (1999) *Durban Metropolitan Open Space System Framework Plan*, Durban Metropolitan Council, Durban

Mathews, F (2001) 'Deep ecology' in D Jamieson (ed) *A Companion to Environmental Philosophy*, Blackwell, Massachusetts

Merchant, C (1992) *Radical Ecology: The Search for a Livable World*, Routledge, New York

Naess, A (1986) 'Intrinsic value: Will the defenders of nature please rise?' in M E Soule (ed) *Conservation Biology: The Science of Scarcity and Diversity*, Sinauer Associates, Sunderland, MA

Naess, A (1998) 'The deep ecological movement' in M E Zimmerman, J B Callicott, G Sessions, K J Warren and J Clark (eds) *Environmental Philosophy: From Animal Rights to Radical Ecology*, Prentice Hall, Upper Saddle River, New Jersey

Roberts, D C (1990) *An Open Space Survey of Municipal Durban*, unpublished PhD thesis, University of Natal (Durban)

Seed, J, Macy, J, Fleming, P and Naess, A (1988) *Thinking Like a Mountain: Towards a Council of All Beings*, New Society Publishers, Philadelphia

United Nations Environment Programme (UNEP) (1999) *GEO-2000: UNEP's Millennium Report on the Environment*, Earthscan, London

World Commission on Environment and Development (1987) *Our Common Future*, Oxford University Press, Oxford

Chapter 10

Mining Conflicts, Environmental Justice and Valuation

Joan Martinez-Alier

INTRODUCTION

This chapter presents an overview of several historical and contemporary environmental conflicts.[1] The international environmental liability of mining corporations is discussed. Comparisons are made between conflicts in the US and in South Africa which fall under the rubric of the environmental justice movement. Such conflicts raise many questions – the economic valuation of damages is only one of these. Others involve questions such as who has the power to impose particular languages of valuation? Who rules over the ways and means of simplifying complexity, deciding that some points of view are out of order? Who has power to determine which is the bottom line in an environmental discussion?

There is a new tide in global environmentalism that arises from social conflicts about environmental entitlements, the burdens of pollution, the sharing of uncertain environmental risks and the loss of access to natural resources and environmental services. Many such ecological distribution conflicts, whether they take place inside or outside markets, whether they are local or global, come about because economic growth depends on an increased use of environmental resources. In such ecological distribution conflicts, the poor often find themselves fighting for resource conservation and a clean environment even when they themselves do not claim to be environmentalists.

The claims to the environmental resources and services of others, which are differentially empowered and endowed, can be contested by framing the discussion within a single standard of value or across plural values. In this chapter, valuation of damages is discussed as well as international environmental liability. The relations between ecological distribution conflicts and economic valuation are as follows. First, the pattern of prices in the economy will depend on the concrete outcomes to ecological distribution conflicts. Second, ecological distribution conflicts (which often arise outside the market) are not only fought through demands for monetary compensation established in actual or fictitious marketplaces, but they may also be fought out in other arenas. When the study of an ecological distribution conflict reveals a clash of incommensurable values, it helps to develop an ecological economics, which moves beyond the obsession of 'taking nature into account' in monetary terms, and is able to cope with value pluralism.

ENVIRONMENTALISM AVANT-LA-LETTRE: COPPER MINING IN JAPAN

I have chosen copper mining as a starting point, for two reasons. First, by looking at historical cases of environmental conflict, which were not yet represented in the language of environmentalism, we can interpret them as instances of social conflict today. This helps interpret modern day conflicts where the actors are still reluctant to call themselves environmentalists (Guha, 1989). Second, by comparing historical with contemporary conflicts on copper mining, I make the point that copper has not become obsolete (despite aluminum and optic fibre). On the contrary, the frontier for the extraction of copper reaches new territories; this is a persuasive argument against the believers in the 'dematerialization' of the New Economy.

Environmentalists in Japan remember Ashio as the site of Japan's first major industrial pollution disaster. This was a large copper mine not far from Tokyo owned by the Fukurawa Corporation. There was a major workers' riot against conditions in 1907. Japanese social historians have debated whether the riot was 'spontaneous' or organized by ancient brotherhoods. There were also already some 'direct action' socialists in Japan at the time who may have played a role. While, as we shall see, in Rio Tinto in Andalusia in 1888 there was a common front between miners and peasants against pollution, this does not seem to have been so at Ashio. Here, thousands of peasants along the Watarase River fought for decades against pollution from heavy metals, which damaged not only crops but also human health. They also fought against the building of a large sediment basin to store the polluted waters, which led to the destruction of the village of Yanaka in 1907 including its cemetery and sacred shrines.

> *The mine's refinery belched clouds containing sulphuric acid that withered the surrounding forests, and the waste water ... ran off into the Watarase River, reducing rice yields of the farmers who irrigated fields with this water ... Thousands of farming families ... protested many times. They petitioned the national authorities and clashed with the police. Eventually their leader, Tanaka Shozo, created a great stir by directly petitioning the emperor for relief. As environmental destruction reemerged in the 1960s as a major social issue, and popular concern with the impact of pollution intensified, so Ashio's legacy as 'the birthplace of pollution in Japan' has endured...* (Nimura, 1997, p20–21, also Strong, 1977)

Ashio was not unique in the world, and Fukurawa's own publicity remarked that Butte in Montana was a fearful place to live: 'The smelting process has utterly destroyed the beauty of the landscape, evil gaseous smoke has killed all plant-life for miles round about; the streams are putrid with effluent, and the town itself seems buried under monstrous heaps of slag' (Strong, 1977, p67). Such were then the realities of copper mining in America. Ashio in comparison was not so bad except that, unlike Montana, there were thousands of unhappy peasants downstream.[2]

Fukurawa procrastinated taking anti-pollution measures for decades. They profited on the novelty and uncertainty of the chemical pollution in question, and from the closeness between government and business in Japan. In cost-benefit language, it was argued: 'Suppose for the sake of the argument that copper effluent were responsible for the damage to farmlands on either side of the Watarase – the public benefits that accrue to the country from the Ashio mine far outweigh any losses suffered in the affected areas. The damage can in any case be adequately taken care of by compensation' (article in the *Tokio Nichi Nichi Shinbun* of 10 February 1892, in Strong, 1977, p74). In today's parlance, a Pareto improvement means, in the strict sense, that a change such as a new mining project improves somebody's situation, and does not worsen anybody's situation. In this sense, Ashio did not fulfil the criterion. However, a Pareto improvement in a wider sense allows for compensation under the so-called Kaldor-Hicks Rule, so that those better off can (potentially) compensate those worse off, and still achieve a net gain. This was Fukurawa's claim.

Tanaka Shozo (1841–1913), the son of a peasant headman of a village in the polluted area, the leader of the anti-pollution struggles, could not yet have known about cost-benefit analysis and welfare economics. In the 1890s, he became a member of the Diet in Tokyo. He was famous for his fervent speeches and was a man with deeply religious feelings. He was the retrospective father figure of Japanese environmentalism – born more in a tradition of pro-peasant environmental justice (and also of care for urban ecology and concern for forest protection and the water cycle, Tamanoi et al, 1984) than wilderness preservation, although within a national context of industrialism and militarism that put environmentalism on the defensive.

Ashio was certainly not the only cause of early popular Japanese environmentalism. Another instance occurred when 'the Nikko company built its copper refinery on the tip of the Saganoseki peninsula (in Oita Prefecture) in 1917 – local farmers objected strenuously. They feared that the acrid smoke from the refinery would blight the mountains and ruin the mulberry trees, on which their silk industry depended. Ignoring them, the town officials agreed to the refinery. The farmers felt betrayed. The angry farmers swarmed into town and cut through the village leader's house pillars, a tactic (uchikowashi) drawn straight from the Tokugawa period... The police brutally suppressed this protest, beating and arresting 100 participants. Nikko built the mill, and it operates to this day' (Broadbent, 1998, p138).

ONE HUNDRED YEARS OF POLLUTION IN PERU

The Environmentalism of the Poor, and the environmental justice movement, are global cross-cultural movements in all countries including the US (Guha, 2000). The communities in the Central Sierra region of Peru also struggled against mining companies. The Cerro de Pasco Copper Corporation polluted pasturelands in the 1920s and 1930s. Huancavelica supplied mercury to Potosi in the 16th century. Silver had been mined in colonial and postcolonial times. Towards 1900, there was a world boom in copper, lead and zinc mining because of the proliferation of electrical instruments, tools, machines, armaments, railroads. Domestic capitalist miners were making small fortunes.

In 1901 the Peruvian government changed the mining code, allowing the mining of deposits on private property (instead of state property and a regime of administrative concessions) (Dore, 2000, pp13–15). The Cerro de Pasco Corporation, from New York, bought many of the deposits, and started a large-scale, underground mining operation. It first built several small smelters, then in 1922 a big smelter and refinery at La Oroya, the effects of which became a cause célèbre. 'The new smelter polluted the region's air, soil and rivers with arsenic, sulphuric acid and iron-zinc residues' (Dore, 2000, p14). The pastures withered, people became ill. There was a legal case brought against the company by peasant communities and by old and new hacienda owners up to 120 kilometres away. The mining company was forced by the court to buy the lands it had polluted as a form of compensation. Later, when the mining operations and La Oroya smelter became less polluting (at least with respect to the air because of scrubbers, if not with respect to rivers), all this land became a valuable asset to the company, which then started a large sheep ranching business, resulting in border conflicts with nearby communities. The enormous ranch (of about 300,000 ha) was expropriated in 1970 by land reform, and still exists as the SAIS Tupac Amaru, owned by surrounding communities. It is one of the few large-scale sheep ranches in Peru which has not been taken over and split up into individual peasant communities. In the early 1900s, the Cerro de Pasco Corporation initially had difficulties in recruiting skilled labour. It resorted to enganche, a form of debt peonage. As Dore points out (2000, p15), the large-scale pollution caused by the La Oroya smelter contributed to solving the labour shortage, because agricultural yields decreased in the small plots where agriculture is practised at such altitude, and animals died. Peasant labour became available. This was another blessing in disguise.

Mining in Peru was long dominated by the Cerro de Pasco Copper Corporation, but in the 1950s and 1960s the main sites for copper extraction began to move southward towards Cuajone and Toquepala. These are large open-pit mines near Ilo, an extension of the rich deposits of Chuquicamata and other mines in northern Chile. The Southern Peru Copper Corporation owned by Asarco and Newmont Gold has subjected the city of Ilo and its 60,000 inhabitants to water and air pollution for 30 years. The smelter was built in 1969; it spewed almost 2000 tonnes of sulphur dioxide daily. Tailings and slag were discharged without treatment on land and also into the ocean where, it was claimed, 'several kilometres of coastline are totally black' (Ivonne

Yanez, in the ELAN website, 4 October 1996). In this case the actors on both sides are explicitly speaking an environmental language (Diaz Palacios, 1988, Balvin et al, 1995). The Southern Peru Copper Corporation is Peru's major single exporter. The conflict is more urban than it was in the central Sierra and two appeals to international courts have been made. The local authorities presented a successful complaint in 1992 to the (unofficial) International Water Tribunal in the Netherlands in 1992, obtaining its moral support. A class-action suit was initiated at the District Court for the Southern District of Texas, Corpus Christi Division, in September 1995 but it was dismissed after the Peruvian State asked that the case be brought back to Peru. The plaintiffs, on behalf of people from Ilo, most of them children with respiratory illnesses, complained that the pollution from sulphur dioxide had not appreciably decreased in the last years, despite the construction of a sulphuric acid plant (which recuperates sulphur dioxide). The federal court judge decided on 22 January 1996 against admitting the case into the US judicial system on grounds of forum non-conveniens. (This legal principle allows a judge to decline to hear. or to transfer, a case even though the court is the appropriate case because of inconvenient location.)

RIO TINTO AND OTHER STORIES

It was in Huelva, in the southern Spanish region of Andalusia in the 1880s, years before the words environment and ecology became common social coinage, that the first big environmental conflict associated with the name of Rio Tinto took place (Amery, 1974, Ferrero Blanco, 1994). The old royal mines of Rio Tinto were bought in 1873 by British and German interests under Hugh Matheson, first chairman of the Rio Tinto Company. A very large open-pit mining operation was launched. Eighty years later, in 1954, the mines were sold to new Spanish owners, with the original Rio Tinto Company keeping one third interest. This British company Rio Tinto (renamed Rio Tinto Zinc) went on to become a worldwide mining and polluting giant (Moody, 1992) – its name, its business origins, its archive in London, all point to Andalusia. It was here that a massacre by the army, of local farmers, peasants, and syndicalist miners took place. It was the culmination of years of protests against sulphur dioxide pollution.

Historians still debate the number of deaths caused when the Pavia Regiment opened fire on a large demonstration in the plaza of the village of Rio Tinto: 'The company could not find out, and in any case soon decided it was better to play down the seriousness of the whole affair and gave up its attempts to discover the number of casualties, though Rio Tinto tradition puts the total number of dead at between one and two hundred' (Amery, 1974, p207, also Ferrero Blanco, 1993, p83). Historians also debate whether the miners complained only against the fact that excessive pollution prevented them from working on some days (days of manta, ie blanket) and therefore from earning full wages on those days, or whether they complained against pollution per se because of damage to their own and their families' health.

The company was taking out a large quantity of copper pyrites, employing some 10,000 miners. The idea was to sell the copper for export, and also as a by-product the sulphur in the pyrites (used for manufacturing fertilizers). The amount of ore extracted was so large, that in order to obtain the copper quickly, a lot of the sulphur was not recuperated. It was thrown into the air as sulphur dioxide when roasting the ore in a process of open-air calcination, previous to smelting the concentrate. 'The sulphurous fumes from the calcining grounds were a major cause of discontent. They produced an environment that everyone resented, for the pall of smoke which frequently hung over the area destroyed much of the vegetation and produced constant gloom and dirt' (Amery, 1974, p192). Large and small farmers, to whom the company was paying monetary compensation, managed to convince some of the councils from small surrounding villages to forbid open-air calcination in their own municipal territories.

The company successfully plotted (through members of the Spanish Parliament on its payroll) to segregate Rio Tinto as a municipal territory of its own. They used the reasonable argument that population in the mining area had increased a lot. The company was keen to have local municipal officers favourable to it. On 4 February 1888, the immediate cause of the strike was the worker's complaint about the non-payment of full wages on manta days, the demand for the abolition of piecework and the cessation of deducting one peseta weekly from the wage bill to cover the expenses of the medical fund. Maximiliano Tornet, the miners' syndicalist leader and anarchist who had been deported from Cuba back to Spain some years earlier, had managed to make an alliance with the peasants and farmers (and some landowners and local politicians) to create the Huelva Anti-Smoke League. When the army arrived in the plaza it was full of striking miners and peasants and peasant families from the region damaged by sulphur dioxide. There was an argument was going on inside the Rio Tinto town-hall about whether open-air teleras should be prohibited by municipal decree, not only in surrounding villages, but also in Rio Tinto itself. In terms of today's language of environmental management, the local stakeholders (syndicalists, local politicians, peasants and farmers) did not achieve successful conflict resolution, let alone problem resolution. Had the municipality publicly announced a decree against open-air calcination, the tension in the plaza would have diminished and the strike would have been called off. On the other hand, other stakeholders, that is, the Rio Tinto Company and the civil governor in the capital of the province, were mobilizing other resources, namely, arranging for troops to be brought into action. It is not known for sure who first shouted 'fire', perhaps a civilian from a window (Amery, 1974, p205), but the soldiers understood the shout as an order to start shooting into the crowd.[3]

The popular interpretation of this episode in terms of environmentalism became unexpectedly relevant 100 years later, as the village of Nerva, in this same region, struggled against regional authorities over the siting of a large hazardous waste dump (in a disused mine) in the 1990s. Local environmentalists and village officials explicitly appealed to the living memory of the 'year of shots' of 1888 (Garcia Rey, 1997). Meanwhile, sceptics of the popular environmentalism thesis point out that, in 1888, the workers were more worried about wages than about pollution and that the peasants and farmers were manipulated by local politicians. The local politicians wanted to make money from the Rio Tinto Company or had their own disagreements with other politicians at a national level about the treatment of the British company – so conspicuously British that it sported an Anglican Church and a cricket team.[4]

'Retrospective' environmentalism related to mining and air pollution is becoming a staple of social history in many countries. In addition to air pollution, mining also creates considerable water pollution in areas such as the Watarase River in Japan and in Ilo in Peru. The use of azogue, a mercury-containing substance used by the Spaniards in Potosi to amalgamate with silver, and which today is used in Amazonian rivers to amalgamate with gold, is an important source of contamination. Mercury, through consumption of fish, was the source of famous diseases in Japan from Minamata Bay in the 1950s onwards. These were the best-known non-radioactive pollution episodes in the post-1945 history of Japan.

Extraction of copper continued increasing at a rate of about 1.5 per cent per year in the 1990s. In the late 1990s, in the region of Intag (Cotocachi, province of Imbabura) in northern Ecuador, Mitsubishi was defeated by a local non-governmental organization, Decoin, with help from Ecuadorean and international groups, in its plans to start mining for copper. The plan was to relocate 100 families to make way for opencast mining, bringing in thousands of miners in order to extract a large reserve of copper. This is a beautiful and fragile area of cloud forest and agriculture, with a mestizo population. Rio Tinto had already shown interest, but its previous incursions in Ecuador (at Salinas in Bolivar, at Molleturo in Azuay) ended in retreats. A Mitsubishi subsidiary, Bishi Metals, started in the early 1990s with some preliminary work in Intag. After many meetings with the authorities, on 12 May 1997 a large gathering of members of affected communities resorted to direct action.

Most of the company's goods were inventoried and removed from the area (and later given back to the company), and the remaining equipment was burned with no damage to persons. The government of Ecuador reacted by bringing a court case for terrorism (a rare event in Ecuador) against two community leaders and the leader of Decoin but the case was dismissed by the courts one year later. Attempts to bring in Codelco (the Chilean national copper company) to mine were also defeated. When Accion Ecologica from Quito sent one activist, Ivonne Ramos, to downtown Santiago to demonstrate with support from Chilean environmentalists on the occasion of a state visit of the president of Ecuador, she was arrested. The publicity convinced Codelco to withdraw. Accion Ecologica also organized a visit by women belonging to the Intag communities, to copper mining areas in Peru, like Cerro de Pasco, La Oroya and Ilo. The women came back to Intag carrying sad miners' music and lyrics, which became an immediate hit in Intag. These triumphant local women to this day deny that they are environmentalists, or, God forbid, eco-feminists.[5] Today, there are several initiatives for alternative forms of development in Intag, one of them being the export of 'organic' coffee to Japan arranged through environmental networks first contacted in the fight against Mitsubishi. But the copper ore is still there, underground, and the world demand for copper (despite calls for the 'dematerialization' of the economy) keeps increasing.

BOUGAINVILLE AND WEST PAPUA

Copper mining is more successful elsewhere, though not without conflicts. In the island of Bougainville, the Rio Tinto Zinc Company found themselves up against local opposition. Despite the agreement that the company had made with the government of Papua New Guinea, which has sovereignty over Bougainville, enabling Rio Tinto Zinc to exploit the site of what was described as the most profitable copper and gold mine in the world, local peoples protested. Back in 1974, it was reported that 'the natives of Bougainville have stopped throwing geologists into the sea ever since the company [Rio Tinto Zinc] declared itself willing to compensate them for the land it had taken with cash and other material services'. However, it was also reported that monetary compensation was not enough: 'The village communities affected gave the highest importance to land as the source of their material standard of life. Land was also the basis of their feelings of security, and the focus of most of their religious attention. Despite continuing compensation payments and rental fees, local resentment over the taking of the land remains high, and there is strong opposition to any expansion of mining in Bougainville, whether by the existing company, the government, or anyone else' (Mezger, 1980, p195). Finally, the tiny island of 160,000 inhabitants erupted into a secessionist war at the end of the 1980s. We notice here the use of languages such as sacredness, and national independence. We notice also that the language of monetary compensation was used.

Not far from Bougainville, the copper extraction frontier reached West Papua under Indonesia's sovereignty 30 years ago with a copper and gold mine called Grasberg. This mine was owned by Freeport McMoRan from New Orleans. Rio Tinto had an interest in this mine. The plans for the year 2000 were to mine daily 300,000 tons of ore, of which 98 per cent would be dumped into the rivers as tailings. The 'ecological footprint' (see Rees and Westra, Chapter 5) of this operation includes, of course, not only the discarded tailings but also the overburden, that is all the materials removed before reaching the ore. The total copper content to be finally recovered would be nearly 30 million tons of copper, three years of world production, which would come into the market at a rate which would make Grasberg the supplier of nearly 10 per cent of world copper every year. This open cast mine is located in high altitudes, next to a glacier. The deposit originally formed the core of a 4100-metre mountain, and the bottom of the open pit now lies at 3100 metres above level. Water pollution in the near by Ajkwa river has, until now, been the major environmental complaint. Acid drainage will be an increasing problem. The ecology of the island is particularly sensitive, and the scale of this operation is enormous. In 1977, at the initial stages of operation, some Amungme rebelled, and destroyed the slurry pipeline carrying copper concentrate to the coast. Reprisals by the Indonesian army were terrible.

The many complaints against Freeport McMoRan led to an initially unsuccessful class-action suit in New Orleans in April 1996 by Tom Beanal and other members of the Amungme tribe. Beanal (1996) declared:

> *These companies have taken over and occupied our land ... Even the sacred mountains we think of as our mother have been arbitrarily torn up, and they have not felt the least bit guilty ... Our environment has been ruined, and our forests and rivers polluted by waste ... We have not been silent. We protest and are angry. But we have been arrested, beaten and put into containers: we have been tortured and even killed.*

Tom Beanal was reported to have obtained some money from the company for his own non-governmental organization, an attempt at a classic procedure for conflict resolution, but the legal case made some progress in the Louisiana courts in March 1998 on the issue of whether US courts could have jurisdiction over international mining operations. The best-known representative of the Amungme is Yosepha Alomang. She was subjected to detention in horrible conditions in 1994 and was prevented from leaving the country in 1998 when she wanted to attend a Rio Tinto's shareholders' meeting in London.[6]

Some of Freeport's shareholders have been publicly concerned about the liabilities incurred by the company in Indonesia. Henry Kissinger is a director of Freeport. The company was deeply involved with the Suharto regime, having given shares in the company to relatives and associates of the ex-President. Freeport is also the biggest source of tax revenue for Indonesia. Which line will the new Indonesian government, and also the separatist movement in West Papua (Organisasi Papua Merdeka, OPM), take towards plans by Freeport (and Rio Tinto) to expand the extraction of copper and gold ore? The OPM has staged ceremonies raising the Papuan flag in the last 30 years, answered violently by the Indonesia Army and by Freeport's security forces (one famous instance took place on Christmas Day of 1994 at Tembagapura, a locality near the Grasberg mine). Will claims for an ecological debt to be paid by Freeport McMoRan be made, not through a private class-action suit brought by indigenous tribes but, as a result of an Indonesian governmental action, an international replica of a Superfund case in the US? Attempts to obtain indemnities for international externalities caused by transnational corporations outside their legal country of residence are interesting ingredients in the calculation of the many environmental liabilities which the North owes to the South, the sum of which would amount to a large ecological debt.

The Indonesian state had an authoritarian regime (or less politely, a capitalist dictatorship) from the mid-1960s until the end of the 1990s. The circumstances in West Papua (or Irian Jaya, this being at the time the official Indonesian name) with both a very rich mine and a separatist movement, provided reasons for a heavy military presence. It would be a cruel joke to say that a suitable environmental policy (implementing the 'polluter pays principle') would have allowed externalities to be internalized into the price of exported copper and gold. Environmental economists forget to include the distribution of political power in their analysis. Some of them even believe in their touching innocence that environmental damages arise because of 'missing markets'. In fact, externalities should be seen in general not as 'market failures' but as 'cost-shifting successes' (Martinez-Alier and O'Connor, 1996, 1999). The language of indigenous territorial rights (whose official acceptance would be a novelty in Indonesia), and even the stronger language of a separate national Papuan identity (which is historically relevant, since West Papua was annexed by Indonesia after the departure of the Dutch), may be used now, after the end of the dictatorship, in order to fight the human and environmental disaster caused by the world's largest gold mine and the third largest copper mine.

In international conflicts of a purely political nature, such as disputes between states over a strip of useless territory, the reaching of a peace agreement and the drawing of a new frontier make both the conflict and the problem disappear. Sometimes – as in the last 20 years with the threat by CFCs to the ozone layer, or the problem with transboundary sulphur dioxide emissions in Europe – agreements are reached which lead to regimes that attempt to solve both the conflict and the problem. On the other hand, in many other environmental cases, solving the conflict is not equivalent to solving the problem. In order to advance towards problem resolution, what is needed is not conflict resolution, but conflict exacerbation.

A FEW GOLD MINING CONFLICTS

Gold is sometimes produced together with other metals such as copper, but is often the primary objective of a mining enterprise. Gold mining is similar in a sense to shrimp farming, or to the extraction of a tropical wood like mahogany. Its demand is directly driven by consumption. About 80 per cent of all gold that is dug out of the ground ends up as jewellery. Is consumption, however, the real driver of the current economy? Are not changes of techniques the real drivers of capitalism, and are they not introduced in production rather than consumption, because of the pressures of competition and profits? Moreover, could not enough consumption to maintain production levels be secured already by the incomes gained in relatively dematerialized activities – a Seattle economy without Boeing? Interesting but premature questions, because the economy is not dematerializing. Consumption has a life of its own, it is not determined by the necessity to sell production. The economy is driven by the profit-rate. Conspicuous consumption or the wish to obtain positional goods also drives the economy. Hence the use of increased incomes in order to buy more and more gold, a habit of the human species in which the East and the West truly meet.

The price of gold still makes it profitable to open new mines. Gold lasts a long time but the existing stock of gold in the world, counting also the central banks' reserves, does not seem to satisfy humankind's desires. There is pressure to open new mines and not to substitute for gold. Why do the central banks not sell the gold they have? Gold mining is particularly destructive, both when it is small-scale (eg the garimpeiros in Brazil) and when it is large-scale by corporations such as Placer Dome, Newmont, Freeport, Rio Tinto or Anglo-American. Gold mining leaves behind enormous 'ecological rucksacks', as well as pollution such as mercury and cyanide.

The participants at the Peoples' Gold Summit in San Juan Ridge, California, which was held on 2–8 June 1999, asked for a moratorium on the exploration for gold. The manifesto resulting from this meeting begins:

> *Life, land, clean water and clean air are more precious than gold. All people depend on nature for life. The right to life is a guaranteed human right. It is, therefore, our responsibility to protect all of nature for present and future generations. Large-scale gold mining violently uproots and destroys the spiritual, cultural, political, social and economic lives of peoples as well as entire ecosystems. Historic and current destruction created by gold mining is greater than any value generated.[7] Commercial gold mining projects are located mainly on indigenous lands. By violating their land rights, mining companies deny the right to life of those indigenous peoples, whose relationship to land is central to their spiritual identity and survival. We need to support the self-determination of indigenous peoples and the recovery, demarcation and legal recognition of campesinos, tribal and indigenous peoples' lands … Large-scale and small-scale, toxic chemical-dependent gold mining damages landscapes, habitats, biodiversity, human health and water resources. Water especially is contaminated by cyanide, acid mine drainage, heavy metals and mercury from gold mining. Additionally, the hydrological cycle is changed and water sources are grossly depleted by pumping water from aquifers.'*

There is not enough space here to describe in detail the conflict in Peru, at the time of writing, between the Yanacocha mine in Cajamarca (where Atahualpa met Pizarro) and local communities that belong to the Federacion de Rondas Campesinas. The gold mine is owned by Newmont, and also by a local company, with a 5 per cent share belonging to the International Finance Corporation of the World Bank. In Venezuela, under the government that preceded President Hugo Chavez, Decree 1850 of 1997 tried to open up the 3 million ha forest reserve area of Imataca to gold mining. A movement arose which consisted of the sparse local indigenous Pemon population, some environmental groups such as Amigransa (the friends of the Gran Sabana), some anthropologists and sociologists and some members of Parliament. They all used different languages (from Indian demonstrations in the streets of Caracas to legal appeals to the Supreme Court) in the service of the same cause. They managed, for the time being, to stop mining in Imataca. The environmental commission of the Chamber of Deputies of Venezuela appealed to the Supreme Court against Decree 1850, quoting a figure between US$7000 and US$23,000 per ha for the restoration of the vegetable cover affected by oil exploitation. This was a useful, if moderate, figure used to calculate the large environmental liabilities that gold mining, with its toxic effects and large ecological rucksacks, implies.[8]

ENVIRONMENTAL JUSTICE

'Environmental justice' is not an expression taken, as one would expect, from philosophy or ethics but from environmental sociology and the study of race relations. In the US since the late 1980s and early 1990s it has come to mean an organized movement against environmental racism. It is also relevant for Indian reservations in the US, particularly in the context of uranium mining and nuclear waste. Indeed, 'environmental justice' could subsume historic conflicts about sulphur dioxide, the Chipko and Chico Mendes cases, the current conflicts surrounding the use of carbon sinks, the conflicts of oustees from dams, the fight for the preservation of rainforests or mangroves for livelihood and many other cases around the world which sometimes have to do with 'racism' and sometimes not.

The environmental justice movement in the US (Bullard, 1990,1993, Pulido, 1994, 1996, Bryant and Mohai, 1992, Bryant, 1995, Sachs, 1995, Gottlieb, 1993, Szasz, 1994, Schwab, 1994, Westra and Wenz, 1995, Dorsey, 1997, Faber, 1998, DiChiro, 1998, Camacho, 1998, Taylor, 2000) is quite different from the two previous environmentalisms in this country, namely, the efficient and sustainable use of natural resources (in the tradition of Gifford Pinchot), and the cult of wilderness (in the tradition of John Muir). As a self-conscious movement, Environmental Justice fights against the alleged disproportionate dumping of toxic waste or exposure to different sorts of environmental risk in areas of predominantly African-American, Hispanic, Native American or low income populations.

The language employed is not that of uncompensated externalities but rather the language of race discrimination, which is politically powerful in the US because of the long civil rights struggle. In fact, the organized environmental justice movement is not an outgrowth of previous currents of environmentalism but rather an outgrowth of the civil rights movement. Thus, in the developing world in the 1980s, the main socio-environmental question was whether an indigenous, independent environmentalism of the poor existed. This question was first theorized in India and, later, in Latin America and Africa, because of episodes of defence of common property resources against the state or the market (Guha and Martinez-Alier , 1997, 1999). Meanwhile, in the US the question was whether the buoyant mainstream environmental movement would deign to consider the existence of 'environmental racism', whether it could accept and work with 'minorities' who were mainly concerned with urban pollution.

There are many cases of local environmental activism in the US by 'citizen-workers groups' (Gould et al, 1996) outside the organized environmental justice movement, some with 100 years' roots in the many struggles for health and safety in mines and factories. Perhaps struggles and complaints against pesticides in Southern cotton fields, and certainly in the fight against toxic waste at Love Canal in upstate New York led by Lois Gibbs (Gibbs, 1981, 1995), who also later led a nation-wide 'toxics-struggles' movement showing that poor communities would not tolerate any longer being dumping grounds (Hofrichter, 1993), were part of this growing movement. In the 'official' environmental justice movement are included celebrated episodes of collective action against incinerators (because of the uncertain risk of dioxins), particularly in Los Angeles, led by women (see Conclusion). Also in the 1980s, other environmental conflicts gave rise to groups such as People for Community Recovery in South

Chicago (Altgeld Gardens), led by Hazel Johnson; and the West Harlem Environmental Action (WE-ACT) in New York, led by Vernice Miller. In 1989, the Southwest Network for Economic and Environmental Justice (SNEEJ), led by Richard Moore, was founded, with its main operations in Albuquerque, New Mexico, out of grievances felt by Mexican and Native American populations.

In October 1991 the First National People of Colour Environmental Leadership Summit took place in Washington, DC, the Principles of Environmental Justice were proclaimed and the movement for Environmental Justice became well known. President Clinton's Executive Order 12898 of 1994 on Environmental Justice was a triumph for this movement. It directed all federal agencies (though not corporations or private citizens) to act in such a way that disproportionate burdens of pollution do not fall on low-income and minority populations in all territories and possessions of the US. Thus, both poverty and race were taken into account, but nothing was said about impacts outside the US. Happy is the country where 'low-income' people are regarded as a minority (alongside or overlapping with racial 'minorities')!

The insistence on 'environmental racism' is sometimes surprising to analysts from outside the US. In fact, some foreign academics refuse to acknowledge the racial angle, and have boldly stated that: 'If one were asked to date the beginning of the environmental justice movement in the US, then 2 August 1978 might be the place to start. This was the day when the CBS and ABC news networks first carried news of the effect of toxic waste on the health of the people of a place called Love Canal' (Dobson, 1998, p18, see also Dobson, Chapter 4). However, the Love Canal activists, led by Lois Gibbs, were not people of colour, they were white, as such categories are understood in the US, and therefore were subject only to metaphorical, not real 'environmental racism'. Other non-US academics agree with the interpretation that Environmental Justice is in the US a movement against 'environmental racism'. I also agree. Thus, the seminal moment (Low and Gleeson, 1998, p108) came in 1982 in Warren County, North Carolina.

Of course, one could also argue that the world environmental justice movement started long ago at a 100 dates and places all over the world. For instance, in Andalusia in 1888, when miners and peasants at Rio Tinto were massacred by the army. Or when Tanaka Shozo threw himself in front of the Emperor's carriage with a petition in his hand. Or, in the US, not in North Carolina, but in Wisconsin during the 1970s and 1980s as alliances of Indian tribes and environmentalists struggled against mining corporations (Gedicks, 1993). Or, the many other struggles of resistance by Native Americans, from Canada to Tierra del Fuego. Which will be considered the worldwide environmental justice first? Possibly Chico Mendes' assassination day, Ken Saro-Wiwa's, or perhaps the day the *Rainbow Warrior* was sunk by the French secret services in New Zealand, and its Portuguese cook died? Or when Karunamoi Sardar died defending her village in Horinkhola, Khulna, Bangladesh, on 7 November 1990?[9]

The self-conscious US environmental justice movement of the 1980s and 1990s shifted the whole discussion about environmentalism in the US away from preservation and conservation of Nature towards social justice. It also destroyed the NIMBY image of grass-roots environmental protests by turning them into NIABY protests (not in anyone's backyard). Also, it expanded the circle of people involved in environmental policy by practising 'post-normal science' in the 'popular epidemiology' movement. In the US, legislation against racism (such as Title VI of the Federal Civil Rights Act of 1964) forbids discrimination based on race. However, in order to establish the existence of racism, it is not sufficient to prove that environmental impact is different (for instance, that lead in children's blood level is different according to racial background). It must also be shown that there is an explicit intention to cause harm to a minority group. The uncertainties of environmental risk (for instance, dioxin), and the statistical difficulties in separating racial and economic factors in toxic waste location decisions have given rise to a rich practice of 'popular epidemiology' (Novotny, 1998). It might be difficult to prove that race more than poverty correlates with toxic waste, but if it is convincingly shown, then the chances of redress are high. Lay persons gather scientific data and other information, and they also process the results offered by official experts in order to challenge them in cases involving toxic pollution, a clear case of 'extended peer review'.

By emphasizing 'racism', the movement for Environmental Justice also emphasizes incommensurableness of values. This is its greatest achievement. If I pollute a poor neighbourhood, by applying the polluter pays principle (PPP), I may compensate the damage. This is more easily written than done, because, how much is human health worth? In which scale of value? Nevertheless, the PPP implies that a worsening ecological distribution is in principle compensated by an improving of economic distribution. The objective is to make pollution expensive enough so that its level will decrease by a change in technology or by a lower level of pollution production. Whatever the objective, the principle implies a single scale of value. Now, the same problem phrased in terms of 'environmental racism' becomes a different problem. I can inflict damage to human dignity by using a racial insult or by racial discrimination. Paying a fine does not entitle me to repeat such conduct. There is no real compensation. Money and human dignity are not commensurate.

Bullard, who is both an academic and an activist, realizes the potential of the environmental justice movement beyond 'minority' populations, and asserted in 1994: 'Grass-roots groups, after decades of struggle, have grown to become the core of the multi-issue, multi-racial, and multi-regional environmental justice movement. Diverse community-based groups have begun to organize and link their struggles to issues of civil and human rights, land rights and sovereignty, cultural survival, racial and social justice and sustainable development… Whether in urban ghettos and barrios, rural "poverty pockets", Native American reservations or communities in the developing world, grass-roots groups are demanding an end to unjust and non-sustainable environmental and development policies…'.[10] Notice the clear awareness that environmental justice is functional to sustainability and that it concerns poor people everywhere, including developing world communities, that is, billions of people. Low-income 'people of colour' are a minority in the US but they are certainly a majority in the world at large.

There are some ecological distribution conflicts in the world (the European conflicts on nuclear risks as expressed at famous fights in Gorleben or Creys-Malville (see Blowers, Chapter 3), or the European conflict against US 'hormone beef' and transgenic crops, or the current conflict on the Three Gorges dam in China, for instance), for which the analysis and resolution of the 'environmental racism' metaphor is not useful. On the other hand, we could retrospectively apply 'environmental racism' to one of the many forms of racism that the Spaniards showed in America, by imposing a terrible load of mercury poisoning on indigenous workers in silver mines (Dore, 2000). Environmental racism is often a useful language for conflicts that have been fought, up to now, under the banner of indigenous territorial rights. Activists and lawyers in the class action suit against Texaco in Ecuador, blamed Texaco in advertisements in US newspapers in 1999 for 'environmental racism'. Profiting from the publicity against Texaco because of a court case for internal racism against black employees in the US (settled out of court in 1997 for US$176 million), sympathizers for the Ecuadorian plaintiffs placed an advertisement in the *New York Times* (23 September 1999) which stated: 'The lawsuit alleges that in Ecuador, Texaco dumped the poisonous water produced by oil drilling directly onto the ground, in nearby rivers, and in streams and ponds. The company knowingly destroyed the surrounding environment and endangered the lives of the indigenous people who had lived and fished there for years. These are people of colour, people for whose health and well-being Texaco shows only a cavalier disregard … It's time that Texaco learns that devaluing the lives and well-being of people because of the colour of their skin is no longer acceptable for any American company'. Notice that this language, so effective in the US, was not used when the case started in 1993. Moreover, it would be problematic though not impossible to apply it to Texaco's successor, Petroecuador, which has used similar technology damaging not only indigenous people but also average mestizo Ecuadorian settlers. Perhaps 'internal colonialism' (Adeola, 2000) could be used against Petroecuador, as against the Nigerian authorities, while 'racism' could be reserved for Texaco (or Shell, in Nigeria, see Agbola and Alabi, Chapter 13).

INTERNATIONAL ENVIRONMENTAL LIABILITIES

Environmental conflicts in South Africa are often phrased in the language of environmental justice (Bond, 2000, see also Roberts, Chapter 9). Thus, a conflict in the late 1990s which continues at the time of writing, places environmentalists and local populations against a project near Port Elizabeth. The project is the development of an industrial zone, a new harbour and a smelter of zinc for export. Billinton, a British firm, owns the project that would guzzle up electricity and water at cheap rates while poor people in the area cannot get the small amounts of water and electricity they desperately need, or at the very least must pay increasing rates under current economic policies. The Billinton project has costs in terms of lost tourism revenues because of the threats to a proposed national elephant park extension nearby, beaches, estuaries, islands and whales (Bond, 2000, p47). There are also costs in terms of the displacement of people from the village of Coega. This point was emphasized in a letter sent by the Southern Africa Environment Project to Peter Mandelson, the then British Secretary of State for Trade and Industry: 'We are writing on behalf of those who have historically lacked the capacity to assert their rights and protect their own interests but who now seek to be heard and to call to the attention of the international community the injustice that is now about to be inflicted upon them'. The life of the people of Coega was already full of memories of displacements under the racist regime of apartheid. Although Billinton could no longer profit from the lack of voice of the people under apartheid, it was alleged that it now sought 'to take advantage of the region's desperate need for employment to enable construction of a highly polluting facility that would never be allowed adjacent to a major population centre in the UK or any other European country'.[11]

A small improvement in the economic situation of the people would be obtained at high social and environmental cost. The displacement of people, along with increased levels of sulphur dioxide, heavy metals, dust and liquid effluents are a few of the consequences of this project. An appeal was made to the British Minister to take into account the Organisation for Economic Co-operation and Development (OECD)'s guidelines for multinational enterprises which has included a chapter on environmental protection since 1991. Unfortunately the chapter consists only of recommendations which the authorities cannot enforce directly. The Minister was asked in any case to exercise his influence upon Billinton informally.

The environmental impacts left behind by the apartheid regime are now surfacing. There are large liabilities to be faced. Best known is the asbestos scandal, which includes international litigation initiated by victims of asbestosis against British companies, particularly Cape. Nearly 2000 persons asked for compensation because of personal damages as a result of Cape's negligence in supervising, producing and distributing asbestos products. The lawyers argue that Cape was aware of the dangers of asbestos at least from 1931 onwards, when Britain asbestos regulations were introduced. Nevertheless production continued in South Africa with the same low safety standards until the late 1970s. Medical researchers have found that 80 per cent of Penge's black miners (in Northern Province) who died between 1959 and 1964 had asbestosis. The average age of the victims was 43. Cape operated a mill for 34 years in Prieska, Northern Cape, where 13 per cent of workers' deaths were attributed to mesothelioma, a very painful asbestos-related cancer. Asbestos levels in this mill in 1948 were almost 30 times the maximum UK limit. There are other cases in South Africa of asbestos contamination, by companies such as Msauli and GEFCO, at locations such as Mafefe, Pomfret, Barberton, Badplass (Felix, in Cock and Koch, 1991).

Contaminated abandoned mines and asbestos dumps must now be rehabilitated by the post-apartheid South African Government. Simultaneously, court cases were started against Cape in the UK. The House of Lords (in its judicial capacity) ruled until July 1999, when the judgement was reversed, that such cases could be heard in London rather than in South Africa. British companies could be sued in British courts. Against GATT-WTO doctrine, the asbestos court case and similar ones, if successful, would show that international regulation is required not only about the safety and quality of the final products, but also on the process of production and its side-effects. When regulation failed or was non-existent, and when effective protest was impossible because of political repression, there are retrospective liabilities to be faced. The courts will perhaps institute little by little a sort of international Superfund obligations for the transnational companies. True, the South African apartheid state was blind to damage to black workers. The asbestos and mining companies most probably fulfilled internal South African laws in regard to safety, wages and taxes. Nevertheless, they should be held accountable for the 'externalities' that they have left behind. Given the chance, workers and their families would have complained, not so much because they were environmentalists but because their health was threatened. The law firm which represents the asbestosis victims (Leigh, Day) also brought actions in London for damages to workers at Thor Chemicals in KwaZulu-Natal on behalf of victims of poisoning by mercury and on behalf of cancer victims from Rio Tinto's Rossing uranium mine in Namibia.[12]

In April 1990 massive concentrations of mercury had been detected in the Umgeweni River near the Thor Chemicals' Cato Ridge plant. This was reported in the national and international press. Thor Chemicals imported mercury waste into South Africa, partly through Cyanamid, an American company. South African environmental groups, mainly Earthlife under Chris Albertyn's leadership, allied themselves with the Chemical Workers Industrial Union, the local African residents under their chief, and also white farmers from the Tala Valley who had already endured a bad experience of pesticide spraying from the neighbouring sugar industry. A true 'rainbow' alliance, which also incorporated US activists against the Cyanamid plant in question, complained against such 'garbage imperialism' or 'toxic colonialism' by asking: 'Why did Thor, a British company, decide to build the world's largest toxic mercury recycling plant on the borders of KwaZulu in a fairly remote part of South Africa? Why not build it closer to the sources of the waste mercury in the US or in Europe?' (Crompton and Erwin, in Cock and Koch, 1991, pp82–84).

Actually, 'the practice of exporting hazardous wastes for disposal in developing countries has been described as environmental injustice or environmental racism on a global scale' (Lipman, 1998). The Basel Convention of 1989 forbids the export of hazardous waste from rich countries except for recovery of raw materials or for recycling. It was complemented on 25 March 1994 by a full ban negotiated at a meeting in Geneva on all exports of hazardous waste from the 24 rich, industrialized countries of the OECD. The agreement was reached over the opposition of the richest countries which received from Greenpeace, in this context, the name of the Sinister Seven. Some defections inside the European Union (Denmark, and later Italy) helped an alliance among China, Eastern European countries and in general all Southern poor countries in order to close, at least in theory, the 'recycling' loophole of the initial 1989 convention though which 90 per cent of the waste was flowing. Thus, pending ratification and domestic implementation of this agreement, and assuming that Article 11 of the Basel Convention (which allows for bilateral or multilateral hazardous waste exporting agreements provided they comply with 'environmentally sound management') is not abused, then a sad chapter of industrialization would be closed. Rich countries would not be able to exploit the weaker regulations of poorer countries to avoid their own responsibility for minimizing waste. Clearly, the issue is far from over. The pressure for the export of toxic waste still increases, although the Basel Convention has had a positive effect. This is the context in which it was announced that nearly 3000 tons of Taiwanese toxic waste from the group Formosa Plastics had been dumped in a field in the port of Sihanoukville in Cambodia in November 1997. Taiwan is not a party to the Basel Convention. The waste was scavenged by poor villagers, many of whom later complained of sickness; one died quickly (Human Rights Watch, 1999). The logic of Lawrence Summers' Principle still remains compelling.[13]

The perception of risk changes with time, sometimes because scientific research produces clear results, and sometimes because, on the contrary, scientific uncertainties cannot be dispelled, and a feeling of danger creeps in. Then the question is asked: Who is responsible for cleaning up the (newly perceived) mess, for paying indemnities or making reparations? How to assign environmental liabilities, granting that restoration may be impossible when irreversible damages or deaths are involved. Thus, the Superfund legislation in the US is supposed to achieve the cleaning up of hazardous waste sites (chemical dumps, mine tailings, brownfields), which are called 'orphaned' when no existing corporation or private citizen or public body is responsible. In this exceptional case, the burden of proof lies with polluting companies rather than with the polluted citizens or with the regulatory agency. Companies have to prove against EPA allegations that no risk of damage exists from the waste they have abandoned. Nuclear waste is excluded however from the Superfund legislation, which arose in the late 1970s (at the end of President Carter's administration). Its official name is Comprehensive Environmental Response Compensation and Liability Act (CERCLA). As in Europe after the Seveso alarm (dioxin release from a chemical firm near Milan), in the US after the Love Canal scandal near Niagara Falls in upstate New York, there was a feeling that something should be done to remedy damage, and to make future damage costly by imposing strict norms of private or public liability. Superfund may also be interpreted as a government response to the first stirrings of the environmental justice movement. Clean-up operations under Superfund are financed by special charges on the oil and chemical industries, when the sites are 'orphaned'. When the companies are identified and still active, they have to pay for the clean-up. The EPA must not act in an 'arbitrary and capricious' manner but it has no obligation to prove that there is actual damage, only that there is a risk of damage. Critics of Superfund point out that the costs are too high compared to benefits, including administrative costs, and that the communities near the waste sites cleaned up do not always benefit economically because the improved environmental situation is countered by the adverse environmental image.

Notice that there is no international Superfund to which appeals can be made, should common-law judicial actions against Texaco, Freeport, MacMoRan, Dow Chemical, Cape or the Southern Peru Copper Corporation fail.

After listing a number of cases in the US in which indemnities have been paid by corporations, such as the Exxon Valdez incident, a Venezuelan journalist asked himself in January 2000: 'Being Venezuela [is] a country dominated by the oil and mining industries, the question is, which is the *pasivo ambiental* (ie environmental liability) of all this oil and mining activity in our country?'[14] It is fascinating to watch the diffusion of the term pasivo ambiental in a mining and oil extraction context in Latin America as one writes this chapter. Hector Sejenovich, from Buenos Aires, was perhaps the first economist to use this term when he calculated the environmental liabilities from oil extraction in the province of Neuquen, Argentina. The Argentinian Minister for the Environment Oscar Massei was quoted on 6 February 200 (journal *Rio Negro*, online) as saying that regional incentives to oil companies in Neuquen may not include flexibilization of environmental standards. The government, he added ominously, had in its possession the study made for UNDP which evaluated the *pasivos ambientales* from oil exploitation in Neuquen at US$1 billion. In Peru, a new law project was submitted to Congress in 1999 (project no 786) creating an National Environmental Fund – as sort of internal Global Environmental Facility (GEF) as a congressman put it. The fund would finance environmental research, would improve and restore the environment and would promote ecological agriculture. Its economic resources would come mainly from a percentage of the revenue from the privatization of state enterprises. After complaining about the environmental deterioration in the last decades because of mining and fisheries, after commenting on increasing desertification and deforestation in the country, congressman Alfonso Cerrate quoted the unsuccessful case of the privatization of CENTROMIN (the state firm which was the successor to the Cerro de Pasco Copper Corporation). The environmental problems of Centromin must have been a factor in the lack of buyers at the auction. The question was, 'who will pay for the ecological debt? Who will assume the environmental liability (*pasivo ambiental*) accumulated throughout the years by CENTROMIN and other state firms?'

In Chile, new legislation on liabilities after closing mines was under discussion in 1999 and 2000. The Sociedad Nacional de Mineria was aware that environmental standards should improve, that there was a danger of being internationally accused of ecological dumping, and it was in favour of applying international environmental standards adapted to national realities. 'En el tema del pasivo ambiental [with respect to environmental liability]', it added, discussions were proceedingbut the general feeling in the industry was that the state should assume environmental liabilities.[15] The Bolivian vice-minister of mines, Adam Zamora, referring to the pollution in the Pilcamayo river (that flows down from Potosi towards Tarija and eventually Argentina), increased by the bursting of a tailings dyke at Porco belonging to Comsur, had said in 1998: 'la neuva politica estatal minero-metalurgica tiene como responsabilidad remediar los pasivos ambientales originados en la actividad minera del pasado [the new State policy on minerals and metallurgy incorporates the responsibility of mitigating the environmental liabilities caused by mining in the past]' (*Presencia*, 16 June 1998). In fact, environmental liabilities in Potosi reach back to the 16th century.

CONCLUSIONS

Driven by consumption, the throughput of energy and materials in the world economy has never been so large as today. Paradoxically, increases in eco-efficiency sometimes leads to increased demands of material and energy, because their costs diminish (the Jevons Effect). Externalities will not decrease, on the contrary, they increase because of the growth of the world economy. We are certainly not in a 'post-material' age. Externalities (ie cost-shifting) must be seen as part and parcel of the economy, which is necessarily open to the entry of resources and to the exit of residues. The appropriation of resources and the production of waste result in ecological distribution conflicts, which give rise to a worldwide environmental justice movement which in fact started many years ago. The Environmentalism of the Poor and the movement for Environmental Justice (local and global), has grown from the complaints against the appropriation of communal environmental resources and against the disproportionate burdens of pollution, may help to move society and economy in the direction of sustainability.

Activists in the US environmental justice movement have emphasized the links between the increasing globalization of the economy and environmental degradation of habitats for many of the world's peoples. Robinson (1999) has argued that:

> *In many places where Black, minority, poor or Indigenous peoples live, oil, timber and minerals are extracted in such a way as to devastate eco-systems and destroy their culture and livelihood. Waste from both high- and low-tech industries, much of it toxic, has polluted groundwater, soil and the atmosphere. Environmental degradation such as this, and its concomitant impact on human wealth and welfare, is increasingly seen as violation of human rights.*

She continues:

> *As mining, logging, oil drilling and waste-disposal projects push into further corners of the planet, people all over the world are seeing their basic rights compromised, losing their livelihoods, cultures and even their lives. 'Environmental devastation globally and what we call 'environmental racism' in the US, are violations of human rights and they occur for similar reasons.*[16]

The management and resolution of such ecological distribution conflicts requires cooperation between business, international organizations, NGO networks, local groups and governments. Can this cooperation be based on common values and on common languages? I argue that this is not always the case, that whenever there are unresolved ecological conflicts, there is likely to be not only a discrepancy but incommensurability in valuation (Faucheux and O'Connor, 1998, Funtowicz and Ravetz, 1994, Martinez-Alier, Munda and O'Neill, 1998, 1999, Martinez-Alier and O'Connor, 1996, 1999, O'Connor and Spash, 1999). Environmental conflicts are expressed as conflicts on valuation, either inside one single standard of valuation, or across plural values. In other words, 'semiotic resistance' (O'Connor, 1993, Escobar, 1996, p61) to environmental abuse may be expressed in many languages. To see in statements about human rights, indigenous rights, sacredness, culture, livelihood, a lack of understanding or an a priori refusal of the techniques of economic valuation in actual or fictitious markets, indicates a failure to grasp the existence of value pluralism. Different interests can be defended either by insisting on the discrepancies of valuation inside the same standard of value, or by resorting to non-equivalent descriptions of reality, ie to different value standards.

Should legislation require dispersed minerals to be concentrated again to their previous state and the dispersed overburden to be restored, it would indeed change the pattern of prices in the economy. Beyond economic values, choices on the use of 'natural capital' involve decisions about which interests and forms of life will be sustained, which will be sacrificed or abandoned. A common language of valuation is not available for such decisions. For instance, it can be stated that while humans have different economic values they all have the same value in the scale of human dignity. When we say that someone or something is 'very valuable' or 'not very valuable', this is an elliptical statement (which requires the further question, in which standard of valuation? (O'Neill, 1993). For policy, what is needed is not cost-benefit analysis but rather a non-compensatory multi-criteria approach able to accommodate a plurality of incommensurable values (Munda, 1995, Martinez-Alier, Munda and O'Neill, 1998, 1999).

While conventional economics looks at environmental impacts in terms of externalities which should be internalized into the price system, one can see externalities (following Kapp) not as market failures but as cost-shifting successes which nevertheless might give rise to environmental movements (Leff, 1995, O'Connor, 1988). Such movements will legitimately employ a variety of vocabularies and strategies of resistance, and they cannot be gagged by cost-benefit analysis or by the cost-effectiveness approach. In conclusion, conflicts on the access to natural resources or on the exposure to environmental burdens and risks, may be expressed:

- Inside one single standard of valuation. How should the externalities caused by a firm be valued in money terms, when demanding compensation in a court case? How could an argument for conservation of a natural space be made or contested, in terms of the number and biological value of the species it contains, or in terms of its net primary production? An appeal to the particular experts is appropriate here.
- Through a value standard contest or dispute, that is a clash in the standards of value to be applied, as when losses of biodiversity, or in cultural patrimony, or damage to human livelihoods, or infringements on human rights, are compared in non-commensurable terms to economic gains from a new dam or from a mining project or from oil extraction. There is a clash in standards of valuation when the languages of environmental justice, indigenous territorial rights or environmental security, are deployed against monetary valuation of environmental risks and burdens. Non-

compensatory multi-criteria decision aids or participatory methods of conflict resolution are more appropriate for this second, common type of situation, than the mere appeal to the disciplinary experts (O'Connor and Spash, 1999, p5).

Thus, ecological distribution conflicts are sometimes expressed as discrepancies of valuation inside one single standard of value (as when there is a disputed claim for monetary compensation for an environmental liability), but they often lead to multi-criteria disputes (or dialogues) which rest on different standards of valuation. What is 'the price of oil?' – asked Human Rights Watch in a report on the Niger Delta. 'Todo necio/ confunde valor y precio' – agreedAntonio Machado long ago.

NOTES

1   Since May 2000, when it was presented at the Royal Geographical Society/Institute of British Geographers Conference 'Towards Sustainability: Social and Environmental Justice' at Tufts University, this chapter has been expanded into a book that Edward Elgar will publish in 2002 with the title *The Environmentalism of the Poor – a Study of Ecological Conflicts and Valuation.*
2   Butte has been known as the "richest hill on Earth" in Montana local lore and history, an honour which belongs not to Butte's copper but to Potosi's Cerro Rico's silver. Butte recently has earned the more dubious distinction of being the Environmental Protection Agency's geographically largest 'Superfund' clean-up site, a legacy of mining history' (Finn, 1998, p250, fn. 8). Butte used to belong to the Anaconda Company, which bought from Guggenheim the Chuquicamata mine in Chile, possibly the largest copper mine on earth. No Superfund for Chuquicamata, nor for Potosi.
3   Ferrero Blanco (1994, p214) lists the articles of the Criminal Code which were infringed. There were no judicial pursuits in Spain or in Britain.
4   Sceptics also point out correctly that in Aznalcollar, a village inside the polluted area of 1888, the miners of Bolliden clamoured in 1999 for 'their' mine to reopen, against middle-class environmentalists from Seville and Madrid. Bolliden is a Swedish-Canadian company whose tailings dike broke down in 1998 contaminating with heavy metals 10,000 ha of irrigated agriculture (where cultivation has been discontinued), and threatening the Doñana national park in the delta of the Guadalquivir.
5   Accion Ecologica (Quito) and Observatorio Latinoamericano de Conflictos Ambientales (Santiago de Chile), A los mineros: ni un paso atras en Junin-Intag (Quito, 1999).
6   Survival for Tribal Peoples (London), Media Briefing May 1998, 'Rio Tinto critic gagged'.
7   I would myself frame the issue in terms of incommensurableness of values rather than the value being larger or smaller.
8   *The Economist*, 12 July 1997, p30, *El Universal* (Caracas), 3 August 1997, pp1–12.
9   'Horinkhola and the surrounding villages have been declared a '"Shrimp-Free Zone", and every November 7, thousands of landless peasants gather here in a show of solidarity with this community's resistance against the shrimp industry' (Ahmed, 1997, p15).

10 R Bullard, Directory. People of Colour Environmental Groups 1994–1995, Environmental Justice Resource Center, Clark Atlanta University, Georgia.

11 Available at http://www.saep.org, letter from Norton Tennille and Boyce W Papu to Peter Mandelson, 7 September 1998.

12 Ronnie Morris, 'UK court demolishes double standards', *Business Report*, 4 March 1999, and subsequent information downloaded from http://www.saep.org. A UN report stated in 1990 that the Rossing uranium mine in Namibia was 'a theft under the law and must be accounted for when Namibia becomes independent'.

13 Internal World Bank memo, as reported in *The Economist*, 8 February 1992, under the title 'Let them eat pollution'. This has become a favourite text for the environmental justice movement.

14 Orlando Ochoa Teran, *Quinto Dia*, 18 January 2000, relayed by J C Centeno through the Environment in Latin America discussion list (ELAN at CSF). Some of us have been struggling for years to introduce the equivalent expression, deuda ecologica.

15 Danilo Torres Ferrari, 'Los avances de la normativa sobre Cierre de Faenas Mineras', *Boletin Minero* (Chile), 1122, June 1999.

16 Deborah Robinson, 'Environmental devastation at home and abroad: The importance of understanding the link', 1999, http://technicalassistance.com/ipu.htm

REFERENCES

Adeola, F O (2000) 'Cross-national environmental injustice and human rights', *American Behavioral Scientist*, 43(4), pp686–706

Agarwal, A and Narain, S (1991) *Global Warming: A Case of Environmental Colonialism*, Centre for Science and Environment, Delhi

Ahmed, F (1997) *In Defence of Land and Livelihood. Coastal Communities and the Shrimp Industry in Asia*, Consumers' Association of Penang, CUSO, InterPares, Sierra Club of Canada, Ottawa and Penang

Amery, D (1974) *Not on Queen Victoria's Birthday: The Story of the Rio Tinto Mines*, Collins, London

Anderson, M R (1996) 'The conquest of smoke: Legislation and pollution in colonial Calcutta' in D Arnold and R Guha (eds) *Nature, Culture and Imperialism: Essays on the Environmental History of South Asia*, Oxford University Press, Delhi

Arnold, D and Guha, R (eds) (1996) *Nature, Culture and Imperialism: Essays on the Environmental History of South Asia*, Oxford University, Delhi

Balvin, D, Tejada Huaman, J and Lozada Coastro, H (1995) *Agua, Mineria y Contaminacion: El Caso Southern Peru*, Labour, Ilo

Barham, B, Bunker, S G and O'Hearn, D (1994) *States, Firms and Raw Materials: The World Economy and Ecology of Aluminum*, University of Wisconsin Press, Madison

Baviskar, A (1995) *In the Belly of the River: Tribal Conflict Over Development in the Narmada Valley*, Oxford University Press, Delhi

Beckenbach, F (1996) 'Ecological and economic distribution as elements of the evolution of modern societies', *Journal of Income Distribution*, vol 6 (2), pp163–91

Beinart, W and Coates, P (1995) *Environment and History: The taming of Nature in the USA and South Africa*, Routledge, London and New York

Beanal, T (1996) Speech at Loyola University, New Orleans, 23 May. See: http://www.corpwatch.org/issues/PID.jsp?articleid=987

Bond, P (2000) 'Economic growth, ecological modernization or environmental justice? Conflicting discourses in post-apartheid South Africa', *Capitalism, Nature, Socialism*, vol 11 (1), pp33–61

Broadbent, J (1998) *Environmental Politics in Japan: Networks of Power and Protest*, Cambridge University Press, New York

Brosius, J P (1999) 'Comments to A. Escobar, after nature: Steps to an anti-essentialist political ecology', *Current Anthropology*, vol 40 (1)

de Bruyn S M and Opschoor J B (1997) 'Developments in the throughput-income relationship: Theoretical and empirical observations', *Ecological Economics*, vol 20, pp255–68

Bryant, B (ed) (1995) *Environmental Justice: Issues, Policies and Solutions*, Island Press, Washington, DC

Bryant, R and Bailey, S (eds) (1997) *Third World Political Ecology*, Routledge, London

Bryant, B and Mohai, P (eds) (1992) *Race and the Incidence of Environmental Hazards*, Westview, Boulder

Bullard, R (1990) *Dumping in Dixie: Race, Class and Environmental Quality*, Westview, Boulder

Bullard, R (1993) *Confronting Environmental Racism: Voices from the Grassroots*, South End Press, Boston

Bunker S (1996) 'Raw materials and the global economy: Oversights and distortions in industrial ecology', *Society and Natural Resources*, vol 9, pp419–29

Camacho, D E (ed) *Environmental Injustices, Political Struggles. Race, Class and the Environment*, Duke University Press, Durham and London

Carrere, R and Lohman, L (1996) *Pulping the South: Industrial Tree Plantations and the World Paper Economy*, Zed, London

Cleveland, C and Ruth, M (1998) 'Indicators of dematerialization and the materials intensity of use', *Journal of Industrial Ecology*, vol 2, pp15–50

Cock, J and Koch, E (eds) (1991) *Going Green: People, Politics and the Environment in South Africa*, OUP, Cape Town

Costanza, R (ed) (1991) *Ecological Economics: The Science and Management of Sustainability*, Columbia University Press, New York

Dembo, D, Morehouse, W and Wykle, L (1990) *Abuse of Power – Social Performance of Multinational Corporations: the Case of Union Carbide*, New Horizons Press, New York

Diaz Palacios, J (1988) *El Peru y su medio ambiente. Southern Peru Copper Corporation: una compleja agresion ambiental en el sur del pais*, IDMA, Lima

DiChiro, G (1998), 'Nature as community: The convergence of environmental and social justice' in M Goldman (ed) *Privatizing Nature: Political Struggles for the Global Commons*, Pluto, London

Dobson, A (1998) *Justice and the Environment: Conceptions of Environmental Sustainability and Dimensions of Social Justice*, Oxford University Press, Oxford

Dore, E (2000) 'Environment and society: Long-term trends in Latin American mining', *Environment and History*, vol 6, pp1–29 (previous version in Spanish in *Ecologia Politica*, 7, 1994)

Dorsey, M (1997) 'El movimiento por la Justicia Ambiental en EE.UU. Una breve historia', *Ecologia Politica*, vol 1, pp23–32

Downs, A (1972), 'Up and down with ecology: The issue-attention cycle', *Public Interest*, vol 28, Summer

Draisma, T (1997) *Mining and Ecological Degradation in Zambia: Who Bears the Brunt When Privatization Clashes with Rio 1992?*, Environmental Justice and Global Ethics Conference, Melbourne, October (revised version, August 1998)

Dryzeck, J S (1994) 'Ecology and discursive democracy: Beyond liberal capitalism and the administrative state' in M O'Connor (ed) *Is Capitalism Sustainable?*, Guilford, New York

Ekins, P, Max-Neef, M (eds) (1992) *Real-life Economics: Understanding Wealth Creation*, Routledge, London

ELAN website at: http://csf.colorado.edu/forums/elan/96/oct96/0010.html

Epstein, B (2000) 'Grass-roots environmentalism and strategies for social change', *New Social Movements Network*, updated 28 February, http://www.interwebtech.com/nsmnet/docs/epstein.htm

Erickson, J D and Chapman, D (1993) 'Sovereignty for sale: Nuclear waste in Indian country', *Akwe:kon Journal*, Fall, pp3–10

Erickson, J D, Chapman, D and Johny, R E (1994) 'Monitored retrievable storage of spent nuclear fuel in Indian country: Liability, sovereignty, and socio-economics', *American Indian Law Review*, University of Oklahoma College of Law, pp73–103

Escobar, A (1996) 'Constructing Nature, Elements for a post-structural political ecology' in R Peet and M Watts (eds) *Liberation Ecologies*, Routledge, London

Faber, D (ed) (1998) *The Struggle for Ecological Democracy: The Environmental Justice Movement in the United States*, Guilford, New York

Faucheux, S and O'Connor, M (eds) (1998) *Valuation for Sustainable Development: Methods and Policy Indicators*, Edgar Elgar, Cheltenham

Ferrero Blanco, M D (1994) *Capitalismo Minero y Resistencia Rural en el Suroeste Andaluz: Rio Tinto 1873–1900*, Diputacion Provincial, Huelva

Finn, J L (1998) *Tracing the Veins: Of Copper, Culture and Community from Butte to Chuquicamata*, University of California Press, Berkeley

Friedman, J and Rangan, H (eds) (1993) *In Defense of Livelihood: Comparative Studies in Environmental Action*, UNRISD, Kumarian Press, Hartford, CT

Funtowicz, S and Ravetz, J (1991) 'A new scientific methodology for global environmental issues' in R Costanza (ed) *Ecological Economics: The Science and Management of Sustainability*, Columbia University Press, New York

Funtowicz, S and Ravetz, J (1994) 'The worth of a songbird: Ecological economics as a post-normal science', *Ecological Economics*, vol 10 (3), pp189–96

Gadgil, M and Guha, R (1995) *Ecology and Equity: The Use and Abuse of Nature in Contemporary India*, Routledge, London

Garcia Rey, J (1997) 'Nerva: No al vertedero. Historia de un pueblo en lucha', *Ecologia Politica*, 13

Gedicks, A (1993) *The New Resource Wars: Native and Environmental Struggles Against Multinational Corporations*, South End Press, Boston

Ghai, D and Vivian, J M (eds) (1992) *Grassroots Environmental Action: People's Participation in Sustainable Development*, Routledge, London

Gibbs, L M (1981) *Love Canal: My Story*, State University of New York Press, Albany

Gibbs, L M (1995) *Dying from Dioxin: A Citizen's Guide to Reclaiming our Health and Rebuilding Democracy*, South End Press, Boston

Goldman, M (ed) (1998) *Privatizing Nature: Political Struggles for the Global Commons*, Pluto, London

Gottlieb, R (1993) *Forcing the Spring: The Transformation of the American Environmental Movement*, Island Press, Washington, DC

Gould, K A, Schnaiberg, A and Weinberg, A (1996) *Local Environmental Struggles: Citizen Activism in the Treadmill of Production*, Cambridge University Press, New York

Greenpeace (1988) *International Trade in Toxic Waste*, Greenpeace, Brussels

Greenpeace (1994) *The Database of Known Hazardous Waste Exports from OECD to Non-OECD Countries*, 1989–94, Greenpeace, Washington, DC

Guha, R (1989) *The Unquiet Woods: Ecological Change and Peasant Resistance in the Himalaya*, University of California Press, Berkeley (revised edition 1999)

Guha, R (2000) *Environmentalism: A Global History*, Longman, New York

Guha, R and Martinez-Alier, J (1997) *Varieties of Environmentalism: Essays North and South*, Earthscan, London

Guha, R and Martinez-Alier, J (1999) 'Political ecology, the environmentalism of the poor, and the global movement for environmental justice', *Kurswechsel* (Vienna), Heft 3, pp27–40

Hays, S (1959) *Conservation and the Gospel of Efficiency: The Progressive Conservation Movement 1898–1929*, Harvard University Press, Cambridge, MA

Hays, S (1998) *Explorations in Environmental History*, University of Pittsburgh Press, Pittsburgh, PA

Hecht, S and Cockburn, A (1990) *The Fate of the Forest: Developers, Destroyers and Defenders of the Amazon*, Penguin, London

Hofrichter, R (ed) (1993) *Toxic Struggles. The Theory and Practice of Environmental Justice*, New Society Publishers, Philadelphia, PA

Hornborg, A (1998) 'Toward an ecological theory of unequal exchange: articulating world system theory and ecological economics', *Ecological Economics*, 25 (1), pp127–136

Human Rights Watch (1999) *The Price of Oil: Corporate Responsibility and Human Rights Violations in Nigeria's Oil Producing Communities*

Human Rights Watch (1999) *Toxic Justice: Human Rights, Justice and Toxic Waste in Cambodia*, New York

Jackson, T and Marks, N (1999) 'Consumption, sustainable welfare, and human needs – with reference to UK expenditure patterns between 1954 and 1994', *Ecological Economics*, vol 28, pp421–41

Keil, R et al (eds) (1998) *Political Ecology: Global and Local*, Routledge, London

Kuletz, V (1998) *The Tainted Desert: Environmental and Social Ruin in the American West*, Routledge, New York

Kurien, J (1992) 'Ruining the commons and responses of the commoners: Coastal overfishing and fishworkers' actions in Kerala state, India' in D Ghai and J M Vivian (eds) *Grassroots Environmental Action: People's Participation in Sustainable Development*, Routledge, London, pp221–58

Leff, E (1995) *Green Production: Toward an Environmental Rationality*, Guilford, New York

Lipman, Z (1998) 'Trade in hazardous waste: Environmental justice versus economic growth', Conference on Environmental Justice, Melbourne, Http://www.spartan.unimelb.edu/au/envjust/papers

Lohman, L (1991) 'Peasants, plantations and pulp: The politics of eucalyptus in Thailand', *Bulletin of Concerned Asian Scholars*, 23 (4)

Lohman, L (1996) 'Freedom to plant. Indonesia and Thailand in a globalizing pulp and paper industry' in M J G Parnwell, and R L Bryant (eds) *Environmental Change in South-East Asia: People, Politics and Sustainable Development*, Routledge, London and New York

Low, N and Gleeson, B (1998) *Justice, Society and Nature: An Exploration of Political Ecology*, Routledge, London and New York

Mallon, F (1983) *The Defense of Community in Peru's Central Highlands*, Princeton University Press, Princeton, NJ

Martinez-Alier, J (1991) 'Ecology and the poor: a neglected issue in Latin American history', *Journal of Latin American Studies*, vol 23 (3), pp621–40

Martinez-Alier, J (1995) 'Political ecology, distributional conflicts, and economic incommensurability', *New Left Review*, 211

Martinez-Alier J and Hershberg, E (1992) 'Environmentalism and the poor', *Items*, vol 46(1) March, Social Sciences Research Council, New York

Martinez-Alier J and O'Connor, M (1996) 'Ecological and economic distribution conflicts' in R Costanza, O Segura and J Martinez-Alier (eds) *Getting Down to Earth: Practical Applications of Ecological Economics*, ISEE, Island Press, Washington, DC

Martinez-Alier J and O'Connor, M (1999) 'Distributional issues: An overview' in J Van den Bergh (ed), *Handbook of Environmental and Resource Economics*, Edward Elgar, Cheltenham

Martinez-Alier, J, Munda, G and O'Neill, J (1998) 'Weak comparability of values as a foundation for ecological economics', *Ecological Economics*, vol 26, pp277–86

Martinez-Alier, J, Munda, G and O'Neill, J (1999) 'Commensurability and compensability in ecological economics' in M O'Connor and C Spash (eds) *Valuation and the Environment: Theory, Methods and Practice*, Edward Elgar, Cheltenham

McCully, P (1996) *Silenced Rivers: The Ecology and Politics of Large Dams*, Zed, London

McDonald, D (ed) (2000) *Environmental Justice in South Africa*, James Currey, London

McNeill, J R (2000) *Something New Under the Sun: An Environmental History of the Twentieth-Century World*, Norton, New York

Mezger, D (1980) *Copper in the World Economy*, Heineman, London

Mikesell, R F (1988) *The Global Copper Industry*, Croom Helm, London

Mol, A (1995) *The Refinement of Production: Ecological Modernization Theory and the Chemical Industry*, Van Arkel, Utrecht

Mol, A (1997) 'Ecological modernization: Industrial transformation and environmental reform' in M Redclift and G Woodgate (eds) *The International Handbook of Environmental Sociology*, Edward Elgar, Cheltenham

Moody, R (1992) *The Gulliver File – Mines, People and Land: A Global Battleground*, Minewatch-WISE-Pluto Press, London

Morehouse, W and Subramanian, M A (1986) *The Bhopal Tragedy: What Really Happened and What it Means for American Workers and Communities at Risk*, a preliminary report for the Citizens Commission on Bhopal, Council on International and Public Affairs, New York

Munda, G (1995) *Multicriteria Evaluation in a Fuzzy Environment: Theory and Applications in Ecological Economics*, Physika Verlag, Heidelberg

Naredo, J M and Valero, A (1999) *Desarrollo Economico y Deterioro Ecologico*, Argentaria-Visor, Madrid

Nimura, K (1997) *The Ashio Riot of 1907: A Social History of Mining in Japan*, Duke University Press, Durham and London

Norgaard, R B (1990) 'Economic indicators of resource scarcity: A critical essay', *Journal of Environmental Economics and Management*, vol 19, pp19–25

Novotny, P (1998) 'Popular epidemiology and the struggle for community health in the environmental justice movement' in D Faber (ed) *The Struggle for Ecological Democracy: The Environmental Justice Movement in the United States*, Guilford, New York

O'Connor, J (1988) 'Introduction', *Capitalism, Nature, Socialism*, 1

O'Connor, M (1993) 'On the misadventures of capitalist nature', *Capitalism, Nature, Socialism*, vol 4(3), pp7–40

O'Connor, M (ed) (1996) 'Ecological distribution', special issue of the *Journal of Income Distribution*, 6(2)

O'Connor, M and Spash, C (eds) (1999) *Valuation and the Environment: Theory, Methods and Practice*, Edward Elgar, Cheltenham

O'Neill, J (1993) *Ecology, Policy and Politics*, Routledge, London

Opschoor, J B (1995) 'Ecospace and the fall and rise of throughput intensity', *Ecological Economics*, vol 15(2), pp137–40

Painter, M and Durham, W (eds) (1995) *The Social Causes of Environmental Destruction in Latin America*, University of Michigan Press, Ann Arbor

Parikh, J K (1995) 'Joint implementation and the North and South cooperation for climate change', *International Environmental Affairs*, vol 7 (1), pp22–41

Peet, R and Watts, M (eds) (1996) *Liberation Ecologies*, Routledge, London

Princen, T (1999) 'Consumption and environment: Some conceptual issues', *Ecological Economics*, vol 3, pp347–363

Pulido, L (1994) 'Restructuring and the contraction and expansion of environmental rights in the US', *Environment and Planning A*, vol 26, pp915–36

Pulido, L (1996) *Environmentalism and Economic Justice: Two Chicano Struggles in the Southwest*, University of Arizona Press, Tucson

Robleto, M L and Marcelo, W (1992) *Deuda Ecologica*, Instituto de Ecologia Politica, Santiago, Chile

Rocheleau, D et al (eds) (1995) *Feminist Political Ecology*, Routledge, London

Sachs, A (1995) *Eco-justice: Linking Human Rights and the Environment*, Worldwatch Institute, Washington, DC

Saro-Wiwa, K (1995) *A Month and a Day: A Detention Diary*, Penguin, London

Schnaiberg, A et al (1986) *Distributional Conflicts in Environmental Resource Policy*, Edward Elgar, Aldershot

Schwab, J (1994) *Deeper Shades of Green: The Rise of Blue-collar and Minority of Environmentalism in America*, Sierra Club Books, San Francisco

Shabecoff, P (2000) *Earth Rising: American Environmentalism in the 21st Century*, Island Press, Washington, DC

Strong, K (1977) *Ox Against the Storm – A Biography of Tanaka Shozo: Japan's Conservationist Pioneer*, Paul Norbury, Tenterden, Kent

Stroup, R L (1997) 'Superfund: The shortcut that failed' in T L Anderson (ed) *Breaking the Environmental Policy Gridlock*, Hoover Institution Press, Stanford

Szasz, A (1994) *Ecopopulism: Toxic Waste and the Movement for Environmental Justice*, University of Minnesota Press, Minneapolis

Tamanoi, Y, Tsuchida, A and Murota, T (1984) 'Towards an entropic theory of economy and ecology – beyond the mechanistic equilibrium approach', *Economie apliquee*, vol 37, p279–94

Taylor, B R (ed) (1995) *Ecological Resistance Movements: The Global Emergence of Radical and Popular Environmentalism*, Sate University of New York Press, Albany

Taylor, D (2000) 'The rise of the environmental justice paradigm', *American Behavioral Scientist*, vol 43 (4) January

Wapner, P (1996) *Environmental Activism and World Civic Politics*, State University of New York Press, Albany

Wargo, J (1996) *Our Children's Toxic Legacy: How Science and Law Fail to Protect Us from Pesticides*, Yale University Press, New Haven and London

Von Weizsäcker, E, Lovins, A B and Lovins, L H (1997) *Factor Four: Doubling Wealth, Halving Resource Use (The New Report to the Club of Rome)*, Earthscan, London

Wenz, P (1988) *Environmental Justice*, State University of New York, Albany

Westra, L and Wenz, P (1995) *Faces of Environmental Racism: Confronting Issues of Global Justice*, Rowman and Littlefield, Lanham, MD

World Resources Institute, Wuppertal Institut et al (1997) *Resources Flow: The Material Basis of Industrial Economies*, World Resources Institute, Washington, DC

Chapter 11

Women and Environmental Justice in South Asia

Anoja Wickramasinghe

INTRODUCTION

Social discourses on women's rights and feminist geography (ies) have pointed out the necessity of providing equal opportunities for men and women in every aspect of life in order for social advancement to take place. The wide spectrum of issues of development concern that have been discussed at various levels suffer either from a lack of integrity, or inadequate attention is given to women's rights to manage local environments. Add these issues to the politics of environmental resource management in South Asia, and there has been no drastic change either in the local environments themselves, or in the lives of women, particularly those who manage the local environment for their livelihood and security.

In this regard, South Asia as a geographical region provides a wide range of scenarios on controversial issues of development concern. While efforts for promoting women in environment and development have been put forward by various interventionists, women's status as managers of local environments has either deteriorated, or women have been pushed further towards marginal environments.

Concentrating on these issues, this chapter presents, with some examples, a typology of land-based livelihood systems of South Asian women. These land-based systems have been subjected to serious deterioration as a result of external forces. The adverse repercussions that have taken place in the lives of women in South Asia and their land-based livelihood systems are associated with the conventional social stratification, unequal distribution of power, globalization, trade liberalization, adoption of capital-intensive technology and the acquisition of authority over local resources by state agencies, and more strongly, by the lack of women's citizenship rights to local environments. The implications for environmental politics, particularly from the perspective of empowering women, are serious because the conventional gender gaps have been re-endorsed in the newly created development paradigms where unequal distribution of power over local resources has been established.

The justification for women's claims to local environments, as presented in this chapter is associated with two conditions. The first is related to their citizenship rights, and the second is related to their right to shape the sustainability of local habitats to which they have contributed over generations, and from which they gain security over livelihood sources. The arguments put forward in this chapter, enrich the debate on women's environmental citizenship and the arguments for feminist environmentalism. The final section of the chapter is on who decides the sustainability of local environments, emphasizing the need for setting new local agendas for the future of social and environmental justice for women in the region. It also shows that the claim for environmental justice for women whose livelihood systems are local resource-based will not be achieved as long as stratification in the conventional social, economic and political systems alienate them.

PERSPECTIVES

The academic discourse, particularly the feminist discourse on women and the environment, has introduced a new paradigm to conventional environmentalism, where people have been placed in the periphery (see Roberts critique of anthropocentrism in Chapter 9). The livelihood system of women in the South Asian region, which has traditionally been oriented towards subsistence, demonstrates the reasons why the linkages between women and the environment need to be understood more deeply than at our current, superficial level. The local environment, with marked diversities in biophysical and social contexts, has provided physical space for women to become the producers of subsistence and for them to become the actors equipped with the technology and authority to deal with multiple production systems. Women's work relations have been part of day to day life, irrespective of the conditions under which they work. The extent to which modern development interventions have been able to recognize the grounded realities for the betterment of women's livelihoods in South Asia, and the sustainability of the local environment, has remained an unanswered question.

The grounds for feminist environmentalism originated in South Asia, in association with women's work as providers of basic survival needs such as food, fuel, water and also as the practitioners of herbal medicine. The issues of overcrowding, population growth and environmental degradation on the one hand, and distribution of resources, and unequal access and control over them on the other, have stimulated a wider policy debates over the years. The process has been directly associated with the women's movement in general, which has extended its roots across the world with the declaration of 1975 as the International Year of Women by the United Nations (UN) at that year's conference on Women and Development. The follow-up declaration of the UN Decade of Women (1976–1985) by the United Nations, the second UN

Conference in Nairobi to mark the end of the decade, and the UN Conference on Women and Development held in Beijing in 1995 have stimulated research on women and environmental issues. It is important to connect this process with the development of the 'environmental movement', a parallel process that has stretched across the world. The publication of the 'Limits to Growth' by the Club of Rome in 1972, which raised an awareness of environmental degradation and the UN Conference on the Human Environment held in Stockholm in 1972; the World Commission on Environment and Development in 1987; and the UN Conference on Environment and Development held in Rio de Janeiro in 1992, are the landmarks of this parallel process.

Feminism and the feminist movement on the one hand, and environmentalism and the environmental movement on the other, have introduced a paradigm of feminist environmentalism. Conceptually, this has stimulated an in-depth analysis and investigation into women's visions and practical approaches to the sustainable management of their local environments. The stimulus has been provided by women in agrarian economies around the world, and it evolved around their roles in sustaining the environment and its resources for social, ie community, sustainability. In conceptualizing the linkage between women and the environment, eco-feminism has been in the forefront. South Asia, with its strong cultural context, has been in the centre of eco-feminism (Shiva, 1989, Mies and Shiva, 1993), and it was found that rural women's struggles to safeguard the environment, such as the 'Chipko Movement' were pivotal in this process. The maintenance of the resource base of the livelihoods of rural South Asian women has turned the focus to matters pertaining to the management of land, forests, soil, water, etc, for which they have gained centuries of knowledge and experience as well as capabilities. The feminist discourse, which has emerged in connection with women and the environment as nature, cannot be substituted for this due to the diversity in the real context. For instance, in dealing with the issues of women and the environment, Sontheimer (1991) has placed attention on land, forests, water, etc, as areas that women deal with in everyday life. Dankelman and Davidson (1988) have made it clear that women are capable of managing resources and that they perform the tasks of the managers of water, forests, etc. Cuomo (1998) in her critique of feminism and her justification for preferring 'ecological communities', and Buckingham-Hatfield (2000), have furthered the discussion on how women's work in the South is directly linked with the local environment, where technology and services are not fully developed to provide water and energy, etc, which are essentials for sustenance.

Women's work in South Asia has evolved, and is centred in the local environment, therefore the ways in which environmental degradation impacts upon women have been brought to our attention by many authors. In this context, Shiva's work on *The Violence of the Green Revolution* (1991) in which ecology, health and development have been brought together; Agarwal's (1986) book on *Cold Hearths and Barren Slopes*; and Sims' (1994) compilation on *Women, Health and Environment* reveal some of the issues that have been dealt with since 1980s. An analysis of the contexts and issues put forward by the many other scholars and activists are beyond the scope of this chapter. However, it is important to note that the perspectives and cases that have been studied have enabled activists to put the linkages between women and the environment in the laps of policy-makers.

South Asia as a geographical region, despite the internal variations within and between countries, has been seriously affected by environmental degradation, resource depletion, deforestation and modernization, and by a reduced per capita land area. The majority of women in South Asia, particularly those who are in rural areas, have been pressurized under these conditions, as well as their underprivileged position in their local environments, and society generally.

IMBALANCES AND INEQUALITIES

The prerequisites for seeking environmental justice for women in South Asia are not merely matters of structuring the rationale(s) for women's claims, but also of highlighting other issues that undermine women's status. Women have never been able to secure equal opportunities and have remained secondary citizens in the social hierarchy. For example, as revealed in the Human Development Reports of the United Nations Development Programme (1995, 1997, 1999), there are disparities between men and women in all countries and thereby, in regional scenarios, in terms of many aspects of development (ie adult literacy, education enrolment rates, real Gross Domestic Products (GDP) per capita, and earned income share (see Table 11.1)). These figures show that women have had fewer opportunities or have to deal with various constraints impinging upon them in achieving a status comparable to that of men. Existing inequalities point to the relatively low status that women have in society. The GDP per capita for women, for example, is extremely low and is often less than half, or one third of that of men. In Sri Lanka it is nearly 41 per cent; it is 38 per cent in India; 30 per cent in Pakistan; 54 per cent in Nepal; 51 per cent in Bhutan and 58 per cent in Bangladesh. Accordingly, the earned income of men and women differs remarkably. The reasons for the low GDP per capita for women and the reporting of women as economically less active in all these six countries is determined by their low environmental status.

Rates of literacy, educational enrolments, etc, have either been inadequate or have not contributed to elevating their status in the local environment, although progress has been reported over the years. There were no indicators to measure the status of women in the local environment because it has been partly influenced by ascribed gender relations. Ironically the relatively high rate of illiteracy and the low level of education enrolment of women in five of the countries, Sri Lanka excepted, has enabled policy-makers to ignore women's rights to local environments. In status-oriented societies in Asia, women's low achievements and gender ideology often impinge upon their rights and opportunities to negotiate with policy-makers and earn social recognition. The consequence has been the continued concentration of women as passive workers in local environments without the power of control in their workplace environment. The difference in the distribution of earned income between men and women is outstanding in all countries. It is extremely high in Pakistan and

Table 11.1 *Gender inequalities in human development in six countries in the South Asia region*

Variable	Sri Lanka		India		Pakistan		Nepal		Bhutan		Bangladesh		South Asia		All developing countries	
	W	M	W	M	W	M	W	M	W	M	W	M	W	M	W	M
Life expectancy at birth 1997 (years)	75.4	70.9	62.9	75.4	65.1	62.9	57.1	57.6	62	59.5	58.2	58.1	63.1	62.3	66.1	63.0
Adult literacy (%)	87.6	94.0	39.4	66.7	25.4	55.2	20.7	55.7	30.3	58.1	27.4	49.9	38.6	65.0	62.9	80.0
Ed. Enrolment ratio (%)	67.0	65	47	62	28	56	49	69	10	14	30	40	44	60	55	64
Real GDP per capita (US$)	1452	3545	902	2389	701	2363	763	1409	985	1940	767	1320	950	2606	2088	4374
Earned income (%) of share	34.5	65.5	25.7	74.3	20.8	79.2	33	67	–	–	23.1	76.9	–	–	31.7	68.4

Source: UNDP, 1999

Bangladesh, with nearly 59 per cent and 53 per cent differences respectively, where women's right to own land is restricted. The situation in India is marginally better, but relatively unsatisfactory when compared with Sri Lanka, where customary practices do not prevent a woman's right to own land.

Amazingly, a well-established imbalance between the distribution of work, and women's share of the labour force, has been reported at various enumerations. An in-depth analysis provides evidence for the greater enrolment of women in farm production work, and also in forest management. Yet, their share in the adult labour force is 34 per cent in Sri Lanka, 31 per cent in India, 24 per cent in Pakistan, 40 per cent in Nepal, 39 per cent in Bhutan and 42 per cent in Bangladesh (see Table 11.2). These generalized figures neither show the women's share in agricultural output nor do they reveal the rates of women's activity, or input in reality. In South Asia, women's economic activity rates are low, and are in the range of 29 to 44 per cent in these six countries. Raju and Bagchi (1993), while compiling an academic discourse on women and work, have stated that the underreporting of women's work is a serious issue. The inherent skew in this picture, due either to underreporting or misconception, has resulted in the marginalization or alienation of women from their local environment. More seriously, it has weakened their claims to local environments, both for space and livelihood resources. The female economic activity rate never gets close to 100 per cent of the male economic activity rate. For the South Asia region as a whole, it is only 51.7 per cent with the range of variation between 40 to 77 per cent for individual countries. It is 55.4 per cent of the male economic activity rate for Sri Lanka, 50.3 per cent in India, 40.3 per cent in Pakistan, 69.6 per cent in Nepal, 66.7 per cent in Bhutan and 77.2 per cent in Bangladesh. The volume of data synthesized in the UN Human Development Reports reveals that despite the efforts made across the region to adopt strategic means to promote women as producers, the Female Economic Activity Index points out the shortfalls. The Index, worked out using 1985 as the baseline, and taking its value as 100, shows that since then, in Sri Lanka it has gone up to 151.3, Pakistan 123.3 and Bangladesh 107.8. In India it is 95, Nepal 98.3, and 94.7 in Bhutan (see Table 11.2). The question of women's rights to the local environment under these circumstances needs re-examining from the perspectives of women's empowerment and sustainability of the local production base, the land.

The regional and in-country scenarios support women's claims for management of the local environment as equal partners in economic activity, and in the labour force. While challenging the shortfalls of male-dominated systems in the region, academic discourses stimulate forums for discussion. However, the justification for women to have an equal share of opportunity is not a strong one, because they have not gained recognition as equal partners in the subsistence economies in the region. It also points to the fact that the conventional ideology of underreporting women's work, and under-valuing women's work, as Raju and Bagchi (1993) and many others have noted, continues as it did when it started about three decades ago. This means that greater environmental justice for women in South Asia should have a strong case, with empirical evidence to strengthen the arguments that are capable of challenging the conventional norms and established discriminatory attitudes.

Table 11.2 *Data related to women's relative status in six countries in the South Asia region*

Variable	Sri Lanka	India	Pakistan	Nepal	Bhutan	Bangla-desh	South Asia	All developing countries
Female economic activity rate (1997)	30.5	29	20.8	37.9	38.8	44.4	29.1	39.3
As a per cent of male (1997)	55.4	50.3	40.3	69.6	66.7	77.2	51.7	68.0
Female Economic Activity Index (1985=100)	151.3	95.0	123.3	98.3	94.7	107.8	99.4	111.3
Female unpaid family workers (1990—1997)	53	—	33	61	—	71	—	—
Labour force (1990—1993)	41	38	28	—	—	47		
Women's share in adult labour force (1994)	27	24	13	32	32	41		
Per cent labour force in agriculture (1990)	48	64	52	94	94	65		

Source: UNDP, 1995, 1997, 1999

CONTEXTUAL DIVERSITY AND ENVIRONMENTAL ISSUES

The countries of the South Asia region show evidence of some commonalties in terms of stage of development and socio-economic background. The first feature found across the region is the rural nature of the majority of its population. The second is the lower position attained by these countries in human development, gender development and in gender empowerment. The third is the high dependence on agriculture when compared with all other sectors of the economy. These features, coupled with high population densities in the region, indicate the high pressure on environmental resources. South Asian nations will have to deal with these issues if they are to move into a phase of sustainable development.

The inherent environmental diversity of the South Asia region, with its complex mixture of ethnicity, religion and culture, highlights the contextual diversity that one has to deal with in seeking environmental justice for women. The positioning of women at various hierarchical levels has been embodied in cultural practices, which vary tremendously across South Asia. On the surface, the situation ranges from strict cultural limits enforced in predominantly Muslim countries like Pakistan and Bangladesh, to the relatively lenient practices in South India, and particularly in Sri Lanka. Also on the regional scale, the strictures of segregated behavioural patterns in gender-specific physical space can be found in Pakistan, Bangladesh and also Nepal at one extreme, while in Sri Lanka and the southern states of India, the sexual division of labour is widespread.

It can be argued that eco-feminism originated, evolved and was moulded from dependence on the land and local resources. The habitats in which women's livelihood systems have evolved are diverse in terms of spatial distribution of farms, forests, trees, etc. Many elements have either been fragmented or devastated, or replaced with modern interventions. To a greater extent, the environmental issues of women's concerns are common across the region, and require some negotiation and collective action.

The biophysical, agro-ecological and demographic features of the countries in South Asia have contributed to geographically distinct regions with direct and indirect implications for development and livelihood systems. The level of labour input and management strategies varies from the mountain regions of Nepal and India to the lowlands of Bangladesh. The distinct variations within and between the countries in women's environmental linkages are socio-ecological. Ecologically, the local habitats within which women live frame the nature of work relations. This regional phenomenon with its high dependence on local environments, points to the fact that at any scale, future transitions in the environmental scenario presented, should be a women-centred one to reduce the established adverse impacts upon women.

The women–environment linkage systems are rather complex and are influenced by many factors. These include quality and quantity of resources available to women, local livelihood systems, policy, legislation, customary practices, and so on. These tend to vary spatially due to the variation in local contexts. The intra-country variations are greater in India, due to its size, cultural and environmental diversity. Between 1951 and 1991 the population of South Asia has increased from 440 million to 1100 million. The countries in the region, except Bhutan are densely populated and marked with continuing deforestation (see Table 11.3). The increased population pressure on resources has reduced the space and resources available for women. As one would have expected, the pressure on agricultural land have increased tremendously, where expansion of land under agriculture has not been possible to keep pace with the growth of the rural population. Between 1965 and 1990, there was nearly 60 per cent growth in the rural population, whereas the growth in the spatial extent of agricultural land was about 5.1 per cent. Resource exploitation and expansion of agriculture onto fragile areas has promoted droughts, desertification and salinization. The reduction in soil fertility, water, biomass production and biodiversity has pressurized women and their livelihood more than any other sector.

The deforestation that has continued at unprecedented rates during the 20th century has raised some serious issues. The major challenge has been to sustain production without letting the marginal lands, which have been converted from forests to agricultural production, turn into deserts. With agricultural expansion,

Table 11.3 *General statistics related to land and land use in six countries in the region*

	Bangladesh	Bhutan	India	Nepal	Pakistan	Sri Lanka
Land area ('000 ha)	13,017	4700	297,319	13,680	77,088	6463
Pop. Density (1995) per '000 ha	9252	349	3147	1602	1823	2840
Domesticated land as a per cent of land area	79	9	61	32	34	36
Land use (1991–1993) Cropland ('000 ha)	9703	133	169,547	2354	22,890	19.3
Permanent pasture ('000 ha)	600	272	11,533	2000	5000	439
Forest & Woodland ('000 ha)	1896	3100	68,330	5750	3470	2126
Other ('000 ha)	818	1194	47,909	3756	45,728	1995
Population (million) 1997	122.7	1.9	966.2	22.3	144	18.3
Annual growth of labour force (1991–2000)	2.9	1.4	1.9	2.5	3.0	1.7
Forest & woodland (per cent of total, 1992)	14.5	54.5	23	39.1	5.3	32.5
Arable (per cent of total 1992)	69.5	2.9	57.1	17.2	27.4	29.5
Irrigated land (per cent of total 1992)	34.3	25.4	27	36.1	81.0	28.9
Annual deforestation ('000 ha 1980–1989)	8	1	1500	84	9	58
Per cent of deforestation 1980–1989	0.9	0.1	2.3	4.0	0.4	3.5
Reforestation ('000 ha per year 1980–1995)	17	1	138	4	7	13
Rural population (millions, 1990)	90	–	621	17	76	13
% change of rural population (1965–1990)	63.6	–	57.2	70	76.2	44.4
Extent of agricultural land (million ha)	9.2	–	169.5	2.6	20.8	2.0
% change between 1964–1990	2.2	–	4.4	44.4	8.3	11.1

Source: UNDP, 1995, 1997, World Resources, 1992–1993, Peiris, 2000

despite the variations between countries which are in the range between 8000 to 1,500,000 ha per annum (except for Bhutan), nearly 1.7 million ha of forest is lost annually. This is serious because of the extremely low recovery rate, where annual reafforestation is limited to about 0.2 million ha. The implications of these changes are yet to be explored locally as well as regionally. The human cost of such trends on the lives of women who use the forest directly, as a primary or secondary source of basic materials, is high. They depend directly on fuelwood and water rejuvenated through natural processes in their day to day life. The spatial differences in the availability of these resources within the sphere of women's access, and the distance to the resources, have structured diurnal patterns in women's time allocation (Wickramasinghe, 1997b), and the nature of their contacts with the resource. Depletion of the sources of forest products, fuelwood and water has increased the time that women have allocated on procuring and portaging.

The problems emerging in the areas under crop production are also serious due to land degradation. It has resulted in losing potentially useful land and has also caused the loss of sources of products, income and food. Women's concern over these has stimulated them to become the local conservationists through their contribution to the protection of biodiversity and domestication. The home-centred garden culture and the biodiversity of home gardens provide some practical examples of this nature.

Women in the South Asia region have been in a privileged position to become environmentalists through their first-hand experience. The bulk of the population of these countries is rural, and depends heavily on land-based livelihoods. The rural share of the population in 1997 was 77 per cent in Sri Lanka; 73 per cent in India; 65 per cent in Pakistan; 89 per cent in Nepal; 94 per cent in Bhutan; and 81 per cent in Bangladesh. This is also reflected in having a high share of labour in agriculture. It is 49 per cent in Sri Lanka; 62 per cent in India; 47 per cent in Pakistan; 93 per cent in Nepal; 94 per cent in Bhutan; and 59 per cent in Bangladesh. This means that women in these countries who have dominated the predominantly subsistence agricultural economy have had opportunities to become the experts in resource management as part of their day to day life. They have explored technologies to deal with the local situation and have equipped themselves with the indigenous knowledge which has been transferred from one generation to another. One problem found throughout the region is that measures like Gender Development Index (GDI) adopted to indicate the level of women's development has not been able to grasp the position gained by women in their local environments. As a result, such indices have undermined social learning and the real achievements of rural women. It is important to note that indigenous knowledge and technology would have made a change to the present global picture, if provisions had been made. It is important to note that women in South Asia are not in an equal position when compared with others, besides their disadvantaged position when compared to men. Whatever policy measures are adopted, the present position of the countries in South Asia make it clear that the advancements made by women in these countries are inadequate. There are no clear indications as to what extent local environmental and resource issues impinge upon improvements in women's lives.

ENVIRONMENTAL JUSTICE FOR SUSTAINABILITY

The broader issues urging the move towards greater environmental justice are still predominant, and are left unanswered in South Asia. This is due to two reasons. The first is that local systems are male biased and permit men to exercise a greater control over women's labour and resources on which women's livelihoods depend. The second is that the household is assumed to be a homogeneous, male-headed unit with men having the capacity to represent women's concerns and interests. One good example of this nature is the land distribution in the Mahaweli settlements in Sri Lanka. The land titles are given to men, the heads of the family units, in spite of negotiations to introduce a system of joint titles to the allotments, or granting women independent titles. A similar situation was found in promoting participatory forestry in Sri Lanka. This situation symbolically represents the well-established opportunities to exploit women's labour and space. Because of their reproductive responsibilities and concerns, their local initiatives are suppressed, impinging upon wider opportunities they could claim as managers of the local environment.

The problems in the agriculture sector, which is the mainstay of the rural economy and women's livelihoods in South Asia, also supports the debate on feminist environmentalism. Any recession in this sector reflects the serious issues faced by women farmers, both in their productive and reproductive domains. In their day to day contact with local environmental resources, women are exposed to hazardous effects of environmental changes and technology. Women's contacts with 'land', and its water, soil and flora are consistent and regular. The time that women allocate to agriculture or land-based work is three to four times greater than that of their male partners (Mencher, 1993, p99). The importance of agriculture in livelihood security and its centrality in promoting women as responsible environmental strategists needs special emphasis. Women through their investments and inputs to the land, household, and community survival, have contributed to local, regional and national sustainable development. Between 1970 and 1992, the percentage share of agriculture in the GDP has been reduced throughout the region. It has reduced from 67 per cent to 52 per cent in Nepal; 55 per cent to 34 per cent in Bangladesh; 45 per cent to 27 per cent in India; 37 per cent to 27 per cent in Pakistan; and 28 per cent to 25 per cent in Sri Lanka. Between 1965 and 1992, the percentage of the labour force in agriculture has been reduced from 94 per cent to 93 per cent in Nepal; 84 per cent to 59 per cent in Bangladesh; 73 per cent to 62 per cent in India; 60 per cent to 47 per cent in Pakistan; and 56 per cent to 49 per cent in Sri Lanka. The reduction in the share of agriculture and labour, and also the increasing land degradation, desertification and deepening poverty in the region pose serious threats to women, in particular their responsibility over food security. These changes incur an impact on women's labour, income and subsistence crop production. The previous work in Sri Lanka conducted on land-based occupations and management of local resources. Wickramasinghe (1997a, p37), Zwarteveen (1994) and also Zwarteveen and Endeveld (1995) have concluded that women's issues are agrarian issues. Their lack of control over land, technology and service delivery systems is symbolic of displacing the managers of local environments. Throughout history, strategic interventions have attempted to 'exclude' rather than 'include' those who work in the local environment, consistently and regularly. In South Asia, there are three types of subordination. The first is the subordination of women by men in the presence of ascribed gender relations. The second is the subordination of women in the process of development where agriculture has been placed at the edge of development (while facilitating globalization and open economic policies to exploit women as a group of passive workers and cheap labour). The third is the subordination of women by environmentalists, conservationists and extensionists by excluding them, their local knowledge systems, technology, experience, linkages and concerns. This means that 'women's issues' are not just related to their role in the economy, but include their 'capacity of functioning' (Banerjee, 1992, p27).

Women's positions in the local environment are not comparable to that of men due to three factors in South Asia. The first is that the majority of women are '*de facto*' users of the resources. The second is that there are no mechanisms or instruments to guarantee women's rights other than their ascribed status in social institutions such as the family and the community. The third is that state efforts, or the policies of the countries in the region, are focused heavily on economic empowerment through income generation, rather than addressing key environmental issues such as those impinging on women's control over resources and local environments. In addition, the spatial diversity in culture has induced complexities regarding women's rights to resources.

In India, for instance, the National Perspective Plan for Women (Government of India, 1988) has placed emphasis on the issues affecting women's access to land. Similarly the Eighth Five Year Plan (1992–1997) (Government of India, 1991) has made it clear that a prerequisite for improving women's status is to change inheritance laws so that women get an equal share in parental property, inherited or self-acquired. In addition, state governments were asked to allot 40 per cent of the surplus land to women alone and the rest jointly in the names of husband and wife (Agarwal, 1994). A similar process has taken place in Nepal, Bangladesh and in Pakistan where women have been denied equal opportunities to inherit or possess land. Nepal's Eighth Five-Year Plan (1992–1997) (National Planning Commission, 1992) and Pakistan's report on the Sixth Plan (1983–1988) (Planning Commission/Government of Pakistan, 1983) have recognized women's need for land. In Bangladesh no special attention has been given to this issue (Fourth Five-Year Plan 1990–1995, Ministry of Planning, 1995). One striking feature revealed in the literature (Huda, 1997, p294) is that the existence of a strong division of labour between fields and the household has resulted in non-acknowledgement of women's role in agriculture. Since by tradition, women are not involved in the work in the fields, they depend on male kinsmen, sharecroppers or labourers for the operation of land. This dependence on male help often leads to the loss of, or transfer of control of, female-owned land in return for favours or support. This situation has multiple consequences, where women's economic participation is largely determined by patriarchal control, which induces control over property, resources and labour (Begum, 1989, p519).

In Sri Lanka, no rigid rules are in existence. But attention paid to land issues, and also to the environmental issues affecting women's rights to hold land and exercise control over resources, is marginal (National Plan of Action for Women, Sri Lanka, 1993). The notion of the household as one entity, the ideology of men as heads, income earners, producers, decision-makers and managers of the sources of income and resources, and women as home makers and reproducers, has provided grounds for excluding women's rights important policy issues. Women's gender, the predominant conventional ideology, women's roles and responsibilities, state policies, inheritance laws (collectively or in various combinations), all ignore women's rights to land. Enforced participation without rights to decide and claim space, is not only a property issue but also demands a human rights response. Moving beyond the fences of farmlands or private property to state property such as the forest, the need for restoring the rights of the traditional custodians, of which the majority are women, could be raised.

WOMEN AS CONTRIBUTORS

Environmental concerns of rural women, who are often placed in a non-working, economically inactive category, are very diverse. They are influenced by family responsibilities and the ways in which the tasks of household well-being have been placed in the conventional domain of women. Women's contributions are systematic, regular and not accidental. They are regularized through social customs, practices and norms. No comprehensive analysis or a single explanation to cover all variations is possible regarding the environmental context for the division of labour. Women's time, knowledge, experience and skills have been the main inputs to local environments. The gendered patterns that are found in the sexual division of labour in forest-based and farm activities elaborate two points. The first is related to the domination of women's labour, knowledge, and experience over resources and material sources of survival importance. The second is the linkage between subsistence and the environment. Although it is too hard and fast to promote the idea that such linkages are biologically determined, there is evidence to support the importance of women's contribution to intergenerational livelihood and security. Their involvements in safeguarding resources against exploitation, or against individual desires, are the essence of responsible management and intergenerational sustainability.

In these countries the multiplicity of women's responsibilities in rural households and communities divides their time between highland farms, home gardens, waterlogged paddy fields and forests. The input requirements are seasonally fixed. The interseasonal variations in women's time allocation on production is determined by the diversity and composition of the systems. Women farmers are consistently occupied in fields, producing a variety of crops (Wickramasinghe, 1993a, p159) or gathering natural stocks. In the dry zone of Sri Lanka, where household production systems consists of paddy, home garden and chena (plots on slash and burn cultivation), it has been found that 44 per cent of the work in the home gardens is done by women alone, 21 per cent by men alone and 35 per cent jointly. In chenas or swidden fields, 40 per cent of the work is done by women alone, 36 per cent by men alone and 24 per cent jointly. In paddy cultivation, in the waterlogged fields 15 per cent is done by women alone, 52 per cent men alone and 33 per cent jointly. Work in home gardens and chenas, because of their greater diversity and rotational practices, changes frequently so it demands more labour than that in paddy cultivation. Women's home-based occupations are found to be essential to sustain paddy production. In India, women work longer hours and work more regularly on their own farms than do men (Raghuram, 1993, p109). In Chalnakhel village in Nepal (Bhadra, 1997, p74) the share of women's input to farm work is 79 per cent. In Khanmohona, in Bangladesh, the women's input to on-farm activities is in the range of 5 to 10 per cent, and in home-based agriculture related work, in threshing, cleaning and so on, it is about 80 to 90 per cent (Das, 1997, p51).

Production and reproduction as integral parts of women's livelihood systems are well structured, with agro-ecologically significant variations. There are patterns in the distribution of work in space and the allocation of their time in each system. Analysed by evaluating the time that they spend in their workplaces – fields, forests, commons and home gardens, often without payment – favours women's rights over the workplace, and for social justice for their input. Work for the family and community sustenance furthers the debate on women's work for sustainable survival rather than for short term benefits, under which intergenerational connections are formed. The farm is the place for seasonal crop production and a space for women to integrate other elements like medicinal herbs, edible greens, trees and livestock. It is in the same space that women practice all their conservation measures, establish communities of nurturers with nurturing capacities, species with resistance to stress conditions with sensitive ones requiring greater care. A number of in-depth studies conducted in Sri Lanka have affirmed that women's initiatives have made home gardens a refugium of species, diverse in composition, connected with home/family and food security. Their inputs are related to the integration of crops, trees and livestock, and also include indigenous methods of pest and weed management, the use of organic materials for soil improvements and water conservation, etc (Wickramasinghe, 1994a, p37, 1995a, 1995b, p164). These are partly voluntary investments to local environmental management, which enter into the production spheres in the form of time, energy, knowledge, experience and social concerns.

In the rural agrarian economies in South Asia, women's work is a demonstration of the diversified livelihood systems in local environments. They are closely linked with multiple resources, and integrated resource management. Traditionally, time and labour are used to measure the sustainability of women's contribution, but these measures do not explain their interconnectedness and integrity in managing local environments. The frequently changing patterns of women's time allocation and their workplace are essential to assure habitat diversity and the multiple needs of subsistence economies, where no single source is capable of meeting all. In this respect whether women are the primary managers of such systems, or should they be a category of worker for the family, are the questions in hand, to be addressed in policy formulations.

WOMEN AND LOCAL RESOURCES

From the perspective of habitat diversity for livelihood security, local resources and the constituent sub-systems in village ecosystems are a complex phenomenon. The resources are parts of the livelihood system and habitat. The spatial variety of forests, biodiversity and biomass resources have been the traditional ways to cope with seasonal food insecurity. Either these substitute the livelihood sources of the households, where agriculture remains the major source, or agriculture is substituted through seasonal products. The best examples of these material related associations are rooted in the context of non-timber products derived from the natural forests, agroforests, commons, forest plantations, home gardens, reservational areas and shrub lands, depending on the location. Empirical studies in Sri Lanka, in the dry zone, (Wickramasinghe, 1997b, p89, 1997c, p87) and in the forest fringe areas of the Adam's Peak Wilderness (Wickramasinghe, 1995c p81) have revealed that non-timber products of the forests are the main sources of income and subsistence. The linkages are mediated primarily through women. They are engaged in forest-based activities and home-based processing. They are not only the users of forests and gatherers of forest products, but also the ones who convert raw biomass into consumable or marketable products, the knowers of the phenological cycle of the forests, their species and also the trees and other plants in the non-forest lands. Rather similar evidence has been revealed in the studies conducted in India (Saxena, 1987, 1995, 1997, Mukherjee, 1995, p3, p130, Sarin, 1995). The widespread nature of the connections between women's livelihood and biomass sources and resources are shown by researchers in Bangladesh (Das, 1997), and in Nepal (Bhadra, 1977).

Most of the research conducted in the past which has concentrated heavily on 'food, fuelwood, and fodder' has in fact weakened the uniqueness of the contextual relationships. A widely explored aspect here has been the impact of deforestation on domestic energy and women (Wickramasinghe, 1994b, 1993b, p46, Agarwal, 1986, Kumar and Hotchkis, 1989). The consequences of deforestation on domestic energy and thereby women in poor families who are unable to go for alternative energy types, are multiple. Additional time and energy required for gathering and portaging fuelwood are provided by women with adjustments in the domestic domain and by reducing their leisure. The increasing deforestation and the loss of woody biomass have forced the women in Bangladesh and in many parts of India to depend heavily on straw, tree leaves, residues, cow dung and various types of poor quality, non-woody biomass substitutes. Resource or resource utilization is one part of a multifaceted system. Among others, Saxena (1995, p71) has noted that the livelihood needs of poor women are preconditions for the sustainability of natural resources. The extent to which women's needs and the contribution of natural resources, mainly various forms of biomass, could guide future strategies have not been fully explored. Women's forest-based and home-based work on biomass resources have been not fully evaluated either. It has been found that women are the environmental entrepreneurs (Wickramasinghe, 1995d, p44). They carry knowledge on technology to produce edible and medicinal oils using Mee

(Madhuca longifolia) and seeds of Wild Bread-fruit (*Artocarpus nobilis*); and also maintain buffer stocks of food, such as Goraka (*Garcinia cambogia*), Tamarind (*Tamaridus indica*), Jak (*Artocarpus heterophyllus*), Bread-fruit (*Artocarpus altilis*) and Wild Bread-fruit (*Artocarpus nobilis*).

The issues of women's concern in conservation are also associated with genetic diversity. For instance, nearly 84 varieties of plants, wild rice, sesame, pumpkins, melons, etc, that are of multiple potential have been maintained by women (Wickramasinghe, 1997d, p51). With the expansion of modern agriculture such varieties have been displaced. The Green Revolution and modernization have threatened crop production, genetic resources and local control over the systems (Shiva, 1994).

Women should have rights to keep such resources and knowledge under their custody and stewardship. Their rights to secure authority over their livelihood sources, which they have maintained over generations, their rights to claim to the knowledge and technology that various agencies are negotiating for, are the questions to be taken up from the perspectives of women's environmental rights. Women's rights over resources have already been acquired because they are becoming scarce. The alienation of their authority over the resources that they depend upon has created better opportunities for externally driven agencies/strategies to make transformations according to external demands.

There are serious problems in the region associated with water resources. Its provision for household use is conventionally shouldered by women, while the sources and the replenishing systems are controlled by men. For instance, the expansion of seasonal crops like tobacco in the Ma-oya catchment since 1956 has enhanced erosion and reduced infiltration and as a result nearly 72 per cent of the springs have dried out. The options that women have in these areas are cumbersome, and for a pot of water, women have to walk a kilometre in some instances and often use polluted sources for washing and bathing. The seasonal food scarcities under which women's domestic burdens have become critical are part of the problem created during the recent past. These circumstances demonstrate that environmental issues extend beyond the unequal distribution of resources, and encompass the issues of resource ownership, control and conventional norms of the division of tasks.

WOMEN AND THE WORLD OF IMBALANCE

Why is it that women seek environmental justice? Why is it that South Asia is of central importance to the scenario of women and environmental justice? It is clear that environmental justice for women in a region of diverse contexts and complexities, aims to bring about equal citizenship rights without the discriminatory attitudes that have been structured or created over generations in relation to sex, caste, ethnicity, class, wealth or any other qualities. The resolution of problems and the creation of space(s) for all to act against environmental degradation, together with women's day to day role as self-reliant clients, has been impeded by conflicting priorities of women and the state.

The report of the conference on 'Healing the Earth – Women's Strategies for the Environment' held in 1991 in preparation for the 1992 UN Conference on Environment and Development, stated that, 'the world is an island with limited resources. We, as women, are implicitly working towards the recovery of the environment, reclaiming cultural heritage, maintaining diversity in the eco-system, developing appropriate technology and embracing the principle of cooperation and respect for life'. What the environment means to women who depend on it has to be understood in order to introduce a vision with operational mechanisms and to decide on strategic interventions. No single area or sector could deal with the multiple aspects in the linkages between women and the environment because livelihood security depends on the security over the sources of food, energy, income, shelter, water and biomass, amongst other things, and thereby over the habitats that women occupy. The decisions made at upper hierarchical levels in forestry, agriculture, infrastructure, biodiversity, water resources, energy, etc, reflect the state authority over these and the alienation of women from their management. Policy interventions could either promote or marginalize women's local initiatives in all aspects pertaining to environmental management. The rights of the states have stimulated export-oriented, hybrid monoculture, and paved the way for the huge influx of chemical pesticides and fertilizers to the farm space occupied more regularly by women. Should women lead a life as environmental victims or rather as development victims or should they search for another outlet is a question that remains unanswered.

The debate on environmental justice for women in South Asia is based on certain predominant features that have not changed over the years. The ones of paramount importance are related to the following contexts:

- The wide gap that exists between men and women in the South Asian region and the spread of its roots into the unequal distribution of resources and imbalance in access to livelihood sources.
- The imbalance between women's status in the local environment, stable property such as land in particular and their work and occupational patterns.
- The exclusion of women and their linkages with environmental resources in the process of development, particularly in relation to forestry, agriculture, land, water resources, and energy policies.
- The depletion and degradation of local resources which are to be kept under women's custody because of their regular contacts with resources and knowledge, responsibilities and roles for the individual, family and community well-being.
- The lack of recognition given to women as equal citizens whose responsibility for the local environment is equal to, or more important than that of their partners' for sustainability.
- The strength of women's ecological knowledge, their regular and close contacts and their habitat or ecosystem-based occupations.
- Women's special knowledge on the environment, and the indispensable material linkage with sustenance.
- Women's environmental entrepreneurship, which attributes values to local biomass products, and biodiversity.

The essence of the relation between women and nature as explained by Shiva (1997, p62) is that 'women in India are an intimate part of nature, both in imagination and in practice. At one level nature is symbolized as the embodiment of the feminine principle and at another she is nurtured by the feminine to produce life and provide sustenance'. What we find across Asia today is a culture originated through trial and error in the soils of each region which has made it possible to be innovative in the ways and means to sustain life. In this respect, the ways and means explored, introduced and continued by women, are strong. Their rights to innovation, knowledge, technology and local management cannot be alienated by enforcing the rights of others. The women of different habitats, agro-ecological regions and the communities of different religions, ethnicities, and classes in South Asia have a special set of gender relations that are seen practically in their status in the environment.

Although it is not intended to examine the eco-feminist discourse here, a noteworthy feature is in the ways in which linkages are constructed by authors like Shiva (1997), Agarwal (1997, p68) and others, as these have some bearing on the arguments put forward. The symbolic construction of women and nature that eco-feminists have promoted over the years is of greater relevance to women in Asia. The violence against women and against nature is linked both ideologically and materially. Developing-world women are dependent on nature for drawing sustenance for themselves, their families and their societies. The destruction of nature thus becomes the destruction of women's sources of staying alive (Shiva, 1989). The local context of environmental justice for women is critical where women consider themselves as an integral part of the ecosystem or habitat within which they live. The alliance between women and nature has often become essential because of their concern over the local environment and the nurturing nature of their contacts.

ALIENATION, A CONSEQUENCE OF GENDER RELATIONS

South Asia, being diverse both in terms of its cultural and biophysical context, offers a tremendous variability in women's relative positions within households, communities and also within countries. The implications of women's relative positions in these societies as reflected in gender relations are both direct and indirect. The stereotype of men as producers and primary income-earners has vested a greater role in men to manage the resources on which both women's and men's livelihoods depend. The household or the unit of function and its hierarchical structure in the context of the environment is crucial. It is marked with intra-household characteristics, power relations between men and women, division of roles and responsibilities and also with differences in intra-household control over resources. The responsibilities and roles in the hierarchy are promoted by many factors such as religion, ethnicity, economy, ideology, history and culture. The similarities and differences in women's positions, in this context are the result of contextual variability in the region. Moser (1993), with examples from Asia and Africa, explains how gender relations are socially constructed, contextually specific and often change in response to altering economic circumstances.

When analysing the environmental justice of women, and the variety of settings through which gender relations are maintained is taken into consideration, both ascribed and achieved relations are important (Young, 1997, p51). In Asia, women's positions' in networks of kinship vary spatially, and thereby too, their ascribed relations, under which great variability in their rights to land and private property are asserted. The totality of the social organization has created specific lifestyles and livelihood patterns across space in Asia. Agarwal (1994) has explained that it refers to the relations of power between women and men which are revealed in a range of practices, ideas and representations, including the division of labour, roles and resources between women and men, and ascribing to them different abilities, attitudes, desires, personality, traits, behavioural patterns and so on.

In fact, closely interwoven women's environmental linkages that were found in agrarian systems in Asia are more complex due to their preconditioned status in the social hierarchy. This implies that issues extend beyond landlessness or access to the resources, and include gender relations under which their status is defined. The other variable attributed to this aspect is the relatively low rates of women's involvement in economic and political life, which are described as 'achieved gender relations' (Young, 1997). Women are considered victims of environmental degradation due to social imbalances and economic and political backwardness. They are 'environmental refugees' who have lost their local identities in the phase of transformation, and are invisible workers whose contributions are uncounted, unrecognized and undervalued.

Issues of environmental justice for women in South Asia are partly interwoven with the rural women's relationship to the land. In this connection Agarwal (1994), with many critical analyses, has affirmed the importance of having a woman having a field of her own. She has argued that land defines the social status and political power in the village, and that it structures relationships both within and outside the household. The enforcement of gender relations in the allocation of intra-household resources obstructs their power of control over land-based livelihoods. Property ownership and control in the South Asian are symbols of power and status through which women have to proceed. The debate on environmental justice also deals with strong son-preference societies in South Asia (Chan, 1988, Banerjee, 1992). The economic rationale for son preference is strong in Pakistan, Bangladesh, India and Nepal mainly due to the patriarchal marriage system where the bride moves to the bridegroom's household. No such rigid rules are practised in Sri Lanka, though in reality the patriarchal marriage system is dominant and it allows men to hold greater rights to local environmental resources. There is often a transition in women's space and their labour from parental space to the husband's space. Women are considered as secondary or categorized as 'helpers' in the fields. This system has undoubtedly resulted in having women as 'unauthorized' farmers or as a category, without having their own fields. Young women who gain experience and knowledge in family farms are moved to another system, either completely new in terms of production functions or characterized with different labour arrangements.

Women have no assurance over farms as managers or farm food production for subsistence though they play a significant role in farms, home gardens and paddy fields. The same reasons allow men to have control over the fields as the owners of the land and heads of households. Women own less than 1 per cent of the world's land, work on other people's land, have no tenurial security, and gain access through familial or marital rights (Davidson, 1993, p5). The lack of solid legitimate legal rights, other than the greater share of work and longer hours of contact, detract from women coming to the forefront as farm and environmental managers.

CONCLUSION

Social and environmental injustices in South Asia are interconnected. Therefore attempts made in raising the social status of women through welfare delivery have not been able to eliminate the environmental inequalities that have been developed locally over time. The grounded realities discussed in this chapter demand a new era of environmental policy-making and action through which women are empowered to become equal partners in the management of local environments. What women in Asia need are rights, as environmental citizens, to make decisions and act as equal citizens in their own habitats and ecosystems. Equal opportunities and environmental citizenship are vital to encourage and mobilize a wealth of women's knowledge and experience in managing the local environment sustainably. The struggle, under these circumstances, should have two interrelated goals. First, to eliminate all forms of disparities and promote equal status. Second, to promote equal rights, not on customary grounds or by redefining the strength of customary rights, but by securing legal rights to claim authority and partnership as environmental citizens. The rationale for these goals are deeply rooted in women's occupational space, the land and the local habitats, and it penetrates debates on equal rights and opportunities. The imbalances found in local environments are a reflection of the social and environmental exclusion of a crucial sector, who should be at the centre of environmental management in Asia. Women's stewardship over the environment has evolved over generations and therefore they have every right to claim equal opportunities.

Moreover, women as equal citizens should have rights to a healthy environment. Denial of their rights on gender grounds, under which women have become disadvantaged and underprivileged, cannot be addressed without having fundamental social change, and a change in gender relations in particular, and also in the process of environmental planning.

The preceding analysis demands policy attention aimed at ensuring security of tenure and access to resources, and rights to claim and control, because the majority of rural women in South Asia depend on them. The improvement of women's status and well-being is a prerequisite for the maintenance and improvement of the family and community. The range of economic opportunities that women themselves have explored, as environmental entrepreneurs, are sources for promoting conservation mechanisms to people. Women's positions in indigenous knowledge systems, the time that they spend on a day to day basis, has made it clear that women can be the legitimate stakeholders in a given habitat or an ecosystem. Finally, if the aim is to promote sustainable, environmentally responsible societies, then it is important that in

countries in South Asia, as well as in other parts of the world, women's
environmental citizenship should be guaranteed.

REFERENCES

Agarwal, B (1986) *Cold Hearths and Barren Slopes: The Woodfuel Crisis in the Third World*, Zed
   Books, London
Agarwal, B (1994) *A Field of One's Own: Gender and Land Rights in South Asia*, Cambridge
   University Press, Cambridge
Agarwal, B (1997) 'The gender and environment debate: Lessons from India' in N
   Visvanathan et al (eds) *The Women, Gender and Development Reader*, Zed Books, London
   and New Jersey
Banerjee, N (1992) 'Integration of women's concerns into development planning: The
   household factor' in United Nations Economic and Social Commission for Asia and
   the Pacific (ESCAP), *Integration of Women's Concerns into Development Planning in Asia
   and the Pacific*, United Nations, New York
Begum, K (1989) 'Participation of rural women in income earning activities: A case
   study of a Bangladesh village', *Women's Studies International Forum*, vol 12, no 5,
   pp519–28
Bhadra, C K (1997) 'The land, forestry and women's work in Nepal' in A
   Wickramasinghe (ed) *Land and Forestry: Women's Local Resource-based Occupations for
   Sustainable Survival in South Asia*, CORRENSA (Collaborative Regional Research
   Network in South Asia), Peradeniya, Sri Lanka
Buckingham-Hatfield, S (2000) *Gender and Environment*, Routledge, London and New
   York
Chan, M (1988) 'Material consequences of reproductive failure in rural South Asia' in D
   Dwyer and J Bruce (eds) *A Home Divided*, Stanford: Stanford University Press
Cuomo, C J (1998) *Feminism and Ecological Communities: An Ethic of Flourishing*, Routledge,
   London and New York
Dankelman, I and Davidson, J (1988) *Women and Environment in the Third World: Alliance
   for the Future*, Earthscan, London
Das, S (1997) 'Women's contribution to homestead and agricultural production systems
   in Bangladesh' in A Wickramasinghe (ed) *Land and Forestry: Women's Local Resource-
   based Occupations for Sustainable Survival in South Asia*, CORRENSA (Collaborative
   Regional Research Network in South Asia), Peradeniya, Sri Lanka
Davidson, J (1993) 'Women's relationship with the environment' in G Reardon (ed)
   Women and the Environment, Oxfam Focus on Gender 1, Oxfam, UK
Government of India, (1988) *Government of India, National Perspective Plan for Women,
   1988–2000*, Deptartment of Women and Children Development, Ministry of
   Human Resource Development, Delhi
Government of India (1991) *Eighth Five Year Plan*, Ministry of Human Resource
   Development, Delhi
Huda, S (1997) 'Women's property rights in Bangladesh: Effect of religion and custom'
   in A Wickramasinghe (ed) *Development Issues Across Regions: Women, Land and Forestry*,
   CORRENSA (Collaborative Regional Research Network in South Asia), Peradeniya,
   Sri Lanka
Kumar, S K and Hotchkiss, D (1989) 'Consequences of deforestation for women's time
   allocation, agricultural production and nutrition in hill areas of Nepal', Research
   Report No 69, International Food Policy Research Institute, Washington, DC

Mencher, J (1993) 'Women, agriculture and the sexual division of labour: A three-state comparison' in S Raju and D Bagchi (eds) *Women and Work in South Asia: Regional Patterns and Perspectives*, Routledge, London and New York

Mies, M and Shiva, V (1993) *Ecofeminism*, Zed Books, London and New Jersey

Moser, C (1993) *Gender Planning and Development: Theory, Practice and Training*, Routledge, London and New York

Ministry of Planning (1995) *Fourth Five Year Plan*, Dhaka, Bangladesh

Mukherjee, N (1995) 'Forest management and survival needs community experience in West Bengal', *Economic and Political Weekly*, December 9, pp3130–132

National Plan of Action for Women in Sri Lanka, (1993) *Towards Gender Equality*, Ministry of Transport, Environment & Women's Affairs, Colombo, Sri Lanka

National Planning Commission (1992) *Eighth Five Year Plan*, Kathmandu, Nepal

Peiris, G (2000) 'Environmental protection and development: The South Asian challenge' in V A Pai Panandiker and N Behera (eds) *Perspectives on South Asia*, Konark Publishers Pvt Ltd, Delhi, India

Planning Commission/Government of Pakistan (1983) *The Sixth Five Year Plan*, Karachi, Pakistan

Raghuram, P (1993) 'Invisible female agricultural labour in India' in J H Momsen and V Kinnaird (eds) *Different Places, Different Voices: Gender and Development in Africa, Asia and Latin America*, Routledge, London and New York

Raju, S and Bagchi, D (1993) *Women and Work in South Asia: Regional Patterns and Perspectives*, Routledge, London and New York

Sarin, M (1995) *Community Forest Management: Where are the Women*, The Hindu Survey of the Environment, Madras

Saxena, N C (1987) 'Women in forestry', *Social Action*, vol 37

Saxena, N C (1995) 'Gender and forestry', *The Administrator*, vol XI, pp71–92

Saxena, A, Navtiyal, J and Foot, D (1997) 'Analysing deforestation and exploring policies for its amelioration: A case study of India', *Journal of Forest Economics*, vol 3, no 3, pp253–89

Shiva, V (1989) *Staying Alive: Women, Ecology and Development*, Zed Books, London

Shiva, V (1991) *The Violence of the Green Revolution: Third World Agriculture, Ecology and Politics*, Third World Network, Penang, Malaysia

Shiva, V (1994) *Close to Home: Women Reconnect Ecology, Health and Development*, Earthscan, London

Shiva, V (1997) 'Women in Nature' in N Visvanathan et al (eds) *The Women, Gender and Development Reader*, Zed Books, London and New Jersey

Sims, J (1994) *Women, Health & Environment: An Anthology*, WHO, Geneva

Sontheimer, S (1991) *Women and the Environment: A Reader Crisis and Development in the Third World*, Monthly Review Press, New York

United Nations Development Programme, (1995, 1997, 1999) Human Development Report, New York, Oxford University Press

Wickramasinghe, A (1993a) 'Women's Roles in Rural Sri Lanka' in J H Momsen and V Kinnaird (eds) *Different Places, Different Voices: Gender and development in Africa, Asia and Latin America*, Routledge, London and New York

Wickramasinghe, A (1993b) 'The different impact of deforestation and environmental degradation on men and women: Issues for social justice', *Journal of Human Justice*, vol 5, no 1, pp46–57

Wickramasinghe, A (1994a) 'The Kandyan homegardens in Sri Lanka', *Women and Natural Resource Management: A Manual for Trainers in Commonwealth Asia*, Section III, Women for Conservation, Commonwealth Secretariat, UK

Wickramasinghe, A (1994b) *Deforestation, Women and Forestry: The Case of Sri Lanka*, International Books, The Netherlands

Wickramasinghe, A (1995a) 'Homegardens: Habitation rescuing biodiversity', *MPTS News*, vol 4, no 2, Sri Lanka

Wickramasinghe, A (1995b) 'The evolution of Kandyan homegardens: An indigenous strategy for conservation of biodiversity in Sri Lanka' in P Halladay and D Gilmour (eds) *Conserving Biodiversity Outside Protected Areas: The Role of Traditional Agro-ecosystems*, IUCN, Gland, Switzerland and Cambridge

Wickramasinghe, A (1995c) 'Gender patterns in the management of forest resources in Sri Lanka' in E Silva and D Kariyawasam (eds) *Emerging Issues in Forest Management for Sustainable Development in South Asia,* Asian Development Bank, Manila

Wickramasinghe, A (1995d) *Women, Environmental Management and Entrepreneurial Opportunities,* The British Council, Colombo, Sri Lanka

Wickramasinghe, A (1997a) 'Rural women's problems as issues for agrarian reforms' in A Wickramasinghe (ed) *Development Issues Across Regions: Women, Land and Forestry,* Collaborative Regional Research Network in South Asia (CORRENSA), Peradeniya, Sri Lanka

Wickramasinghe, A (1997b) 'Women harmonizing ecosystems for integrity and local sustainability in Sri Lanka' in A Wickramasinghe (ed) *Land and Forestry: Women's Local Resource-based Occupations for Sustainable Survival in South Asia,* Collaborative Regional Research Network in South Asia (CORRENSA), Peradeniya, Sri Lanka

Wickramasinghe, A (1997c) 'Anthropogenic factors and forest management in Sri Lanka', *Applied Geography*, vol 17, no 2, pp87–110

Wickramasinghe, A (1997d) 'Gender concerns in conservation' in G Borrini-Feyerabend and D Dianne Buchan (eds) *Beyond Fences: Seeking Social Sustainability in Conservation*, vol 2, A Resource Book, IUCN, Gland, Switzerland

Young, K (1997) 'Gender and development' in N Visvanathan et al (eds) *The Women, Gender and Development Reader,* Zed Books, London and New Jersey

Zwarteveen, M (1994) *Gender Issues, Water Issues: A Gender Perspective to Irrigation Management,* Working Paper No. 32, International Irrigation Management Institute (IIMI), Colombo, Sri Lanka

Zwarteveen, M and Endeveld, M (1995) *Rural Women's Questions Are Agrarian Questions: A Discussion of the Intellectual and Political Construction of Realities of Rural Women.* Paper presented at the conference on Agrarian Questions: The Politics of Farming Anno 1995, 22-24 May, Wageningen, The Netherlands

Maori Kaupapa and the Inseparability of Social and Environmental Justice: An Analysis of Bioprospecting and a People's Resistance to (Bio)cultural Assimilation

Stefanie S Rixecker and Bevan Tipene-Matua

INTRODUCTION

Current problems implementing sustainability point to fragmentation based upon Western modernity's propensity towards atomism and reductionism in contending with the environmental problematique. Rather than acknowledging the inherent connections between ecological and social health, modern practitioners focus upon ecological bottom lines and continue to maintain the arrogance that humankind has the ability to manage planet Earth. Other peoples and cultures, such as Maori in Aotearoa New Zealand, do not see planet Earth, or Papatuanuku from a Maori worldview, as divisible into separate spheres or dimensions. Ecological and social justice are inseparable; one is inherently entangled with the other. Aotearoa New Zealand, a land with bicultural distinctions through the Treaty of Waitangi/Te Tiriti O Waitangi and multi-cultural populations, provides an excellent example of how a people with a holistic worldview simultaneously use and challenge contemporary institutions to promote and protect a world which aspires to reconnecting social and environmental justice. In order to focus the complexity of neocolonial challenges in Aotearoa New Zealand, this chapter will address the contemporary responses to genetic engineering and bioprospecting evident in the Aotearoa cultural, sociopolitical, and biotechnological landscape.

E rite ai nga tangata ki nga reo manu o te ngahere, ka ketekete te Kaka, ka koko te Tui, ka kuku te Kereru. hakoa te rereketanga o nga reo manu nei, ka tau tonu te wao nui a Tane. Ko te tikanga, tiakina te rereketanga, kia tu te kotahitanga.

People are similar to the bird songs of the forest, the Kaka chatters, the Tui soars, and the Kereru withdraws. Regardless of the diversity of voice emanating from these bird songs, the forest remains settled. Only by nurturing diversity shall unity be achieved. (Maori proverb)

SUSTAINABLE DEVELOPMENT

The 1987 publication and subsequent dissemination of the World Commission on Environment and Development's *Our Common Future*, more commonly referred to as the Brundtland Report, launched a defining moment in global environmental politics and discourse. Not only did it focus attention upon a particular term, sustainable development, but it did so by using a global perspective to connect ecological, economic and sociocultural issues within and across generations. In doing so, the Brundtland Report popularized the concept of sustainable development and challenged nations, corporations and peoples to find multiple ways of addressing the ongoing and frightening growth of ecological degradation.

The 1992 United Nations Conference on Environment and Development (UNCED), or 'Earth Summit', furthered the use and dissemination of sustainable development, or 'sustainability' more broadly defined. This Summit saw the largest collection of nation states and non-governmental organizations (NGOs) come together in parallel conferences to debate and address global environmental problems, or the 'environmental problematique' (Bührs and Bartlett, 1993, Soroos, 1994). Although a phenomenal event and feat of organization in its own right, the conference included and perpetuated severe constraints on equity, democracy and justice. A variety of factors can be linked to these inequities, yet there is no doubt that the pre-existing and institutionalized disparities between governments, nations and peoples contributed to the lack of a fair and equitable platform for dialogue and negotiation (Morphet, 1996). These inequities, institutionalized through the international relations system, premised upon sovereignty and the nation state, ultimately caused (and now continue) the disharmonious relationships of and between the 'haves' and 'have nots' on planet Earth.

The contemporary system of environmental politics enshrined in and legitimated through international and national laws premised upon a Western legal system and modernity (see Blowers, Chapter 3), makes it especially challenging for indigenous peoples to reclaim and control their native homes, cultures, practices and beliefs. The rise of the corporate powers through globalization has added another dimension to the challenge as multinational corporations (MNCs) become major economic and power brokers in decision-making systems previously reserved for national governments and their respective heads of state. Such political changes mean that decision-making with regard to the environment has become increasingly complex and contentious. Thus, indigenous peoples have had to become more resourceful than ever before to secure and protect their cultural and environmental heritage.

In Aotearoa New Zealand, Maori, as the indigenous peoples, have worked steadfastly to address a variety of environmental issues. Despite the increased global attention to environmental and indigenous matters, the nexus between corporate stakeholders, national governments and international institutions has increased the need for Maori vigilance and innovation. This is no more true than in the contemporary arena of biological technology (biotechnology) and genetic engineering (GE). The issues associated with genetic technologies are diverse and complex, but ultimately they cannot be separated along a clear division of environmental and social justice as much rhetoric and debate around 'sustainability' attempts to do. Fundamentally, one form of justice cannot exist without the other. As this chapter will show, the ongoing politics regarding genetic engineering and Maori in Aotearoa New Zealand reflect the cultural complexity and dynamism of a people's resistance to (bio)cultural assimilation.

GEOGRAPHIC LOCATION: TE IKA A MAUI, TE WAKA A MAUI AND TANGATA WHENUA

To be an indigenous person is to be able to trace one's genealogy to those who encountered a land in its unpolluted state. To be indigenous is to be able to access a knowledge base that has its roots in the natural world, a knowledge base that has developed over centuries as a result of many trials and many errors. All peoples can trace their ancestry to such a knowledge system somewhere; all peoples have indigenous connections to somewhere. For Aotearoa New Zealand, the tangata whenua, or people of the land, have come to be known over the last few centuries as Maori.

For Maori, explaining the origins and complexities of the world is simple. Maori started with potential (te po or nothingness). Out of potential came the world of light which was dominated by Papatuanuku, the mother of all things, and Ranginui, the greatest heaven. The first children of Ranginui and Papatuanuku (Heaven and Earth) were the guardians of the natural world: Tangaroa guardian of the sea, Tawhirimatea guardian of the elements and Tane Mahuta, guardian of the forest are three members of a family of over 70. It was Tane Mahuta who fashioned humans from the elements and the first was a woman named Hineahuone. One of the most famous of these early peoples was Maui, and it was Maui who delved into the depths of the great southern oceans and fished up Aotearoa New Zealand.

Thus, Aotearoa is an island nation in the South Pacific. The two main islands are the Fish of Maui – Te Ika a Maui (the North Island) and the Canoe of Maui – Te Waka a Maui (the South Island) as well as a cluster of smaller islands (including Rakiura or Stewart Island). Combined, the islands cover more than 1600 kilometres in length, while their greatest width is only 450 kilometres, and their total area is 'about the same size as the British Isles or Japan (approximately 27 million hectares)' (Ministry for the Environment, 1997, p2.3). Aotearoa's closest and most frequented neighbour is Australia to the west. The nation's population currently stands at 3.6 million, and people with Maori ancestry represent approximately 560,000 (or 15 per cent) of this total (Ministry for the Environment, 1997, p2.3). The major ethnic group is European (79 per cent) while smaller populations are represented by Pacific Island groups (4 per cent), Chinese (less that 2 per cent), Indian (less than 1 per cent) and other (less than 1 per cent) (Ministry for the Environment, 1997, p2.4). This is the context in and from which contemporary biopolitics in Aotearoa flows.

ENVIRONMENTAL VANDALISM AND CULTURAL PLUNDER: AOTEAROA'S ECOPOLITICS

Aotearoa's earliest ecopolitical history is dominated by complex land and natural resource exchanges that defined and determined human relationships and established obligations among and between people and the land (Papatuanuku). Integral to the traditional Maori world was the relationship with Papatuanuku, and all human relationships revolved around and were interdependent upon the natural world.

Traditional Maori definitions of wealth, mana (prestige), and spiritual, physical and intellectual well-being vastly differ from the Western worldview which dominates and is worshipped by many today. For Maori, unfettered consumption, secular lifestyles and individualistic or non-communal decision-making processes were all foreign concepts. All decisions were based on mana (prestige, authority and power) and mana accrued to those who provided for the people and not at all to those who hoarded wealth or looked after themselves. Social and political structures centred on concepts such as whakapapa (genealogy), tapu (sacredness), kaitiakitanga (guardianship obligations to look after the environment) and rangatiratanga (self-determination – the power to control one's destiny). Maori were organized communally as whanau (family), hapu (extended family) and iwi (larger tribal groupings). These were the pillars of a Maori society.

Thus when European colonizers arrived in Aotearoa, Maori society and Maori people were fit and well. Unlike other indigenous peoples throughout the world, many Maori were keen to have European settlers move into their districts as the new technologies and teachings (particularly Christian) they brought were well sought after. Put simply, Maori became intoxicated with the new technological inventions and knowledge that the early settlers brought with them. They took on board Western ways with little or no question, and such a blind embrace of this new technology (particularly guns and alcohol) had devastating impacts. History has shown that Maori experience of Western technology has repeated itself over and over again. Maori have always been the last to benefit from and the first to suffer the adverse effects of Western science and technology.

In 1840 Maori entered into a Treaty with the Crown called the Treaty of Waitangi or Te Tiriti O Waitangi. The Treaty guaranteed to Maori the right to manage their own resources in accordance with their own environmental preferences. The Treaty also guaranteed to Maori their language and culture and the rights and privileges of British citizens. The Treaty guaranteed to the Crown the right to govern over New Zealand and also gave the Crown the right of 'pre-emption' over Maori land sales (ie Maori could only sell their land to the Crown).

The key provision of the Treaty of Waitangi which has been the focus of most of the debate since its signing is article two which states the following:

> *Her majesty the Queen of England confirms and guarantees to the Chiefs and Tribes of New Zealand and to the respective families and individuals thereof the full and exclusive and undisturbed possession of their lands and estates, forests, fisheries and other properties which they may collectively or individually possess so long as it is their wish and desire to retain the same in their possession...* (Kawharu, 1989, p317)

Since the Treaty was signed, Maori have fought, protested, lobbied and died in an attempt to have its provisions honoured and to have past injustices redressed. Current approaches to environmental and social justice issues must be seen in light of the experiences suffered by Maori through colonization. Maori were dispossessed of their lands either through illegal invasion and confiscation, or other encounters, which later became known as 'land sales'. More importantly, they were denied the opportunity to be kaitiaki or guardians of the natural world around them, including people. European diseases, the general breakdown of Maori social structures, the disconnection of Maori from their traditional lands, and the vast amount of environmental degradation that accompanied the colonizers, all contributed to the Maori race almost dying out in the early 1900s.

Although the battle by Maori for social and environmental justice in Aotearoa continues, the 150-year plus effort has resulted in limited Crown attempts to address historical grievances. There has also been limited recognition, in law, that the unique and valuable knowledge of Maori and their right to self-determination must be protected. Such recognition can be found in the 1975 Treaty of Waitangi Act, the 1991 Resource Management Act and the 1996 Hazardous Substances and New Organisms Act. From an international perspective, these three initiatives are considered (by some) to be the most significant and ambitious attempts at environmental management of their time.

The Waitangi Tribunal – the first indigenous environmental and social justice court

The Waitangi Tribunal was set up in 1975 to hear historical grievances after over a century of protest by Maori about how they had not received a fair deal and had suffered at the hands of illegal racist Crown actions and inactions since 1840. Early claims to the Tribunal related to the Crown's failure to ensure traditional Maori food gathering areas were protected from industry and urban pollution. The Manukau Harbour, Kaituna River and Motunui outfall were the first claims and related to the pollution and destruction of sacred and important waterways. As a result of these early claims, the Crown was forced to take responsibility for ensuring Maori could fulfil their obligations to be kaitiaki (guardians) of the environment for future generations. Using their holistic worldview, Maori broadened their focus to social justice issues to include the impacts of European health, education and legal systems that were transported to Aotearoa and imposed upon Maori. There are now over 700 claims to the Tribunal, all of which seek remedies for the marginalization of Maori and their relationship with the natural world as the result of Crown breaches of the Treaty of Waitangi.

A key claim before the Tribunal is the WAI262 claim. This claim relates to the Crown's failure to ensure Maori heritage, knowledge and resources are protected from being exploited by large biotechnology and biopharmaceutical corporations by using intellectual property rights and free trade agreements. For over ten years, Maori have warned of the injustices and potential dangers of biotechnology and genetic engineering, and at the heart of this concern is the impact on Maori cultural and spiritual imperatives. However, the Waitangi Tribunal is not the only institutional vehicle used by Maori. The Resource Management Act, which incorporates Maori to some extent, is also an institutional mechanism with potential for extending Maori self-determination.

The Resource Management Act 1991

In 1991, the Resource Management Act (RMA) was passed. This Act required decision-makers to: have regard to Kaitiakitanga (guardianship rights); recognize and provide for the relationship between Maori and their culture and traditions with their ancestral lands, water, sites, waahi tapu (sacred sites) and other taonga (spiritually important things); and take into account the Principles of the Treaty of Waitangi.

The recognition of the need to protect the relationship between Maori and biodiversity in the RMA was well received by Maori. They saw this legislation providing an opportunity to participate once again in resource management, thereby enabling a means of re-establishing their relationship with the natural world. However, there have been few environmental and social justice outcomes for Maori as a result of the RMA. Maori have come to realize that the Treaty of Waitangi itself provides the template for Maori involvement in environmental management and ensuring that Maori self-determination is protected. As Matunga (2000, p38) argues:

> *The Treaty provided a basis for the evolution of dual environmental planning traditions, one grounded in indigenous Maori traditions, philosophies, principles and practices, and the other in the imported and evolving traditions and practices of an introduced 'Western' planning tradition. The current 'mainstream' environmental planning system has its roots in a colonial discourse which continues to exclude Maori and rejects both the primacy of the relationship between Maori and the Crown and the redistribution of power. Aotearoa/New Zealand has a dual-planning heritage which needs to form the basis for a new paradigm for environmental planning.*

Over ten years since the RMA was passed, Maori continue to struggle to participate fully and have their unique cultural perspectives recognized and provided for. Decision-makers often ignore or dismiss the cultural and spiritual perspectives of Maori and their relationship with the natural world. Additionally, Maori participation is severely restricted by lack of human and financial resources. Most of the Maori resource management units are expected to respond to hundreds of proposals, regarding possible environmental effects, on a voluntary basis. This pressure to respond to the activities of individuals, industries and others to use and abuse the environment has been compounded over the last two years by the rapid increase in proposals to introduce genetically modified organisms into Aotearoa.

Genetic engineering and bioprospecting: Neocolonial biopolitics

Although the latter half of the 20th century saw a number of countries declare their independence, thereby symbolically and structurally decoupling the shackles of colonization, both the blatant and subtle processes of control and oppression associated with colonization remain steadfast in global politics and economics. Attempts by peoples of the 'Fourth World', ie indigenous peoples, to decolonize have been even more challenging because most indigenous peoples were dispossessed of their lands, and their status as full-fledged nations was denied through processes of national and international colonization and exploitation. In effect, they no longer had control of their lands in order to declare independence as other colonized nations did in the 1960s and 1970s. In this respect, indigenous peoples' contemporary experiences of colonization are more aptly described as neocolonial than postcolonial. Irrespective of its descriptor, such disenfranchisement from national and international politics created an ever stronger voice and form of resistance within and across indigenous peoples around the world. Indeed, the indigenous voice thus:

> *… speaks to the narrative of the community of nation-states because the forces that threaten their survival as indigenous nations arise from the rationalization of their exclusion from both state and world community building. To hear the indigenous voice is to hear a critique of modernization as an ideology of both international and national political community-building processes. It is all the more powerful because it is not only an intellectually critical view but a narrative of critical experience.* (Wilmer, 1993, p194)

During the late 20th century, this indigenous narrative of critical experience has been especially vocal in the global politics of biodiversity and genetic engineering. As such, it has been extended to include the 'Biotech Century' (Rifkin, 1998) and its attendant politics of biology (biopolitics), whereby science and economics have united to create seemingly new and innovative products.

These new products of genetic engineering and modern biotechnology include a variety of forms such as pharmaceuticals, foods, enzymes and diagnostic procedures. They are distinguished from other products and creations because they are generated through some form of genetic alteration or manipulation of organic life forms, also known as recombinant-DNA (r-DNA) technology. However in order to access the life forms, and specify which organisms hold the greatest potential for yielding positive outcomes, scientists require access to two things: knowledge and natural resources.

The natural resources, or life forms, usually come from a non-local geographic location, and rarely are the items attained with permission from the local peoples who were the original stewards of the organism. When researchers do seek out local or indigenous knowledge, it is usually with the intent of determining which local flora and fauna have the greatest chance for generating medicinal or pharmaceutical properties which can be marketed. Thus, these processes of natural resource and knowledge extraction are the most recent form of colonization and imperialism and may be referred to as biocolonialism (Hindmarsh et al, 1998).

Extraction of such information and life forms for the purpose of economic profit has been called biological prospecting (bioprospecting). It is regarded as part of the ongoing colonization of local and indigenous peoples around the world, and it has been labelled biopiracy (Shiva, 1997). Where colonizers once exploited the land and degraded the culture, current science and technology now continue this by exploiting and stealing the remaining treasures of indigenous peoples – their flora and fauna, knowledge, spiritual integrity and relationship with the human and non-human world. Ultimately, resistance to biopiracy is:

> … *a resistance to the ultimate colonization of life itself – of the future of evolution as well as the future of non-Western traditions of relating to and knowing nature. It is a struggle to protect the freedom of diverse species to evolve. It is a struggle to protect the freedom of diverse cultures to evolve. It is a struggle to conserve both cultural and biological diversity.* (Shiva, 1997, p5)

As such, indigenous peoples' resistance to biopiracy and their desire to control their own futures is *simultaneously* about social, cultural and ecological justice; they are inextricably linked. Using the modern, environmental lexicon, this struggle is fundamentally about sustainability – the ability of peoples to sustain themselves and future generations by ensuring that the diversity, integrity and life force between all organic life forms is nurtured and protected.

Kaupapa Maori: Resisting (bio)cultural assimilation

To ignore people's knowledge is to almost ensure failure in development (Brokensha, cited in Tipene, 1997, p83)

The Maori renaissance has been based around the revival and enhancement of language, culture and tradition. From this stems confidence, self-belief, self-esteem, and spiritual wellness. Maori are realizing that without their cultural knowledge and spiritual connections to the land and each other they are doomed. Total immersion Maori language schools; Maori universities; Maori health providers and Maori tribal autonomy are all realities that are underpinned by Maori tradition and are seen as the way forward by many. The resurgence of the importance of Maori knowledge, language and worldviews to Maori has developed in parallel with the wider societal obsession and hype about the wonders of Western knowledge and technology (including genetic engineering).

As Suzuki and Dressel (1999, p4) reminds us:

> [w]e repeat like a religious mantra the unquestioned benefits and power of
> science, information and economics, without inspecting the structures and
> methodology on which they are built. Many of these beliefs are insupportable
> and dangerous. For example, the notion that human beings are so clever that
> we can escape the restrictions of the natural world is a fantasy that cannot be
> fulfilled. Yet it underlines much of government's and industry's rhetoric and
> programmes.

Maori no longer believe the hype about how Western science, technology and
economic ideology will meet the needs of local communities. What many non-
Maori do not understand, particularly the scientific community, is that if you
ignore or disregard a people's language and culture, you render them invisible.
Biotechnology advocates see the world through a reductionist, consumption-
driven, ethnocentric mindset and ignore and render irrelevant the cultural and
spiritual values of indigenous peoples. Whether this is done intentionally or not
is irrelevant. What is important is that Maori will never benefit from, participate
in an empowered way, or agree to this technology and will remain suspicious
unless new approaches to recognizing and respecting indigenous and other
worldviews are adopted.

　　Thus, Maori have been proactive about and attentive to such nefarious
neocolonial biopolitics in order to retain their sovereignty and resist (bio)cultural
assimilation. In addition to the long standing attempts to regain their lands and
resurrect their cultural belief systems, various members in the Maori community
have worked tirelessly campaigning for their right to decide their futures within
the realm of biopolitics. Alongside their ongoing defence of tino rangatiratanga,
or self-determination, Maori have contributed to the national dialogue and
politics surrounding the ethics, policies, and implementation of biotechnology.
Contributions have come from within Maori social structures and systems, such
as Te Kotuku Whenua (a grass-roots voluntary environmental group), while
others have emanated from Crown generated structures, such as Ngä Kaihautü
Tikangä Taiao, the Maori body of the Environmental Risk Management
Authority (ERMA). The perspectives of these two groups and others provide
insights into the process and vitality of Maori resistance.

　　For example, in 1993 Maori held a hui (meeting or conference) and created
the Mataatua Declaration on the Cultural and Intellectual Property Rights of
Indigenous Peoples. This declaration highlighted the need and right of Maori
to determine and retain their cultural knowledge and heritage. This heritage
was under further threat from colonization and exploitation due to the
globalization process of the Uruguay GATT rounds and the promulgation of
the World Trade Organization which sought to establish and increase the use
of Intellectual Property Rights (IPRs). Such rights could and would be used to
claim ownership over a variety of intellectual material, including patents derived
from knowledge and natural resources extracted from indigenous peoples. The
creation and use of IPRs became even more contentious when they were
directly linked to biodiversity during various Conference of Parties meetings

regarding the United Nations Convention on Biological Diversity (CBD) (1992), to which Aotearoa New Zealand is a signatory. These meetings focused more upon the property rights and extraction of biological diversity than upon its conservation or the rights of indigenous peoples, as stipulated in Section 8(j) of the CBD, thereby further compromising indigenous peoples' ability to protect their cultural and ecological heritage (Daes, 1993, Shiva, 1997, Tipene, 1997). Thus, the Mataatua Declaration became a means of direct Maori and indigenous resistance to the ongoing colonization of Maori cultural and ecological heritage.

In order to further protect their heritage, Maori laid a claim with the Waitangi Tribunal with regard to indigenous flora and fauna. This claim, referred to earlier as WAI262, stipulates iwi Maori rights to indigenous flora and fauna in relation to international and national issues such as intellectual property rights, biodiversity and genetic engineering. Maori desires to make their own decisions about their heritage and taonga are obvious in the WAI262 claim, and this is further supported by Maori activists and academics who argue that:

> *[i]t is essential that Hapu/Iwi guardianship rights over their assets be recognized and that any over-arching national body respect and uphold those rights... Decisions by Hapu and Iwi can only be decided by their members.* (Mead and Tomas, 1995, p130–132)

Such a view was echoed by the Ministry of Maori Development, Te Puni Kokiri, in relation to Maori and biodiversity policy when they emphasized 'the need for Maori to be directly involved in the development of policy, plans and strategies as they impact on biodiversity' (Te Puni Kokiri, 1994). However, Te Puni Kokiri (1994, p13) also noted the severe barriers to such inclusion by highlighting that 'problems have arisen with inadequate resources for Maori; ineffective departmental and local authority consultation strategies; and a lack of comprehension of Maori aspirations'. For these reasons, and others, Kilvington and Rixecker (1995, p21) argued that it would 'require the *entire fabric* of Aotearoa New Zealand's governance tapestry to enhance coordinated efforts of biodiversity policy development'. Thus, it is not surprising that this proved to be an impossible feat due to the ongoing constraints placed upon Aotearoa's indigenous peoples.

Despite the severe resource and political constraints, Maori still organized events and opportunities for ensuring their voice and politics were enabled in national and international fora. Another example of this is evident in the Indigenous Peoples Roundtable Meeting of 1994, convened by the Maori Congress in Whakatane, Aotearoa New Zealand. The meeting derived 12 recommendations and one overarching final statement which declared:

> *The Indigenous Peoples Roundtable Meeting ... unanimously affirms the natural inherited and inalienable right of all indigenous peoples to their self-determination over their lives, livelihoods, lands, territories and all other gifts/assets of their heritage.* (Mead, 1994, p14)

From this it is evident that Maori not only aspire to protect their own taonga (gifts), but work to support a collective right on behalf of all indigenous peoples to protect and retain their cultural and natural heritage. In this way, Maori work to redefine national and international biopolitics and continue their resistance to biocolonization.

More recently, such efforts were acknowledged and incorporated into more formal, government-related institutions through the formation of Ngä Kaihautü Tikangä Taiao, the Maori body of the Environmental Risk Management Authority (ERMA). The ERMA was created through the Hazardous Substances and New Organisms Act 1996 (HSNO) which came into effect in July 1998. The organization is regarded as an independent, expert decision-making body charged with protecting the health and well-being of communities and the environment with respect to hazardous substances and new organisms (HSNO, 1996). As such, the ERMA is responsible for considering all applications to introduce hazardous substances and new organisms, including genetically modified organisms (GMOs), into Aotearoa New Zealand. Although approximately 35 people administer the HSNO legislation within the ERMA, only eight people, including two Maori, have the power to determine which applications are successful. In considering the applications, a number of criteria must be provided for or be taken into account (Box 12.1).

Clearly, the criteria explicitly require consideration of Maori social and ecological heritage. Thus, it became necessary to enhance dialogue and decision-making structures with Maori, and this was enabled through the creation of Ngä Kaihautü Tikangä Taiao in 1997 which was formalized through a Memorandum of Understanding in 1998. Despite these structures, decision-making with regard to genetically engineered organisms was no easy task, and both the ERMA and Ngä Kaihautü Tikangä Taiao found it difficult and demanding. As Tipene-Matua (2000, p8), the former Senior Policy Analyst for Maori in ERMA, explained:

> [i]n hindsight the first few applications to field-test GMOs in Aotearoa escaped close scrutiny and requirements for further information on Maori issues largely because they were the first, and Maori (including Ngä Kaihautü and myself) were still coming to grips with some of the issues.

Thus, despite the hard work and efforts of some Maori with respect to indigenous rights and ongoing resistance to biocolonization, the complexity and specialized nature of genetic engineering further disenfranchises Maori from fully engaging in their newfound role as contributors to biopolitical decision-making.

A proposal by a Crown owned Scientific Research Institute, AgResearch, in 1998 exemplifies a key case in point. This proposal sought to produce the human myelin basic protein in the milk of a herd of cattle they proposed to raise on the ancestral lands of the local Maori tribe (from whom they leased the land). The environmental arm of the local tribe, Te Kotuku Whenua, was forced to respond to a proposal that breached Maori tradition. They explained that:

> ## Box 12.1 Criteria for Assessing HSNO Applications
>
> Applications for the introduction of Hazardous Substances and New Organisms to Aotearoa New Zealand must consider:
>
> - safeguarding the life-supporting capacity of air, water, soil and ecosystems;
> - the maintenance and enhancement of the capacity of people and communities to provide for their own economic, social and cultural well-being and for the reasonable foreseeable needs of future generations;
> - the sustainability of all native, and valued introduced, flora and fauna;
> - the intrinsic value of ecosystems;
> - public health;.
> - the relationship of Maori and their culture and traditions with their ancestral lands, water, sites, waahi tapu, valued flora and fauna and other taonga;
> - the economic and related benefits; and
> - New Zealand's international obligations.
>
> *Source:* HSNO, 1996

> *For us, all life is sacred and consequently whakapapa is sacred … One of the loudest arguments against genetics and biotechnology is coming from our own Kaumatua (elders), who are saying very clearly that no one should corrupt or interfere with whakapapa (genealogy). The sanctity and respect for whakapapa is to be maintained. Both mauri (life principle) and wairua (spirit) of living things are sacred. The responsibility falls on us to protect the legacy of our future generations and this includes the guardianship of whakapapa.* (Smith and Reynolds, 1999)

After discussion among the local peoples, Te Kotuku Whenua submitted that raising a herd of cattle with inserted human genes on their ancestral lands would cause a serious spiritual imbalance in their community. They submitted that the GM cattle proposal had the potential to cause long term psychological stress on their people that could, over time, manifest in physical illness and perhaps even death. The Maori term for such illness is 'mate Maori'. The ERMA laboured over this application for two years and, despite the warnings from the local Maori, the application was approved.

In September 2000 Nga Kaihautu joined a group of European submitters and the local tribe in an appeal against the decision to allow the GM cattle field test. This is a landmark case that is likely to be appealed to the highest court in Aotearoa New Zealand, as it will set a precedent about how seriously decision-makers take the impacts of science and technology on Maori, their culture and their relationship with the natural world.

The wider Maori community, similar to the non-elite Pakeha populations, was less aware of the multiple challenges stemming from genetic engineering and its various biotechnological offshoots. However, the New Zealand media's attention to genetic engineering burgeoned in 1999, and this led to greater public interest in and understanding of GE issues and concerns. Tipene-Matua (2000, pp8–9) notes how the media attention changed some Maori views by recounting an iwi member's comment that 'we would've looked at this a lot closer and may have changed position in light of all the public debate over genetic engineering lately'. Indeed, positions did change with time, and the most impressive was that of Ngai Tahu, the most powerful and prominent iwi in the South Island. In their case, they 'went from supporting the GM potato and sugarbeet applications to a blanket opposition on all proposals to introduce GMOs into the South Island' (Tipene-Matua, 2000, p10). As such, it appears that the previous hard work and effort undertaken by Maori in the international sphere is being furthered in the national sphere through Ngä Kaihautü Tikangä Taiao, national and regional hui, Runanga emphases and individual efforts. The extent to which such activities will succeed depends partly upon internal and external political and economic factors. However, it also depends upon the strategic organization of Maori. In this respect, a number of recommendations for future research and activities can be made.

FUTURE SCENARIOS: PROTECTING THE LIFE FORCE OF AND FOR FUTURE GENERATIONS

Maori and others throughout Aotearoa have consistently stated that they are not in a position to give informed responses to the great influx of genetically modified organisms in their communities. Research is needed to assess the relevance and impact of GMOs on Maori. Maori have stated that GMOs breach tikanga Maori and may result in a spiritual imbalance causing long term psychological and physical illness (mate Maori). Thus, Maori have sought comprehensive health risk assessments into the potential psychological and physical effects of redefining Maori relationships with the natural world. Two key questions needing urgent attention include: what is the nature and extent of the potential cultural and spiritual impacts on Maori?, and how do potential medical benefits to Maori or wider society influence these effects?

Explicit recognition of the relationship between Maori and valued flora and fauna is given in the HSNO Act, and the importance of indigenous species to both Maori and scientific research means that such recognition is essential. Maori need to debate among themselves about what, if any, are the 'no go zones' – ie what areas of research are likely to be off limits for Maori spiritually? As such, Maori must also have an opportunity to consider and debate such questions as: how can any limits be further defined or best determined; what is the distinction between Maori cultural (spiritual) concerns regarding indigenous and exotic species – ie how does the significance of indigenous species compare to a rabbit/cow or other exotic species? Or is it simply a 'no' to everything?

Biodiversity protection has flourished through the role of Maori women in protecting and enhancing Maori ecological knowledge. The stated benefits of biotechnology deny the value of Maori knowledge and Maori women's roles in sustaining biodiversity as keepers of this knowledge. To date, there has been no gender analysis of the impacts of biotechnology on Maori women. This information is imperative if a full analysis of the impacts on Maori is to be conducted. Key questions include: do current processes in the GMO debate allow for Maori women's voices to be heard?; how can we ensure that a gender analysis of the impacts of GMOs on Maori is conducted in a way that empowers and transfers knowledge to Maori women and communities while not marginalizing the views of Maori women in the process?

The need for research into the potential impacts on Maori, as well as the sociocultural perspectives of all communities in Aotearoa New Zealand, was recognized by the government who established a Royal Commission on Genetic Modification (RCGM) on 8 May 2000. Additionally, the government placed a 12-month moratorium on GMO releases. The Commission determined its own procedures and processes, and in the end attempted to acquire information through a variety of consultation methods. On 27 July 2001 the RCGM Commissioners (Sir Thomas Eichelbaum (Chair), Dr Jacqueline Allan, Dr Jean Fleming and the Rt Rev Richard Randerson) submitted their report to the Governor-General, and Sir Thomas Eichelbaum (RCGM website, 2001) summarized the Commission's processes by stating that:

> *The Commission has consulted widely with the New Zealand community. We have traversed the country twice with our public meeting and hui programmes. We have received over 10,000 written submissions, and held 13 weeks of hearings. Other processes have been a youth day and a public opinion survey. A point to be emphasized however is that we were not asked to conduct a referendum, but, as an independent body, to make an investigation, and report with our views.*

Ultimately, the success of the Commission, as well as any future research, depends upon the methodology and, in particular, the level and degree of genuine, honest participation engendered through the RCGM process. Although Sir Thomas Eichelbaum's summary indicates a relatively high degree of consultation, the processes used, and the extent to which all New Zealanders could genuinely participate, were questioned throughout this period. A number of papers and research projects currently under way will add further analysis of this recent contribution to Aotearoa New Zealand biopolitics, but our preliminary assessment suggests that the *participatory* aspect of the processes was considerably hampered by time constraints, the consultation methods employed and the structure of formalized participation.

Thus, we argue that any research undertaken in this area must adopt a participatory methodology as opposed to a purely empiricist methodology which emphasizes neutrality and objectivity and often involves researchers controlling the entire process. A participatory approach involves spending time and energy with those we propose to research to determine their research needs and priorities (Small, 1982, Smith, 1999). This involves building trust-based relationships, identifying what information is needed from the research participants, and deciding how information will be transmitted. The main issue is one of control. It is not appropriate to further exploit and expropriate knowledge and information from Maori. As Small (1982, p2) reminds us, it is not appropriate to follow the old research model whereby:

> *[t]he administrators want data for their policies: they gather information 'from those who do not make decisions in order to make decisions for them'. Academic researchers generally treat knowledge like a product which they produce for their survival: they 'mine' ideas and information and package it for consumption in books and journals.*

It is not appropriate to exploit peoples for any purpose, including the protection of ecological systems, in the name of ecological justice. This is because ecological and social justice are fundamentally linked, and attaining sustainability is impossible once one separates these forms of justice. Thus, adopting a participatory approach ensures benefits accrue to those Maori being researched as too often research is driven by the needs of regulators, decision-makers, industry, and academics opposed to the needs of communities and, in this case, tangata whenua. Unless a participatory approach research methodology is adopted when assessing the impacts of GMOs on Maori, mistrust and the sense of a continued oppressive colonial exercise among Maori will remain. The project of neocolonial biopolitics will be fuelled rather than ceased.

Maori ways of viewing the world are about integration and holism. Similar concepts are found in the ancient teachings of all indigenous cultures globally. This teaching provides alternatives to a dominant, modern Western paradigm which defines the natural world as something to be dominated and considers prosperity as having material wealth. For Maori, mana (or prestige) resided in those who could provide for others, all living things were viewed as part of the same family as people, and peace and balance were essential components to self-fulfilment. Only when decision-making processes recognize, respect and protect these unique and important cultural perspectives will Maori obtain social and environmental justice.

We have a long way to go.

REFERENCES

Barnett, A and Wintour, P (1999) 'Did UK science blunder cause mad cow disease?', *The Press,* Christchurch, 10 August, p5

Bührs, T and Bartlett, RV (1993) *Environmental Policy in New Zealand: The Politics of Clean and Green?,* Oxford University Press, Auckland

Daes, E-I (1993) *Study on the Protection of the Cultural and Intellectual Property of Indigenous Peoples.* Presented to the Sub-Commission on Prevention of Discrimination and Protection of Minorities and Working Group on Indigenous Populations, United Nations, New York, July

'Dairy industry boosts research' (1999c) *The Press,* Christchurch, 10 August, p5

'Fletcher targets tree genetics' (1999b) *The Press,* Christchurch, 8 April, p7

Garrett, A (1999) 'Brave new world', *New Zealand Herald,* Weekend Life, 5–6 September, pH3

'Genetically modified food: Food for thought', (1999), *The Economist, v*ol 351, no 8124, 19–25 June, pp21–23

Golden, F and Lemonick, MD (2000) 'The men who mapped the genome', *Time,* vol 156, no 1, pp18–30

'Govt to accept gene tests' (1998) *The Press*, Christchurch, 30 November, p9

Hazardous Substances and New Organisms Act 1996, GP Publications, Wellington

Hindmarsh, R, Lawrence, G and Norton, J (1998) 'Bio-utopia: The way forward?' in R Hindmarsh, G Lawrence and J Norton (eds), *Altered Genes: Reconstructing Nature The Debate,* Allen & Unwin, St Leonard's

'Human gene plan for dairy cattle' (1999a) *The Press*, Christchurch, 8 February, p6

Kawharu, IH (1989) *Waitangi: Maori and Pakeha perspectives of the Treaty of Waitangi*, Oxford University Press, Auckland

Kilvington, M and Rixecker, SS (1995) '(Re)locating governance systems in Aotearoa New Zealand: Global to local linkages in biodiversity policy', paper prepared for the Global Biodiversity Forum, Jakarta, Indonesia

Kimbrell, A (1997) (2nd edn) *The Human Body Shop: The Cloning, Engineering and Marketing of Life*, Regnery Publishing, Inc, Washington, DC

Larkin, P (ed) (1994) *Biotechnology: Genes at Work*, CSIRO, Australia

Legat, N (1999) 'GM food: Should we worry?', *North & South,* Issue 161, October, pp38–50

Matunga, H (2000) 'Decolonising planning: The Treaty of Waitangi, the environment and a dual planning tradition' in P A Memon and H Perkins (eds), *Environmental Planning & Management in New Zealand*, Dunmore Press, Palmerston North

Mead, ATP (1994) *Global Indigenous Strategies for Self-Determination, Report of the Maori Congress Indigenous Peoples Roundtable Meeting,* Taonga Pacific Limited, Wellington

Mead, ATP and Tomas, N (1995) 'The Convention on Biological Diversity: Are human genes biological resources?', *New Zealand Environmental Law Reporter,* July, pp127–32

Ministry for the Environment (1997) *The State of New Zealand's Environment,* GP Publications, Wellington

'The Monsanto files: Can we survive genetic engineering?' (1998), Special Issue, *The Ecologist,* vol 28, no 5, pp249–324

Morphet, S (1996) 'NGOs and the environment' in P Willetts (ed), *'The Conscience of the World': The Influence of Non-Governmental Organisations in the UN System*, Hurst & Company, London

Rabinow, P (1996) *Making PCR: A Story of Biotechnology,* University of Chicago Press, Chicago

Rifkin, J (1998) *The Biotech Century: Harnessing the Gene and Remaking the World,* Penguin Putnam Inc, New York

Royal Commission on Genetic Modification (2001) *Report of the Royal Commission on Genetic Modification* (4 vols), Royal Commission on Genetic Modification/Department of Internal Affairs, Wellington

Shiva, V (1997) *Biopiracy: The Plunder of Nature and Knowledge,* Between the Lines, Toronto

Shreeve, J (1999) 'Secrets of the gene', *National Geographic Magazine,* vol 196, no 4, pp42–75

Small, D (1982) 'Participatory research' (unpublished paper)

Smith, C and Reynolds, P (1999) 'Maori genes and genetics: What Maori should know about the new biotechnology', Iwi Law Centre, Whanganui, New Zealand

Smith, LT (1999) *Decolonizing Methodologies: Research and Indigenous Peoples*, University of Otago Press, Dunedin, Zed Books, London

Soroos, MS (1994) (2nd edn) 'From Stockholm to Rio: The evolution of global environmental governance' in N J Vig and M E Kraft (eds), *Environmental Policy in the 1990s,* Congressional Quarterly Press, Washington DC

Suzuki, D and Dressel, H (1999) *Naked Ape to Superspecies,* Allan & Unwin, St Leonard's

Te Puni Kokiri (Ministry of Maori Development) (1994) *Biodiversity and Maori: Te Ara O Te Ao Turoa,* Wellington

'The GE issue' (1999) *Soil & Health,* July, pp1–92

Tipene, B (1997) *Te Kopere, Protecting, Enhancing and Promoting the Intellectual, Cultural and Biological Heritage of Maori,* unpublished Master of Resource Studies thesis, Lincoln University, Aotearoa New Zealand

Tipene-Matua, B (2000) 'GM: A Maori perspective', in R Prebble (ed), *Designer Genes: A guide on GM for all New Zealanders,* Dark Horse Press, Wellington

United Nations Convention on Biological Diversity 1992, Rio de Janeiro, Brazil

Wilmer, F (1993) *The Indigenous Voice in World Politics: Since Time Immemorial,* Sage Publications, Newbury Park

World Commission on Environment and Development (WCED) (1987) *Our Common Future,* Oxford University Press, Oxford

Chapter 13

Political Economy of Petroleum Resources Development, Environmental Injustice and Selective Victimization: A Case Study of the Niger Delta Region of Nigeria

Tunde Agbola and Moruf Alabi

INTRODUCTION

The exploitation of natural resources (eg petroleum) by a capitalist or dependent capitalist state involves two major processes: surplus extraction and surplus transfer. In a federated state surplus transfer implies the mutual movement of resources among the component regions, for example, between the oil producing and non-oil producing regions. Advocates of federalism argue that this will enhance efficient allocation of scarce resources and at the same time fulfil some of the major fundamental objectives of federalism: spatial equity and social justice.

Unlike surplus transfer, surplus extraction might entail environmental degradation. In the federated state of Nigeria, the state not only controls the exploitation of petroleum resources, it also allocates and utilizes proceeds from the resources. Efforts by the Nigerian government to base their allocation formula on population and equality of states rather than on the principle of derivative, implies that the federal government has violated the basis of equity, fair play and social justice on which modern federalism is based. Similarly, the neglect of the Niger Delta region's environment by the state has resulted in environmental injustice. Thus, petroleum extraction with its attendant environmental problems has made local people believe that the state has failed to protect their lives and property from environmental pollution and that pollution costs are being unfairly imposed on them.

Realizing their rights, the people of the Niger Delta have created a series of resistance movements (see Peña, Chapter 7, and Wright, Chapter 6) in order to protect their environment from further degradation. There is a clear link between human rights abuses and environmental degradation, which can be explained through the process of selective victimization. With selective victimization, the Niger Delta region is losing critical resources as well as a healthy environment, thereby exposing residents to hazardous environmental conditions, while the non-oil producing regions which receive the lion's share of the oil revenue are free to live in a healthy setting. Surprisingly, development efforts in Nigeria, including those involving mineral extraction, have not always been as unsustainable as that being visited upon the Niger Delta today. In the past, the interests of all stakeholders in a community were often taken into consideration. The primary aim of this chapter is twofold. First, to present a time-dimensional overview of sustainable development efforts in Nigeria. Second, to put contemporary developments in context by examining the relationship between the political economy of petroleum resource development and the resulting environmental injustice and selective victimization using the Niger Delta region of Nigeria as a case study.

SUSTAINABLE DEVELOPMENT EFFORTS IN NIGERIA: A TIME-DIMENSIONAL OVERVIEW

Various methods have been used to introduce and sustain development efforts in Nigeria. Prior to the colonial period in Nigeria, indigenous knowledge, technology, and practices that resulted from cultural taboos (in the form of societal norms), adages, standards, behavioural rules, traditional ways of life and sanctions were used to manage and limit the impact of human development on the environment. In the southern part of the country for example, traditional religious sanctions facilitated the emergence of 'sacred groves'. These groves were established natural areas protected from human use and are valuable assets for biodiversity.

After the annexation of Lagos in 1863 and the subsequent introduction of colonial rule, emphasis was placed on environmental sanitation rather than environmental protection and virtually all public health and physical planning laws, policies and practices were aimed at this. Consequently, slum clearance, sanitary inspection and a wide range of sanitation related penalties were used as means of social control. However, the formal institutional approach to environmental protection can be said to have commenced during the colonial period with the establishment of the 1901 Forestry Ordinance and the 1917 Colonial Ordinance through which protected areas were established.

From 1960, when Nigeria was granted independence from Britain, until the middle of the 1970s, environmental issues were treated in an exploitative way, similar to that pertaining in colonial times, even by the local elites. Both the First National Development Plan (1962–68), and Second National Development Plan (1970–74), failed to give adequate priority to sustainable development. Thus, environmental issues essentially emerged as welfare and social service issues.

Even where obvious problems were identified, such as deficiencies in the provision of water, the problem of soil erosion, the silting of waterways, the crisis of industrial and domestic waste disposal and overgrazing and the development of forest resources, they were often couched in purely welfarist, public health or economic development terms (NEST, 1991, p283).

Subsequent development plans, 1975–80 and 1980–85, paid superficial attention to issues of sustainable development, especially environmental sustainability. It took the illegal dumping of toxic waste by the 'leper ship' the *Karin B* in Koko in Edo/Delta state for the Federal Environmental Protection Agency (FEPA) to be established in December 1988. The Agency is expected to coordinate and lead the implementation of the new National Policy on Environment. FEPA has also established new noise pollution standards. This is in addition to the regulations in 'Effluent Limitation' and 'Pollution Abatement in Industries and Facilities Generating Wastes'. Recently, efforts to commercialize solid waste management have been initiated by FEPA. State environmental protection agencies were also established in the states of the federation.

In spite of the various national laws, state environmental edicts as well as local government by-laws were also put in place with a view to achieving sustainable development. In Lagos state, for instance, such edicts include Sand, Laterite and Gravel Spillage Decree (No4, 1984), Road traffic (removal) Abandoned Vehicle Edict of (No7, 1984) and the Special Offences Court Edict (No20, 1984). Others include Town and Country Planning Edict (No1, 1986), Environmental Pollution Control Edict (No9, 1989) and Environmental Sanitation Agency Edict (No3, 1992).

In an attempt to promote the achievement of sustainable development, the Nigerian state formulated the National Policy on the Environment in 1989. In an attempt to achieve the broad aim of sustainable development, the specific goals of the National Policy on the Environment as spelt out by the federal government are to:

- secure for all Nigerians high quality environment adequate for their health and well-being;
- conserve and use the environment and natural resources for the benefits of present and future generations;
- restore, maintain and enhance the ecosystems and ecological processes essential for the functioning of the biosphere, to preserve biological diversity and the principle of optimum sustainable yield in the use of natural resources and ecosystems;
- raise public awareness and promote understanding of essential linkages between environment and development and to encourage individual and community participation in environmental improvement efforts; and
- cooperate in good faith with other countries and international organizations and agencies to achieve optimal use of transboundary environmental pollution.

Other environmental policy and guidelines put in place to help achieve sustainable development include the various policy guidelines on:

- energy in Nigeria, which was issued in 1987 to encourage environmentally conscious exploitation and utilization of energy resources;
- wastes, such as the promulgation of the Harmful Waste Decree of 1988; and
- resources, through the preparation of a Natural Resources Conservation Policy which was expected to serve as a measure to safeguard the citizenry from wanton abuse of the environment and industrial activities that are injurious to human habitation and the entire ecosystem.

Long term sustainable development cannot occur in a situation of deteriorating environmental circumstances. Unplanned and unmanaged urban growth can lead to irreversible destruction of the natural resources environment (Bloxom, 1996, p1). It is in line with this idea that the United Nations Human Settlements Programme (UN-Habitat) formed the Sustainable Cities Programme (SCP) as the operational arm of the World Bank/UN-Habitat/UNDP/Urban Management Programme in 1990. The programme now takes place in cities in 12 countries, including Ibadan in Nigeria. The Sustainable Ibadan Project (SIP) as rightly observed by Bloxom (1996, p2) aims to provide communities with an improved environmental planning and management capacity to ensure that the development of Ibadan City meets the needs of the present inhabitants without compromising the prospects of future generations.

The Rio Earth Summit of 1992 also made Nigeria put appropriate institutions, policies and laws in place. In 1992, an Urban Development Bank was established to focus on urban infrastructure and public utilities; National Urban Development Policy for Nigeria was formulated as a policy guideline for urban development and management; and the Nigerian Urban and Regional Planning Decree No88 of 1992 was adopted to attempt to regulate and guide spatial planning at all levels of government. Other policies and guidelines formulated include the National Guidelines and Standards for Environmental Pollution Control in Nigeria in March 1992 and the Promulgation of the Environmental Impact Assessment (EIA) Decree of December 1992.

The EIA Decree was designed as an assurance tool against environmental risk associated with development. EIA in Nigeria is expected to be utilized for activities:

- which may result in specific environmental impacts requiring a more limited and focused environmental impact assessment (eg large-scale agriculture, mariculture, irrigation and drainage projects; dams, reservoirs and other large river basin development projects; major land clearance, reclamation, development and resettlement projects; new transportation networks; oil, gas or water pipelines); and
- which may result in specific environmental impacts requiring a more limited and focused environmental impact assessment (eg small-scale agricultural and other industries' mini-hydropower and other renewable energy projects, etc).

On the other hand, EIA in Nigeria does not apply to programmes:

- with no significant environmental impacts, eg education, public health, family planning, and other social programmes when they do not involve major construction projects; and
- for environmental restoration, protection and improvement or for emergency situations requiring an immediate response (eg natural disaster, industrial accidents or fires).

In order to protect forest and biodiversity, the main policy goal of the federal government of Nigeria is to require sustainable use of forest resources and preserve the many benefits they confer for soil as well as for wildlife habitats and recreation. Nigeria is now committed to extensive reforestation and afforestation programmes with increased support for NGO and community-based tree planting programmes, coordination and management of contiguous forests with activities in forest reserves, tightening controls on fuelwood extraction from reserves and developing more efficient wood-stores and alternative energy sources. The programmes involve afforestation, agro-forestry, integrated natural resources management and sustainable livelihood alternatives for the rural poor in marginal zones.

In terms of biodiversity, the government of Nigeria now gives priority to the expansion of the present network of national parks and reserves to protect all major ecosystems, the strengthening of programmes for assessing and protecting fauna and flora, especially endangered species, and the increase of support for the local and export marketing of sustainably harvested forest and wildlife products.

The proliferation of non-governmental organizations (NGOs) and the activities of pressure groups have contributed their own impetus towards sustainable development efforts in Nigeria in general, and in the Niger Delta in particular. It is these NGOs and pressure groups that are deeply concerned with the vulnerable Niger Delta environment. Some of their activities, such as 'public educational campaigning ... are increasing including the on-going efforts at "greening" the school curriculum by introducing environmental education' (Faniran, 1997, p15). However, coastal zone management plans as well as guidelines for land use and development activities in coastal areas (eg fishing, dredging, construction, oil spills, etc) have not been given adequate consideration by the Nigerian state.

Since 1992 FEPA's role and authority have been well strengthened. For example, the Urban Compliance Unit has been set up within the Inspectorate and Compliance Monitoring Department with a mandate to oversee issues in specific areas including solid waste management, urban sanitation, public drains and sewers, vehicular emission and noise pollution. In addition, the Unit also monitors waste disposal, develops effective management strategies and enforces compliance with FEPA standards and guidelines.

Unfortunately, and despite the haphazard collection of laws regulating the environment, the Nigerian state, through its connivance with the operating companies, has blatantly violated the environmental laws and extracted its natural resources in the most unsustainable manner. This is best exemplified in the Niger Delta and is the focus of subsequent sections of this chapter.

PETROLEUM RESOURCES DEVELOPMENT AND THE ECONOMY OF NIGERIA

Nigeria could be described as an oil-rich federal country with multiethnic composition consisting of about 374 ethnic groups (Otite, 1990). Today, the country is one of the world's largest producers and exporters of crude oil. Ibeanu (2000, p1) has rightly pointed out that Nigeria is the fifth largest producer within the Organization of the Petroleum Exporting Countries (OPEC), and it currently exports about 1 million barrels of oil per day. Shell Nigeria (the Shell Petroleum Developing Company, a subsidiary of Royal Dutch/Shell) produces about 50 per cent of total oil exports.

Petroleum resources development in Nigeria dates back to the 1930s. According to Akintola (1978, p94), exploration for petroleum began as for back as 1937, but actual production did not start until 1958 when 252,000 tonnes were produced. The production as well as export of crude oil in 1958 is a function of the discovery of oil in commercial quantity at Olobiri by Shell D'Arcy, an Anglo-Dutch oil company, in 1956. In addition to Shell, other multinational oil companies have joined the race for oil resources development in Nigeria, albeit, in conjunction with the Federal Government through the Nigeria National Petroleum Development Corporation (NNPC). Prominent among such companies are Mobil, Agip and Safrap (now Elf Aquitaine), Texaco, Amoseas (now Texaco/Chevron) and Total. Thus, the development of oil resources in Nigeria is a joint responsibility of the state, multinational corporations, multilateral organizations and local elites. Oil resource development, according to these actors, means economic growth. Implicit in their actions is the justification of the prevalent neo-liberal belief that economic growth is absolutely good and that its benefits ultimately trickle down to everyone (Cable and Cable, 1995, p107).

As a result of the production and exportation of oil in commercial quantities, Nigeria has an oil-dependent economy. Initially, the contribution of oil ('black gold') to the Nigerian economy remained insignificant. However, with the Arab–Israeli (Yom Kippur) war of 1973–74, oil became and has since remained the major contributor of foreign exchange to the Nigeria economy. American support for Israel during the Yom Kippur War made the Arabs place an embargo on the exportation of crude oil to the US and Europe. This brought about a sudden increase in the price of crude oil and Nigeria was able to increase the export of crude oil to the sanctioned countries – US and Europe. This brought about an increase in the revenue generated by the oil sector. Okunronmu (1993) reported that federally collected revenue increased from N634.00 million in 1970 to N5,514.6 million in 1975. He argues further that as a percentage of the total federally collected revenue, it increased from 26.3 per cent in 1970 to 77.5 per cent in 1975 and maintained its relative contribution until 1986 when it contributed 74.7 per cent. Writing on oil as a source of the Nigerian government's revenue, Ikporukpo (1996, p162) observed that whereas no revenue was derived from this source in 1957, its contribution of only 0.1 per cent in 1958–59 increased to 87 per cent in 1975–76 and has subsequently been consistently between 66 per cent and 84 per cent. The contribution of

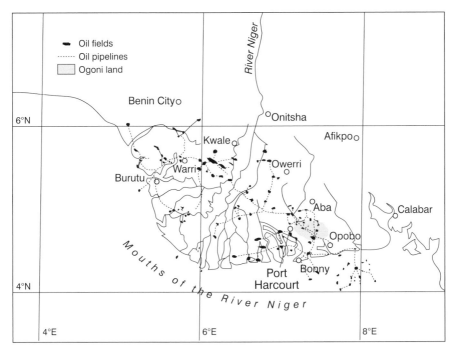

Figure 13.1 *Map of the Niger Delta region of Nigeria*

crude oil to the Nigerian economy has made the UNDP (1997, p11) ascertain that Nigeria remains a mono-cultural economy with oil accounting for 96 per cent of external earnings. All of this oil comes from the Niger Delta region.

The Niger Delta region: An overview

While the political capital of Nigeria is now Abuja, its economic capital still remains in Lagos. More importantly, the Niger Delta represents the 'oil capital' of the country. Thus, 'the region of the Niger Delta, by the Atlantic Coast in the South, is the source of petroleum' (Ikporukpo 1996, p162). The region has a total land mass of about 20,000 km^2 of land, rivers and swamps located between latitudes 4° 30'N–6°20'N and longitudes 5°E–8°30'E. It is bounded solely by the Atlantic Ocean in the south and west, Edo, Anambra and Ebonyi states in the north and Cross-River state and a small international border strip in the east (see Figure 13.1). The Niger Delta region is one of the largest wetlands in the world.

The delta is a vast flood plain built up by the accumulation of sedimentary deposits washed down the Niger and Benue rivers (Moffat and Linden, 1995, p527). It is made up of four ecological zones: coastal barrier islands, mangroves, freshwater swamp forest as well as lowland rainforest. According to Ikporukpo (1996, p162), there are two broad areas of the delta with different drainage characteristics: the outer delta, characterized by its mangrove vegetation, which is more or less a massive swamp dotted with areas of dry land; and the inner delta, made up predominantly of swamp rainforest, which is not as wet.

Of all the 36 states of the federation of Nigeria, only six fall within the delta region. These are Rivers, Delta, Edo, Akwa-Ibon, Cross-Rivers and Imo states. Rivers, Delta, Edo and Akwa-Ibon, account for virtually all of the oil production, with Rivers and Delta state producing about 75 per cent of Nigeria petroleum. This equals over 50 per cent of national government revenue (Moffat and Linden, 1995, p527). The Niger Delta is the home of minorities such as Ijaws, Urhobos, Itsekiris, Efiks, Ibiobios, Ogonis and Ikwerres. These ethnic minorities have very little or no access to national political power. They have been marginalized since independence. Despite its vast oil reserves, the region remains poor (Moffat and Linden 1995, p527). According to Moffat and Linden (1995), GNP per capita in the Niger Delta is below the national average of US$280. The Niger Delta region houses vital industrial installations, including two petrochemical industries, three refineries, a fertilizer industry, a steel complex and major power stations. Other installations include a new liquefied natural gas plant as well as textile and paint manufacturing companies. In spite of these industries, farming, fishing, forest resource collection and agro-cottage activities still constitute the means of livelihood of around 7 million Nigerians who live here. Both staple foods (eg cassava and yam) and cash crops (eg rubber and oil palm) are commonly grown by the inhabitants of the region. These activities, according to Odu (1977), Ikporukpo (1983) and Stanley (1990), have been adversely affected by the widespread incidence of pollution and other forms of environmental damage resulting from petroleum resources development. To corroborate this, Ibeanu (2000, p1) stated that oil spills destroy the freshwater ecosystems, foul farmland, kill animals and endanger human life. Consequently, the actions of the main actors in the petroleum development process (that is, the state, multinational oil companies, multilateral organization and local elites) have multiple effects on the Niger Delta environment.

The destruction of such an environment will likely spell doom for its inhabitants. It is therefore appropriate to examine the political economy of petroleum resources development in Nigeria with particular reference to social and environmental injustice and the reaction of the state to people's resistance to these injustices.

THE POLITICAL ECONOMY OF PETROLEUM RESOURCES DEVELOPMENT IN NIGERIA

Political economists explain social and environmental injustice as the outcome primarily of political and economic relationships and process in the wider society. One vital factor in the recent attempts to develop a political economy approach is the role of the state. The conception of the autonomy school of thought revolves around the idea of state autonomy. This refers to the capacity of the state to act independently of social forces and thus that an arrangement of social forces does not uniquely determine particular state actions (Caporaso and Levine, 1992, p182). This interpretation was rejected, however. From the point of view of Phillips and Williams (1984, p22), the state is either instrumentalist or semi-autonomous within the structural constraints of the capitalist mode of production. Since all ethnic groups in a federation do not have equal access to power, they could not therefore be expected to be protected equally. Thus the power of the state is seen to be limited by the structure of capitalism and this basically supports the interests of the dominant groups.

The opponents of the autonomy school of thought saw the state as an exploitative mediator. Scholars of a Marxist disposition (eg Poulantzas, 1973, Jessop, 1982 and Roemer, 1985) view the state as a tool of group interest. The development of petroleum resources in a developing capitalist economy like Nigeria, for example, is to facilitate the exploitation of the ethnic minorities by the major ethnic groups: the Hausa-Fulani, Igbo and Yoruba. Thus, the state is the guardian which ensures that the inequalities needed for the system of exploitation are maintained (Ikporukpo, 1996, p165). The state, Rothstein (1992) argues, may not have the capacity to act independently.

The notion of access to political capital is a particular entry point for causal explanations of social and environmental injustice in Nigeria. Generally, capital is understood in a broad sense as any stock which is capable of being stored, accumulated, exchanged or depleted and which can be put to work to generate a flow of income or other benefits (Booth et al, 1998). Thus, political capital refers to stocks or reserve of power or private asset, describing the position of individuals or groups of social actors in relation to political institutions (Booth et al, 1998, p69). Unequal access to political capital, therefore, means unequal access to natural capital. Natural capital is defined as the stock of environmentally provided assets (such as soil, atmosphere, forests, water and wetlands) which provide a flow of useful goods or services which may be renewable or non renewable, marketed and non-marketed (Goodland et al, 1998, p73). The development of resources, that is, non-renewable natural capital, has been the sole responsibility of the dependent capitalist state of Nigeria whose dominant class is chosen among the major ethnic groups in the country.

The governance of the Nigerian state has been in the hand of the major ethnic groups in the country (especially the Hausa-Fulani) since independence in 1960. The Hausa-Fulani group in the north of the country has always outmanoeuvred the other two major groups (Yoruba and Igbo) and has always returned the head of government. The northern elites that have controlled the federal government for not less than 35 years deemed it fit to intervene in the development of oil resources with a view to protecting the interests of the north. According to Ikejiafor (1999, p179), there is the fear among this group that an economy that is left totally to the vagaries of market forces would not adequately protect the geo-ethnic interest of the north, and may result in greater domination of the nation's economy by the south. To forestall such an eventuality, the federal government sees it as necessary to maintain a controlling influence in all facets of the nation's economic life. Economic gain, which is associated with the northern elites (and some of their southern collaborators), is another reason for their vested interests in petroleum resources development.

With the majority ethnic groups in power in Nigeria, the power of the state has been used to change the ethnic distribution of rights. Such redistributions have been manifested in the reallocation of the right to petroleum resources and its accompanying revenue. For example, before petroleum resources became the main source of wealth in Nigeria, all the political actors favoured a pro-derivation based revenue allocation principle. However, with oil coming into the economic limelight, the major political actors, coming mainly from the non-oil producing states, decided to change the initial arrangement. Olopoenia (1998, p51) argues that this was possible because a coalition with a majority of the less politically powerful groups offered them an implicit bribe from the oil revenues. According to him, this policy of inter-ethnic corruption of justice in a federal arrangement has had negative developmental consequences for the whole economy. Politics then becomes a zero-sum game, in which the ruling class's (ethnic majority) gains are the residual class's (ethnic minorities) losses. When the politics of resource allocation is played as a zero-sum game, every participant is a loser in the long run. (Olopoenia, 1998, p5). This is the antithesis of environmental justice and sustainable development.

The ostensible aim of petroleum resource development was economic sustainability with a view to achieving economic growth. 'Growth' has always been the aim, and 'economic growth' has been almost synonymous with 'development' (Panos Briefing, 2000, p2). This is peculiar to dependent capitalist economies where environmental resources are seen as not produced or owned by anyone, but common to all. Thus, the free market, which characterized dependent capitalist economies, does not attach any price to environmental degradation. This has contributed to the neglect of the environment where resources are extracted. With its implicit lack of care for its extracting environment, Nigeria can therefore be categorized as a country with an unsustainable economy, as it is liquidating its own natural capital, rather than living off its harvest. Goodland (1994) argues that countries which are truly sustaining themselves rather than liquidating their resources, will be more peaceful than countries with unsustainable economies.

With the allocation of oil revenue based on the principles of equality and size of the state instead of the derivative principle, revenue has not been effectively utilized. It is difficult to achieve a social optimum in this situation. In order to do this, social justice must be taken into account (Willis, 1980, p29). Thus, an allocation of resources is equitable if no individual prefers some other individual's bundle of products to his own (Varian, 1974). A social optimum position can, therefore, be achieved if equity, or social justice, is given adequate consideration in the process of oil revenue allocation. The achievement of the social optimum position in the allocation of oil revenue may be described as an optimum solution to environmental degradation and as a means of achieving environmental justice. Such a unique solution, Willis (1980, p29) argues, may be defined as fair in that it is an allocation that is both equitable and Pareto-efficient.

Petroleum resources development and environmental injustice

According to the US Environmental Protection Agency (EPA) (1995a and 1995b) and Wilkinson (1998, p273), environmental justice is defined as 'fair treatment' such that no one group of people bears a disproportionate share of impacts. Initially, environmental justice called for equity in the distribution of environmental burdens, but more recently (see Faber and McCarthy, Chapter 2, Conclusion) the call has been for productive justice: for clean production so that no one has to suffer. The development of petroleum resources by the Nigerian state and its allied multinational corporations, multilateral organizations and local elites has brought about clear examples of environmental injustice where the vulnerable ethnic minorities of the Niger Delta region bear the heaviest burdens. This is in addition to the burden of economic recession and poverty. Resource development means, for these people, both the loss of critical resources and the hardship of living in degraded settings (Johnston, 1994, p219).

The state's emphasis on capital accumulation with a view to achieving economic growth has made it a largely ineffective agent of social control over environmental degradation caused by oil spills and other related consequences of petroleum resources development. In the course of developing petroleum resources, the natural environment is a commodity. Evidence that the Niger Delta region is on an economic path that is environmentally unsustainable includes agricultural land degradation, flooding and erosion, sea-level rises and habitat degradation. Others include loss of biodiversity, environmental pollution as well as both deforestation and forest degradation. The analogy one can infer from this is that the Niger Delta region, which feeds virtually the whole country, pays the price of petroleum resources development in the form of environmental injustice. Environmental injustice which characterizes the Niger Delta region can also be viewed in terms of an unclean and unsafe/unhealthy environment which causes some people to bear a disproportionate environmental risk, or at least to shoulder more than a socially acceptable minimum level of environmental risk. Hence, environmental injustice implies that the political economy of petroleum resource development has made the people of the Niger Delta region feel that the state has failed to protect their lives and property from environmental hazards and that environmental costs/risks are being unfairly imposed on them. Communities are simply not seeing the benefits from the oil extracted from their territory (Ibeanu, 2000, p8). Other sources of environmental injustice include the building of roads and facilities to service the oil facilities, gas flaring and other waste emissions and oil spills.

Three major causes of oil spills can be identified and these are: equipment failure, human error and sabotage spills. The works of Ramani-Abah (1997), Idonoboye-Obu (1995) and Aprioku (1999) indicate that from the beginning of the first oil spill incident on 17 January 1980 at Funiwa, to the end of 1998, some N12,057.5 billion in damage has been inflicted on the economy of which around N2773.2 billion relate to damage inflicted on the country's infrastructure. Ogbe (1997) estimates that oil spill incidents and volumes occurring between 1980 and 1989 at 679 million barrels from over 2000 incidents (see Table 13.1). As if to corroborate this, the Department of Petroleum Resources (nd) cited in the Niger Delta Environmental Survey (1997) stated that between 1976 and 1996, 2,367,470 barrels of spilled oil were recorded from 4647 spills. For further information, see Table 13.1.

In addition to the oil spills, gas flaring is another major source of environmental injustice in the Niger Delta region. This is because about 90 per cent of the associated gas product of crude oil is flared/wasted, which is at

Table 13.1 *Number of spills and quantity of oil spilled between 1976 and 1996*

S/No	Year	Number of spills	Quantity spilled (barrels)	Quantity recovered (barrels)	Net volume lost to the environment (barrels)
1.	1976	128	26,157.00	7135.00	19,021.50
2.	1977	104	32,879.25	1703.01	31,176,75
3.	1978	154	489,294.75	391,445.00	97,849.75
4.	1979	157	094,117.13	63,481.20	630,635.93
5.	1980	241	600,511.02	42,416.83	558,094.19
6.	1981	238	42,722.50	5470.20	37,252.30
7.	1982	257	42,841.00	2171.40	40,669.60
8.	1983	173	48,351.30	6355.90	41,995.40
9.	1984	151	40,209.00	1644.80	38,564.20
10.	1985	187	11,876.60	1719.30	10,157.30
11.	1986	155	12,905.00	552.00	12,358.00
12.	1987	129	31,866.00	25,757.00	25,757.00
13.	1988	208	9,172.00	1955.00	7,207.00
14.	1989	228	5,956.00	2153.00	3,803.00
15.	1990	166	14,150.35	2785.96	12,057.80
16.	1991	258	108,367.01	2785.96	105,912.05
17.	1992	378	51,187.90	1476.70	49,711.20
18.	1993	453	8,105.32	2937.08	6,632.11
19.	1994	495	35,123.71	2335.93	32,787.78
20.	1995	417	63,677.17	3110.02	60,568.15
21.	1996	158	39,903.667	1183.807	38,719.860
Total		4647	2,369,470.04	549,060.38	1,820,410.50

Source: Niger Delta Environmental Survey (1997), Final Report, Phase1

present the treatment/control process adopted in Nigerian refineries for gaseous emissions. Oily waste as well as water effluents also contribute their own quota, however small, to environmental injustice in the Niger Delta region.

It is in the environmental impacts of oil resources development (which are numerous and interwoven) that injustice manifests itself significantly. Pollution that arises from oil spills destroys marine life and crops, contaminates water formations and streams and renders vast hectares of arable farmland unusable. Problems associated with gas flaring and the high pressure pipelines that criss-cross farmlands include heat radiation and thermal conduction into the environment, deforestation and the destruction of wildlife, production of toxic gases during combustion and generation and dispersal of particulates and incombustible materials like soot into the atmosphere. Others include noise pollution, carbon-dioxide emissions and environmental heat-related impacts as well as the vulnerability of farmland to acid rain. With petroleum resources development being carried out in such an unsustainable manner, the Niger Delta environment has been changed into what may be described as 'wasted homeland'. In a twist of fate, it would seem that what was called 'the white man's grave' in the colonial era, has been turned to the indigenous ethnic minorities' grave of the post-colonial era.

Sometimes, the impact of oil exploitation and associated oil spills may not be limited to the Niger Delta region alone. One of the most important oil spill-incidents is the Mobil-Idoho (1998) spill. This spill spread as far as the Lagos coastlands from Akwam-Ibom where the incident occurred. However, the impact was most severe in the Niger Delta region communities of Olobiri, Opoama-Brass, Fishtown, Agge, Odionia, and Sagana as well as Ezebiri and Koluama, all in the Ijaw-dominated Bbayelsa state. In Delta state, Odimoti and Oguladia were badly affected while in Rivers state, Abalama and Kula were the main victims.

Environmental injustice, human right abuse and selective victimization

Environmental injustice, which emanates from disproportionate environmental degradation, may also be linked directly to the unsustainable mode of petroleum resources extraction in Nigeria. This not only violates the principle of ecological limits and sustainable development, but also the basic principles on which modern federalism is based, that is equity, fair play and social justice. In addition, it violates fundamental human rights since 'a safe, clean environment is a fundamental human right, a moral law' (Hartley, 1995, p285).

The right to health, a decent existence, work, occupational safety and health, an adequate standard of living, freedom from hunger, an adequate and wholesome diet, decent housing, the right to education, culture, equality and non-discrimination, dignity, harmonious development of one's personality, the right to security of person and of family, the right to peace and the right to development are all rights established by existing United Nations covenants (Johnston, 1995, p113). All these depend upon a healthy environment and they represent the ideals that governments strive for in providing for their citizens since they are the basic life requirements that all humans are entitled to.

However, human rights abuses occur as a result of multiple forces resulting from a state's policies and actions. Human rights violations occur in Nigeria as a result of the efforts of the state to gain control of the petroleum resources of the politically and/or geographically marginalized people of the peripheral region of the Niger Delta. Peripheral regions can be a matter of actual geographic distance, or 'cultural' distance, that is, areas populated by less powerful groups (see Blowers, Chapter 3). The environmental crises which have followed unsustainable petroleum resources development in Nigeria are not experienced equally by all the ethnic groups in the country. In spite of all the international and national structures establishing inalienable rights for all Nigerians, the people of the Niger Delta region experience greater environmental hazards and more degraded environments than the inhabitants of the non-oil-producing regions. It is this sociocultural context of selective exposure to hazardous and degraded environmental settings that constitutes a form of human rights abuse. A clear link exists between environmental degradation and injustice, and human rights abuses. According to Johnston (1995, p11), human rights abuse and environmental degradation are linked via the process of selective victimization where pre-existing social conditions result in the loss of critical resources and healthy environmental conditions while others are free to live, recreate, procreate and die in a healthy setting.

Consequently, the existing mode of oil exploitation and extraction favoured by the Nigerian state in the Niger Delta region is responsible for the loss of critical petroleum resources and consequent environmental injustice(s). This induced injustice, which seems to be socially and politically sanctioned against the ethnic minorities of the Niger Delta region, is a glaring example of selective victimization against which the affected people have risen in resistance, through grass-roots opposition.

THE RISE OF GRASS-ROOTS ENVIRONMENTAL MOVEMENTS, THEIR PROTEST MODES AND THE RESPONSE OF THE NIGERIAN STATE

The inability of existing environmental laws or regulations to effectively address the question of environmental injustice has facilitated the rapid development of grass-roots environmental justice movements. Environmental injustice is not peculiar to the developed industrialized Northern countries like the US, where the movement first took shape. It is also a major problem facing the underdeveloped countries of the South. This is because recently the application of more scientific and technological advancements to the development of petroleum resources has increased the magnitude of our impact on the environment. The inadequacy of common law, specifically law of torts, such as nuisance, trespass, strict liability, etc, to cope with problems of assault on the environment, is putting an increasing demand on modern society to evolve statutory environmental laws to cope with the enormous environmental problems of modern technological industrial society (Okorodudu-Fubara, 1998, p16). Environmental law is a system of rules of social control which aim at creating order out of disorder in order to create a sustainable, livable environment. These laws mainly involve defining the relationship between human beings and the world they inhabit and the limits to which we may impact upon the environment through activities such as oil exploitation and extraction.

However, even with the environmental laws and environmental protection agencies in Nigeria today, the issue of environmental injustice still persists in the oil-producing region of the Niger Delta. The popular perception of environmental injustice by local people has thus generated a new form of social control: community-based grass-roots environmental organizations, who now challenge the status quo. The organizations act as informal control mechanisms when the formal mechanism, that is, environmental regulation, has patently failed (Cable and Cable, 1995, p114). The indigenous ethnic minorities now challenge the maldistribution of oil revenue, the distribution of the costs of local environmental degradation and they also demand better protection of their lives as well as their property not only for this present generation but also for future generations. Their focus is therefore justice *and* sustainability.

Thus it is the oil exploration and extraction-induced environmental injustice and selective victimization of the inhabitants of the Niger Delta region which violates fundamental human rights (including 'environmental rights'). This has caused the emergence of a grass-roots environmental justice movement made up of organizations such as the Association of Minority Oil States (AMOS), Movement for the Survival of Ogoni People (MOSOP) and the Ijaw National Council and Movement for the Survival of Ijaw Ethnic Nationality (MOSIEN). The major achievement of these grass-roots environmental justice and rights organizations is that they have been able to tell the whole world that environmental injustice and selective victimization prevail in the Niger Delta region. However, intercommunity rivalry and internal wrangles have also hindered the rate of progress of these organizations. In MOSOP for instance, internal division led on 21 May 1994 to the murder of four prominent Ogonis, who were once active members of the organization, by a crowd of Ogoni youths. The people were Chief Edward Kobani, a one-time Commissioner in the Rivers state government; Chief Albert Badey, a former Secretary to the Rivers state government, and the two Orages: Chief Samuel and Theophilus Orage. Ken-Saro Wiwa (a prominent environmental crusader and a writer who later became the head of MOSOP) and another 12 Ogoni were arrested in connection with the murder of the four Ogonis. In the murder trial, which can best be described as a kangaroo court set up by the then government, Ken-Saro Wiwa and eight other Ogonis ('Ogoni Nine') were sentenced to death and later executed by hanging at the Port Harcourt prison. This took place under the tyrannical rule of the late General Sani Abacha. There were indeed widespread but unconfirmed reports suggesting that he asked for the execution to be filmed and brought to him as a confirmation that the convicts were actually executed (Ibeanu, no date, p20).

In their bid to check the maldistribution of revenue from oil sales, and further degradation of the Niger Delta environment, the various environmental justice organizations, as well as individuals, have engaged in different forms of conflict and protests. As is to be expected when people feel that their families' health is at stake, they sometimes take actions that violate community standards of acceptable behaviour. Notably they engaged in civil disobedience and they occasionally even break the law (Cable and Cable, 1995, p108). Different types of conflict/protests have been used by these various grass-roots organizations and ethnic minorities against perceived injustice and selective victimization. Ikporukpo (1996, p196) identified four such protest modes: advocacy-based protest, protest marches and civil disobedience, sabotage, and armed struggles. The most widespread form of protest are advocacy-based protests through which the environmental justice movements in the Niger Delta have been able to draw the attention of the federal government as well as other well-meaning individuals who believed in social and environmental justice. For example, both the AMOS and MOSOP 'pursue a policy of non-violence through the publications of facts and figures on the perceived injustice' (Ikporukpo, 1996, p172).

In one of such publications, titled, "Ogoni's Bill of Rights", it was observed that:

> *The search for oil has caused severe land and food shortages in Ogoni, one of the most densely populated areas of Africa … that Ogoni people lack education, health and other social facilities. That it is intolerable that one of the richest areas of Nigeria should wallow in abject poverty and destitution.*
> (MOSOP, 1992, p10–11)

With reference to protest marches and civil disobedience, there is hardly an oil-producing community in the Niger Delta that has not at one time or another engaged in these forms of protests. A common characteristic of such protest is that they start as protest marches and end up as a blockade of oil-production facilities and locations (Ikporukpo, 1996, p173). However, in terms of repercussions and violence, the protests at Umuechem (Rivers state), Eket (Akwa-Ibom state) and Bonny (Rivers state) were the most striking. In the Umuechem incident of 1990, the police virtually declared war on the village because of alleged violence by the villagers on the police. Several people, including the traditional ruler, were killed.

Sabotage of oil exploitation as well as disruption of oil distribution, especially through pipelines, is a very important form of protest in the Niger Delta region. Different reasons were advanced for this act of vandalism. OMPADEC (1996) identified three major reasons for sabotage: first, that compensation for damage was inadequate, not paid or did not reach the community; second, that individuals cut lines on their property to obtain compensation in excess of the cost of the actual damage; and third, that individuals disrupted production to force companies to provide amenities to their community.

What should however be noted is that sabotage as a form of protest has not only aggravated environmental degradation in the Niger Delta region, it has claimed several lives. A case in point was the Jesse petroleum pipeline disaster, which occurred on 17 October 1998 at the village of Atiwor, near the main town of Jesse. The disaster area is located midway between Jesse town and Jesse junction, off the Benin-Warri highway. According to Erhivwode (1999), the leakage of the oil pipeline was allegedly linked to the activities of saboteurs who, after their exploits, could not conceal the leakage. In the ensuing blaze, it was estimated that about 1600 persons lost their lives while scooping fuel on that day (NDES 1999, *The Guardian*, 23 October 1998). The majority of the victims of the fire came from oil-producing communities such as Jesse itself, Oghara, Sapele and Mosogar, in addition to certain villages located in the immediate environs of Jesse Town. A similar disaster, which claimed about 300 lives, occurred during March 2000 in Umuejije in Abia state of Nigeria. Thus sabotage as a means of protest has not only caused environmental damage, but considerable loss of life. It has also made the state, as well as their collaborators in the course of petroleum resources development, suffer extra financial loss through repairs and rehabilitation.

Armed struggle as another type of protest is not unheard of in the Niger Delta region. As far back as 23 February 1966, a group of Niger Delta youths under the leadership of Isaac Boro declared a certain part of the Niger Delta region equivalent to the present southern part of Rivers and Delta states as the Niger Delta People's Republic. This was with a view to ending what the leader of 'the twelve day revolution' (Boro, 1982) called a situation where 'petroleum … is being pumped out daily from … (the) veins' of the people of the area. As a means of achieving their broad goal, 'the pipelines connecting Olobiri (the first oil well in the country) and Ughelli on the one hand and Port-Harcourt (a refinery location) on the other were blown up' (Ikporukpo, 1996, p174).

More recently, the Egbesu war of 1998–1999 illustrates a typical armed conflict/struggle between the state and the oil-producing people of the Niger Delta, especially the Ijaw youths. In a grand convention of the Ijaw youths, which took place in the town of Kaiama on 11 December 1998, the Kaiama Declaration was made. This declaration gave oil companies until 30 December 1998 to withdraw from Ijawland. In the ensuing conflict between the state and Ijaw youths in Bayelsa state (a predominantly Ijaw state), a virtual siege existed between December 1998 and early January 1999. The second Egbesu war between the 'Egbesu boys' and the state was triggered when military men in Yenagoa, the capital city of the newly created Bayelsa state, confronted some Ijaw youths said to be participating in a cultural festival. Regular sieges of Ijaw communities or towns such as Odi, Yenagoa and Kaiama by military operatives have made these towns sustain heavy human and material losses.

From the foregoing discussions, it is apparent that, apart from the advocacy-based protest, all three other forms of protest *cannot* be regarded as sustainable in human or environmental terms. Thus, persistent cycles of protest in the Niger Delta region have not only impacted negatively on human life, but have also accelerated the pace at which environmental injustice and selective victimization is gaining ground in the Niger Delta. What should, however, be borne in mind is that these forms of protest are not meant to address the twin issues of environmental injustice and selective victimization alone. They are also an avenue to challenge the political status quo and deep-seated emnities. Grass-roots environmentalists have begun to see that the patterns revealed in environmental conflict and the ensuing protest are reflective of the broader inequities of economic and political power in Nigerian society. The result is that the concept of environmental injustice emerges and this has guided the environmentalists to raise questions, not only about public health, but also about political power in general. The simplest answer to environmental injustice is 'environmental justice'. Thus, there is the need to evolve some means of achieving broader sustainability through the use of social, economic and even politically sustainable solutions, which recognizes the principles of justice.

CONCLUSIONS AND RECOMMENDATIONS

The conclusion one can draw from this chapter is that the political economy of petroleum resources development has not been beneficial to *all* people in the Federal Republic of Nigeria. While the major ethnic groups that occupy the non-oil-producing areas are free to live in a relatively healthy environment, the indigenous ethnic minorities of the Niger Delta, the oil capital of Nigeria, face serious environmental injustices and selective victimization both directly and indirectly at the hands of the state and its collaborators. Environmental justice, social justice and ultimately sustainability, can only be achieved in Nigeria in an atmosphere of good governance.

This entails more than 'government'. The realms of governance include not only the state, but also civil society organizations. Hence, sustainable development in Nigeria, especially in the context of petroleum resources, should not be the responsibility of the state and the private sector alone. It should also integrate the efforts of others with an interest in governance, especially Third Sector, or civil society organizations. Governance should enable the vulnerable groups of the Niger Delta region to take part in decision-making that has to do with oil resources development which amounts to a greater sharing of power. This is necessary because the residents of the contaminated communities of the Niger Delta region face serious obstacles in gaining a voice in decisions that affect them, despite their organized grass-roots environmental groups, because they typically lack the political, legal and scientific resources that their opponents have.

For the achievement of social justice, the principle of allocating oil revenue based on the principle of equality of all states and size of the state should be looked into. Justice and equity go hand in hand and both imply fairness. If the state is to be fair in dealing with the Niger Delta people, then the allocation of revenue should not be based on two criteria alone. It should also include the principle of derivation if only to ameliorate the degraded environment of the area in question. This is why the oil-producing states thanked the present political administration for giving them 13 per cent derivation from oil proceeds to provide infrastructure and to redress the previous imbalances in the region. The reconstruction of the Niger Delta region should also be given adequate consideration by the new democratic government.

The promulgation of the Niger Delta Development Commission (NDDC) by the present democratic regime is a step in the right direction. However, efforts should be made to research the failure of similar Commissions that have existed in the past, with a view to preventing the newly proposed NDDC from collapsing when it is finally established.

The sustainable development of human capital through youth development and empowerment should also be uppermost in the minds of the beneficiaries of petroleum resources. Failure to create such enabling and empowering environments will make it impossible for the marginalized people of the Niger Delta to benefit from economic and social progress. Various communities in the Niger Delta should be allowed to develop in ways that improve the quality of their lives, reduce inequality and restore degraded environments. The federal government should encourage national and international NGOs to go into the area and assist in community-oriented programmes that will positively engage the youth in productive ventures. When many of them are employed, the level of poverty will be reduced and there will be purposeful community cooperation. In the final analysis, the goal is to create a sustainable, just and egalitarian society, but this is clearly a function of a far greater commitment to sustainability, social and environmental justice. This should be the goal of the Nigerian government for all areas of the federation.

REFERENCES

Akintola, F (1978) 'Mineral and energy resources' in J Oguntoyinbo, O Areola and M Filani (eds) *A Geography of Nigerian Development*, Heinemann Educational Books Limited, Ibadan

Aprioku, I (1999) 'Collective response to oil spill hazards in the eastern Niger Delta of Nigeria', *Journal of Environmental Planning and Management*, vol 42 (3), pp389–408

Bloxom, W (1996) 'The sustainable cities programme and the sustainable Ibadan project', *Ibadan EPM News, An Environmental Planning and Management Newsletter of the Sustainable Ibadan Project*, vol 1 (1) first quarter, April

Booth, D, Holland, J, Hentshel, J, Lanjouw, P and Herbert, A (1998) 'Participation and combined methods in African poverty assessment: Renewing the agenda', *Report Commissioned by the UK Department for International Development for the Working Group on Poverty and Social Policy, Special Programme of Assistance for Africa*

Boro, I (1982) *The Twelve Day Revolution*, Umeh Publisher, Benin City

Cable, S and Cable, C (1995) *Environmental Problems: Grassroots Solutions: The Politics of Grassroots Environmental Conflict*, St Martins Press, New York

Caporaso, J A and Levine, D P (1992) *Theories of Poitical Economy*, Cambridge University Press, Cambridge

Erhivwade, F A (1999) 'Detailed report of the Jesse Oil Pipeline fire disaster', Ministry of Information and Culture, Port Harcourt

Faniran, A (1997) 'Geography, environment and (sustainable) development: A case for integrated discipline', extracts from the Fellowship Lecture delivered at the 40th NGA Conference, held at the Department of Geography, Bayero University, Kano, 5 May

Goodland, R (1994) 'Environmental sustainability: Imperative for peace' in D S Graeger (ed) *Environment, Poverty, Conflict*, International Peace Research Institute (PRIO), Olso

Goodland, R (1996) 'The concept of environmental sustainability (ES)' in A Porter and J Fittipaldi (eds) *Environmental Methods Review: Retooling Impact Assessment for the New Century*, The Press Club, North Dakota

Goodland, R, Lemons, J and Westra, L (eds) (1998) *Ecological Sustainability and Integrity: Concepts and Approaches*, Kluwer Academic, Dordrecht and Boston

Hartley W (1995) 'Environmental justice: An environmental civil rights value acceptable to all world views', *Environmental Ethics*, vol l7 (13), pp277–89

Ibeanu, O (2000) 'Oiling the friction: Environmental conflict management in the Niger Delta, Nigeria', *PECS News: A Population, Environmental Change and Security Newsletter*, vol 2 (1), p1 and p8

Ibeanu, O (no date) 'Insurgent civil society and democracy in Nigeria: Ogoni encounters with the state, 1990–1998', *Research report for ICSAG Programme of the Centre for Research and Documentation* (CRD), Kano

Idoniboye-Obu, B (1994) 'Compensation for ecological disturbances and personal losses', paper presented at the Conference on marine pollution control at the River State University of Science and Technology, Port-Harcourt

Ikejiofor, U (1999) 'The God that failed: A critique of public housing in Nigeria, 1975–1995', *Habitat International*, vol 23 (2), pp177–88

Ikporukpo, C (1983) 'Petroleum exploitation and the socio-economic environment in Nigeria', *International Journal of Environmental Studies*, vol 2l, pp193–293

Ikporukpo, C (1996) 'Federalism, political power, and the economic power game: Conflict over access to petroleum resources in Nigeria', *Environment and Planning C: Government and Policy*, vol l4, pp159–77

Jessop, B (1982) *The Capitalist State*, New York University Press, New York

Johnston, B (1994) 'The abuse of human environmental rights: Experience and response' in B Johnston (ed) *Who Pays the Price? The Sociocultural Context of Environmental Crisis*, Island Press, Washington, DC

Johnston, B (1994) 'Environmental degradation and human rights abuse' in B Johnston (ed) *Who Pays the Price? The Sociocultural Context of Environmental Crisis*, Island Press, Washington, DC

Johnston, B (1995) 'Human rights and the environment', *Human Ecology*, vol 23 (2), pp111–23

Moffat, D and Linden, O (1995) 'Perception and reality: Assessing priorities for sustainable development in the Niger River Delta', *AMBIO: A Journal of the Human Environment*, Royal Swedish Academy of Science, pp527–38

Movement for the Survival of Ogoni People (MOSOP) (1992) *Ogoni Bill of Rights*, Saros International Publishers, Port-Harcourt

Niger Delta Environmental Survey (NDES) (1997) *Final Report*: Phase 1

NDES (1999) *Report on our visit to the Burst Crude Oil Pipeline at Ekakpamne*

Nigerian Environmental Study/Action Team (NEST) (1991) *Nigeria's Threatened Environment: A National Profile*, A NEST Publication, Ibadan

Ogbe, M (1997) 'The impact of oil exploration and exploitation on the Niger Delta', *mimeo*

Okorodudu-Fubara, I (1998) *Law of Environmental Protection Materials and Text*, Caltop Publications, Ibadan

Olopoenia, A (1998) 'A political economy of corruption and under-development' Faculty Lecture Series No10, delivered at the University of Ibadan, Ibadan

OMPADEC (1996) Files and documents relating to environmental pollution from oil spill accidents, 1994–1997, OMPADEC, Port Harcourt

Otite, O (1990) *Ethnic Pluralism and Ethnicity in Nigeria*, Shaneson, Ibadan

Phillips, D and Williams, A (1984) *Rural Britain A Social Geography*, Basil Blackwell, Oxford

Panos Briefing (2000) 'Economics for ever building sustainability into economic policy', *Panos Briefing*, No38, March

Poulantzas, N (1973) *Political Power and Social Classes*, New Left Books, London

Ramani-Abah Consultant (1997) 'Compensation practice in Nigeria: A review of current trends', *Technical Report 13*, Ramani Abah Consultant, Port-Harcourt

Roemer, J (1985) *Analytical Marxist*, Cambridge University Press, Cambridge

Rothstein, B (1992) 'Social justice and state capacity', *Politics and Society*, vol 20, pp101–26

Stanley, W (1990) 'Socioeconomic impact of oil in Nigeria', *Geojournal*, vol 22, pp67–79

United Nations Development Programme UNDP (1997) *Human Development Report Nigeria 1996*, UNDP, London

US EPA (1995a) *Environmental Justice Strategy: Executive Order* 12898 (EPA/200-R-95-002), Office of Environmental Justice, EPA, Washington, DC

US EPA (1995b) *Environmental Justice, 1994 Annual Report*: Focusing on Environmental Protection for all People (EPA/200-R-95-003), Office of Environmental Justice, EPA, Washington, DC

Varian, H (1974) 'Equity, envy and efficiency', *Journal of Economic Theory*, vol 9, pp63–91

Wilkinson, C (1998) 'Environmental justice impact assessment: Key components and emerging issues' in A Porter and J Fittipaldi (eds) *Environmental Methods Review: Retooling Impact Assessment for the New Century*, The Press Club, North Dakota

Willis, K (1980) *The Economics of Town and Country Planning*, Granada Publishing Limited, London, Toronto, Sydney and New York

Chapter 14

Environmental Protection, Economic Growth and Environmental Justice: Are They Compatible in Central and Eastern Europe?

Alberto Costi

INTRODUCTION

The concept of environmental justice has developed both at national and international levels. The international community has addressed the issue of the unequal sharing of costs and benefits of environmental protection through such principles as sustainable development, intergenerational equity and common, but differentiated responsibility in managing the global environment (Costi, 1999, pp315–16). At the national level, considerable debate in the US followed the publication of studies demonstrating the discrepancies between the environmental burden suffered by economically and racially disadvantaged minorities and their exposure to greater environmental hazards, and the satisfactory level of environmental protection in areas inhabited by white middle-class communities (Weintraub, 1995, pp567–70). Despite the development of a body of literature on the subject however, the environmental justice movement within Western Europe remains relatively weak.

Whereas numerous studies have discussed the ongoing economic and political transition in Central and Eastern Europe (CEE)[1] and its impact on society (Pickles and Smith, 1998), few works have exhaustively addressed the concept of environmental justice. Elements conducive to environmental justice may be inferred from the recognition of constitutional and political rights in CEE countries: non-discrimination on racial, social, religious, ethnic, sexual, political or other grounds, freedom of expression and of association, the right to a healthy environment and the right of access to information. CEE countries have also been eager to ratify an increasing number of international environmental and human rights instruments and to implement their key legal principles within the domestic legal system (EBRD, 1999). Most of the preconditions for the development of environmental justice, however, have not yet materialized.

The title of the chapter reflects the dilemma confronted by CEE countries ten years on in their difficult reform path. Whereas an excessive protection of the environment might hinder the development of the economy, unsustainable economic growth could jeopardize the environment. The restructuring of the economy imposes a heavy toll on the state budget, favours the privatization of state enterprises mainly to the benefit of foreign investors and induces a culture of consumerism. The transition process also imperils the environment by an overexploitation of natural resources, an increase in the production of waste and the building of a more sophisticated road infrastructure over fertile lands, forests and other ecosystems (Beckmann, 2000).

The aim of this chapter is to demonstrate that the development of environmental justice is crucial if any attempt to accommodate environmental protection and economic growth in CEE is to succeed. The first part examines the reasons for the lack of genuine interest of Communist governments for environmental issues and citizens' concerns for a healthy environment, and the legal and policy framework put in place by the new governments to remedy the situation and promote environmental protection. The second part analyses the reasons for the failure of the new policies to effectively impact on the environment in this politically, economically and socially difficult transition period. This leads me to propose in the third part, a workable model of environmental justice in CEE, reconciling environmental protection and a socially acceptable form of economic development. In the Conclusion, it is argued that citizens are at the centre of the reforms and should play a key role in the decision-making process for the true development of sustainability and environmental justice.

PAST AND CURRENT ENVIRONMENTAL PROTECTION IN CEE

It is admitted that environmental protection in CEE predates the adoption of national environmental instruments in Western Europe (Weiner, 1988). Communist governments adopted groundbreaking environmental impact assessment procedures, planned national environmental policies to be administered by central agencies and set up a series of economic instruments to complement what appeared on paper as tough 'command and control' legislation. Laws were enacted, very often adopting stricter standards than those in place in North America, and the rest of Europe (Szejnwald Brown et al, 1998). Following the United Nations Conference on the Human Environment in Stockholm in 1972, the protection of the environment became a new medium through which CEE countries attempted to prove the superiority of Communist regimes over the West. Comprehensive environmental laws were implemented in most CEE countries. Constitutional provisions imposed duties on states and individuals to protect the environment. The rise of the environmental movement did not meet much resistance from the authorities and expanded in the 1980s, often under the benevolent hand of the state (Bowman and Hunter, 1992, p927). A number of rare species as well as unique ecosystems and habitats still survive in CEE as a result of the efforts of Communist governments to promote the protection of the environment as state policy (Beckmann, 2000).

It is, however, with widespread degradation of the environment that state Socialism is usually associated. The use of natural resources for armament and heavy industry and the frenetic development of gigantic projects, as well as the few examples of successful environmental protection, were all grounded on ideology and subjugated to the realization of the socio-economic plan (Szacki et al, 1993, pp11–13). The over-consumption of energy sources as a result of subsidized prices and the lack of consideration for health conditions in the exploitation of natural resources led to extensive air and water pollution. The centralized nature of power left no space for the participation of local authorities in the decision-making process while the division of environmental responsibilities among many ministries with diverging interests prevented the creation of a coordinated strategy. Environmental legislation remained largely non self-executing and ineffective in practice since state control of economic, social and environmental policies subjected the implementation and enforcement of environmental law to more pressing needs (Jendrocka, 1996, p373).

State Socialism masked the lurking danger of environmental pollution. The programme of the ruling Communist Party identified ecological problems as a primary social concern, a form of unique safety valve through which social discontent was acceptable. The qualification of the environment as a social rather than as a political issue enabled mass conservationist movements, such as the Czech Brontosaurus Society and the Hungarian Ornithological Association, to espouse an ecological agenda. As a result, the environment became more or less the only legitimate basis for opposition to the Communist regime. In reality, environmental movements were tolerated only as long as they remained apolitical. They were intentionally kept atomized and isolated (Jancar-Webster, 1993, pp198–99). Media censorship limited access to information and prohibited any debate on the actual detrimental effects of governmental policy on the environment (Bowman and Hunter, 1992, p926). Environmental standards were marginalized and their compliance only symbolically pursued.

By 1980, the impact of pollution on the natural and human environment became apparent. Dependence upon raw materials for heavy mining and armament industries stripped areas of natural beauty and produced dangerous wastes and fallout, so graphically illustrated by the Chernobyl disaster in Ukraine in 1986. The prolonged delay in disseminating details of the catastrophe to other countries, especially those affected directly by the radiation, generated a tide of fear as to the effects of the use of nuclear energy, and focused public attention on the problems of contamination and pollution in general (Szacki et al, 1993, p14). Some CEE regions were undoubtedly ecological disaster areas and extensive environmental degradation was affecting human health. The lack of reaction by the authorities to the deteriorating living conditions in the ill-fated Black Triangle, a region covering parts of the then Czechoslovak Republic, the German Democratic Republic and Poland, proved that the health condition of workers and their families was not a material consideration for decision-makers (REC, 1994). For Communist governments, environmental protection policies were not an important part of economic and social life: they did not fit with the model of Socialist command economy; accordingly, the authorities did not provide mechanisms for protecting the individual right to a healthy environment (Klarer et al, 1999a, p18).

The severe degradation of the environment and the ensuing health and social hazards explain, at least to a certain extent, the quiet revolutions of 1989–1990. The sentiment of injustice caused by the absence of a consultation process and effective participatory rights of citizens cemented opposition forces and weakened the state. Twenty thousand people gathered in 1989 before the parliament in Budapest for the first time since the 1956 uprising to protest against the construction of the Gabcikovo-Nagymaros Dam. The construction of a dam across a historic and picturesque bend on the Danube, under the impulse of the Soviet authorities, symbolized for many the regime's arbitrary rule. The peaceful movement of protest against the dam provided an important indicator of public strength and potential power in the face of the unresponsive regime (Jancar-Webster, 1993, p193). Five thousand people marched in the streets of Sofia, Bulgaria, in November 1989, to voice their disapproval against pollution, the first public protest in 40 years (Bowman and Hunter, 1992, p926). The old regimes collapsed because the targets predetermined by the command economy were set with no regard to human or environmental costs. The incompetence of the authorities in responding to environmental problems and in providing adequate information antagonized citizens and confirmed the incapacity of the Communist party to live up to their expectations (REC, 1994).

The collapse of state Socialism had far-reaching effects on national policies. There was much hope for widespread political and economic reforms, with particular emphasis on the protection of human rights and the environment. Assistance and support in various shapes and forms provided by the US, the European Union (EU) and international financial institutions (IFIs) generated an enthusiastic response from these emerging democracies with economies in transition (Manser, 1998). In the West, reforms were considered essential to bring political stability to the region, to eliminate security and environmental threats and to rebuild economic systems ravaged by the sudden collapse of the Council of Mutual Economic Assistance. The EU, the US and IFIs prescribed a strict diet of leaner spending policies, greater exercise of market mechanisms and the development of democratic practices. In terms of environmental protection, Western attention focused on the importance of curing past ills and preventing further degradation.

The aspiration to the status of recognized liberal democracies with market economies became the leitmotif of CEE countries in their efforts to gain respectability in the new world order. With limited financial means and daunting challenges facing them, CEE governments were forced to balance state obligations contracted in the international arena with pressing domestic needs (EBRD, 1999). At the same time, society was now undergoing some fundamental changes. The state, individuals, interest groups and industry were forced to redefine their respective roles. Drawing upon lessons of 40 years of state Socialism, the social costs of reforms would only be acceptable if governments endeavoured to adopt and meet reasonable economic and environmental targets. Reforms, to be successful, needed to be undertaken as part of a comprehensive strategy.

The turning point came when trade and cooperation agreements signed before 1989 were replaced with association agreements (also referred to as Europe Agreements) between the EU (then the EC, or European Community) and CEE countries. The Europe Agreements were designed to liberalize trade, harmonize legislation with EU law and help CEE countries prepare for eventual membership. Supported by substantial financial assistance under the PHARE programme,[2] all ten CEE states keenly embarked on the hopeful journey to EU accession. The EU, at the Copenhagen European Council of June 1993, adopted accession criteria[3] and at the Essen European Council in December 1994, approved the pre-accession strategy. Between 1994 and 1996 all CEE countries formally applied for EU membership. According to the pre-accession strategy, candidate states need to meet the accession criteria and adopt the bulk of EU legislation, known as *acquis communautaire*, through approximation and transposition. The European Commission (EC) has highlighted the importance attributed to the restoration of the environment (EC, 1998). It has also expressed reservations as to the institutional capacity of CEE countries to carry out the necessary environmental reforms given the stark contrast between the sophistication of EU law, and the poor state of the environment in CEE countries (EC, 2000). In 2001 the European Commission noted some improvements in the adoption of appropriate environmental legislation in CEE countries while stressing the need for them 'to further strengthen administrative, monitoring and enforcement capacity, in particular in the field of waste, water and chemicals' (EC, 2001).

It is not difficult to imagine why, in these circumstances, governments and environmental movements alike see the protection of the environment as a key ingredient in the success of the transition process. Financial assistance from the European Bank for Reconstruction and Development (EBRD) and the World Bank for a variety of projects has become conditional to the undertaking of prior environmental impact assessments. Inspired by American and Western European models, CEE governments have undertaken the task of enforcing existing norms, replacing outdated legislation and restructuring the economy along sustainable and ecological lines. The signature and ratification of various international and regional instruments, the participation in the Rio 'Earth Summit' in 1992 and the harmonization efforts to comply with EU standards have stimulated the introduction of a new approach to environmental protection. Thus, CEE countries have integrated environmental concerns into their development policy (Klarer et al, 1999b, p29).

Since 1990 a new legal environment has progressively emerged. The regulation of the environment is now conveyed through comprehensive laws providing for horizontal instruments and principles applicable to most sectors of activities: environmental impact assessment, polluter pays principle, prevention and precautionary measures (EBRD, 1999). In addition to the development of substantive principles of environmental protection, procedural rules have materialized, providing a system of citizen access to environmental information, public participation in respect of economic development at the national and local levels and access to justice (REC, 1998, p15). A web of burgeoning sectoral legislation covers air and water pollution, waste, noise, conservation and other specific problems. Environmental strategies in the form of national programmes set out priority objectives to guide public authorities. Provisions in civil and criminal law are also used as robust means to remedy past damage and deter future harmful activity. Permit systems and other administrative authorizations help manage the exploitation of resources (Costi, 1999, pp305–6). The recognition of a right to a healthy environment in most CEE constitutions is another reminder of the importance of the respect for environmental concerns (Sands, 1995, pp223–24).

The promotion of an environmental agenda and the management and enforcement of environmental policies require important financial resources. Governments have increased the use of economic instruments such as emission charges, user fees and eco-taxes to feed environmental funds in order to finance clean-up operations and environmental programmes and to fill the efficiency gap arising from the limitations of regulatory provisions (Klarer et al, 1999a). From early on in the transition, a variety of schemes for the allocation of liability for past pollution and environmental responsibility for current and future operations have been designed in an attempt to cover the costs of restoring the environment without deterring the influx of foreign investments (Bowman and Hunter, 1992, pp965–70).

The reform process launched a decade ago boasts a commitment to the consideration of environmental concerns in the restructuring of the economy and to the promotion of environmental protection through legislative, economic and administrative processes. CEE countries recognize the need to prioritize international and regional cooperation, to become actively involved in the protection of the environment and to value natural habitats and ecosystems as integral parts of the national cultural heritage. The far-reaching character of the reforms has been fuelled to a great extent by the efforts of CEE governments to enact legislation in conformity with the list of priorities specifically mentioned in the Europe Agreements. CEE countries have, therefore, transposed most EU environmental standards within their national legal systems (EC, 2000).

RESULTS AND CHALLENGES OF ENVIRONMENTAL REFORMS

Reforms have enabled CEE countries to implement the principles of market economies and liberal democracy. The privatization of the economy has required the establishment of favourable investment conditions. CEE governments have tried to draft clear and accessible laws in an attempt to generate respect for the rule of law and a secure environment for investments. Increasing efforts are also being made towards the development of political institutions and decentralization. The democratization of the political process and the decentralization of power have permitted the emergence of an open debate on social, political and environmental issues. Citizens and interest groups may question governmental policies (Stec, 1998). Local authorities are involved in the decision-making process on matters of local interest, including the environment (REC, 1994). The dependency of CEE countries on Western assistance has made them more sensitive to critique and has facilitated their cooperation with other governments to protect endangered habitats and species. The quest for closer international ties has engendered economic, political and scientific integration and replaced the distrust and piecemeal nature of East–West relations of the Cold War era.

In so far as the environment is concerned, the rate of success of reforms can only be measured in terms of their actual impact on fauna, flora, air, water and human health. Generally speaking, it cannot be denied that the transition is generating an improvement in the state of the environment in CEE. Improvement in production techniques, importing less polluting goods, energy prices more closely reflecting market costs have all contributed directly or indirectly to a better environment. So has the restructuring of the economy by accelerating the closure of former state enterprises now considered economically or technologically non-competitive (Klarer et al, 1999a, pp16–17).

Nevertheless, after a decade of transition, environmental improvements remain limited. The enormous costs linked to the transition, the incomplete character of political and social transformation, the distorted nature of market transition as well as behavioural patterns reminiscent of the Communist era, help explain the mixed results of the reforms and the preference of CEE governments for economic development over environmental protection in practice.

Cost factors linked to the transition

More effective environmental policies require improved implementation and enforcement practices by governments. They also demand a restructuring of the industrial infrastructure involving the acquisition of cleaner technology and energy efficient processes, as well as an undertaking to clean up the inherited environment. In 1998 it was believed that investments of over ECU 120 billion were required to meet environmental obligations for EU accession in the area of drinking water, air pollution and waste disposal (EC, 1998, p4) (as of 1 January 1999 the EURO replaced the ECU at the conversion rate of 1:1). To meet such expenditures CEE countries should invest annually up to 4 per cent of their gross domestic product (Klarer et al, 1999a, p17). Financial needs are enormous and, under present circumstances, they would have to draw mostly on the already limited resources of the state.

It is beyond the scope of this chapter to analyse the intricacies of economic and social transition in CEE countries. For our purposes, it will suffice to note that new stressful living conditions combined with pressures to shrink state debt and to reduce budget deficits have caused standards of health care, social services as well as life expectancy to fall (EBRD, 1999). Criminality, corruption and unemployment rose significantly, especially in the first few years of the transition.

The economically dictated terms of transition adversely affected each aspect of daily life (EBRD, 1997). Even with foreign aid, CEE countries continue to struggle to find the necessary resources to cover governmental expenses and cannot easily free funds to invest in the environment. Simply stated, the authorities face a difficult choice. Raising taxes further might endanger foreign investment and encourage the growth of the informal economy. An alternative would be to free budgetary resources by further reducing health, social and welfare services, thereby further threatening living standards.

Ambitious environmental standards require additional financial and human resources to monitor and enforce the law. The severe social and economic constraints burdening governments of CEE countries cast a shadow on the importance of environmental policy. It follows that adequate policies and efficient regulation of the environment are either subjugated to, or compromised with, other political, economic and social interests.

Incomplete character of political and social transformation

In recent years free elections and liberalized markets have been branded as the best illustrations of a deeply rooted transformation in CEE countries. For most Western business and political analysts, a successful transition to democracy and sustained progress in economic restructuring lead naturally to a general improvement in social well-being – including the environment. Contrary to expectations, it appears now that the transition process has not automatically led to a vibrant economy and better living standards in most CEE countries (Meth-Cohn, 1999). This is explained by the fact that institutional reforms are lagging behind. The causes are manifold. Reformers include the old political guard, former foes of the Communist governments, and the younger breed of politicians often emerging from the old student or environmental movements. All have developed their own political agenda and party politics blur the long term vision of the state. Framework and sectoral legislation confers wide discretionary powers on the government to adopt decrees relating to the environment. This limits the participation of elected representatives, prevents public debate and fosters secrecy in the decision-making process. Domestic and international pressures dictate the strategy of the government and render difficult the coordination of ministerial activities amid competing interests and scarce financial resources. Ministers in charge of economic and financial affairs have traditionally commanded more respect than those in charge of environmental matters. The former are jealous of their prerogatives and leave limited space for the integration of environmental considerations in economically related decisions. For instance, in the early 1990s, the then Czechoslovak Finance Minister Vaclav Klaus stated that 'environmental issues are the "icing on the cake" of economic development' (Bowman and Hunter, 1992, p922), indicating the limited interest of authorities for the improvement of the environment. Ministers responsible for the environment often play a restricted role in the decision-making process, work with a very tight budget and rely on limited local technical expertise. This hinders their capacity to tackle problems adequately. Structural changes have also proven more difficult to implement than expected. Many of the civil servants with experience in public management were either removed with the advent of the reforms or attracted towards the more lucrative and prestigious private sector. The implementation of technical standards has been detrimentally affected by the lack of experience of new officials, their limited motivation due to meagre wages and the inadequacy of monitoring systems (Casalino, 1995, pp251–52).

Defective institutional reforms affect the implementation of environmental policy. The level of environmental fines is set too low. The high inflation rate in some of the countries under consideration further erodes the real value of fines. Collected fines are insufficient to fund environmental programmes, to compensate individuals and communities affected by pollution and to cover the cost of clean-up operations. For instance, in Bulgaria, the value of environmental fines in 1997 was estimated at US\$0.34 million. In 1995 collection efficiency was believed to be at around 50 per cent of the pollution fines levied throughout that year (Klarer et al, 1999b, p18). Although collection efficiency levels appear to have considerably improved thereafter, it remains that the levels of environmental fines and their collection efficiency are inadequate in the face of the real cost of pollution control. The lack of financial resources also affects local authorities, who are primarily responsible for implementation, monitoring and enforcement of environmental policy. Political restructuring has ensured decentralization of some powers to local authorities, but a transfer of fiscal powers has not supplemented this (Pavlinek, 1996). EU concerns regarding the capacity of CEE governments to implement the environmental *acquis* are legitimate (EC, 2001).

Distorted market transition

Environmental reforms have been based on the understanding that the greater part of the financial burden associated with the protection of the environment would be gradually supported by the emerging private sector. The underlying rationale is that an efficient taxation regime would generate state revenues. The adoption of economic instruments would put a price tag on pollution and modify the behaviour of the industry and individuals alike. Foreign and domestic investors would assume the costs involved in cleaning up the land surrounding the sites of privatized enterprises. To be operational, however, this scenario requires well-functioning markets and sound business practices.

The problem is that in general, CEE countries have been unable to efficiently provide the institutions and services necessary for markets to function smoothly (EBRD, 1999). Meanwhile, private actors in the emerging economic infrastructure are not yet in a position to fully compete according to market mechanisms. The high costs of complying with, and the limited capacity of the state to enforce the law, have led to the emergence of an informal economy (EBRD, 1997). Registered enterprises have often used the same assets to produce output for both the formal and the informal economy, the difference lying in misreported or undeclared activity rather than *proprio* illegal activity. This part of the unofficial economy at least seems to be particularly sensitive to the cost differential between official and unofficial activity, that is, the cost involved in complying with tax and other regulations (EBRD, 1997). At this stage of the transformation process, it is essential that the implementation of environmental market instruments be accompanied by more effective tax and regulatory enforcement mechanisms. Otherwise, market instruments will be inefficient since they will merely result in further increasing the share of the informal economy and make surveillance, control and monitoring all the more difficult and costly. Moreover, participants in the official economy will suffer disproportionately from the costs associated with implementation measures because the burden of market instruments imposed on declared and correctly reported activities will need to be heavier in order to compensate for revenues missing, due to non-compliance by the informal sector. Competition will also be distorted at the expense of properly functioning enterprises, thereby causing further inequities (Costi, 1999, pp319–20). As long as laws are not implemented efficiently and equitably, the industry will be unwilling to meet the compliance costs of market regulations as this would adversely affect their competitive position.

Remnants of behavioural patterns of the Communist era

Political and economic changes have altered the political landscape and the fabric of society. Soon after the changes, freedom from Communism was interpreted as freedom from constraints altogether. The imposition of restrictions upon individuals was labelled as incompatible with civil liberties, even when this was in the common interest. The emergence of a rather undefined rights discourse in the region stressed the justification for private action and downplayed any obligations that could limit it. An attitude of *chacun pour soi* and laissez-faire soon replaced the spirit of solidarity that marked the early days of transition (Sajó, 1996, pp149–50). In reality, building a more equitable society in a region that historically has experienced only very few sporadic episodes of democratic rule requires more than elections on a regular basis. It calls for a profound metamorphosis capable of cutting through behavioural patterns.

It will take time to change beliefs and attitudes inherited from the Communist era. Forty years of Socialist ideology have deeply impressed societies in CEE countries. The most striking feature is social inertia. Individuals remain very much dependent on already existing structures. Those in powerful positions oppose any changes that could jeopardize their privileges (Grapska, 1994, p216). The lack of a sense of public duty or responsibility encourages disregard for the law and induces little respect for contracted obligations, opening the door to any justification to achieve one's ends (Sajó, 1996, pp149–50). From the citizen's viewpoint, defying legislation that is not implemented in practice remains the easy option. The limited acceptance of responsibility clearly endangers the application of the polluter pays principle and the state reverts to its traditional role as the provider of environmental services. This results from the inability in shifting the previous role of individuals as mere recipients of law and policy to partners in reforms. States might have ratified numerous treaties and even implemented international obligations establishing a process for citizens and non-governmental organizations (NGOs) to obtain environmental information, to comment on proposed governmental decisions, to develop and submit their own positions and to have access to justice to review administrative decisions.

Nevertheless, public participation continues to be limited on the whole. As mentioned earlier, the broad delegation of powers to the executive results in the retention of a model of strong executive supremacy, typical of Socialist ideology. Public consultation is still sometimes set aside when more pressing interests attract the attention of the government. Examples include the construction of nuclear plants at Mochovce (Slovak Republic) in the early 1990s and at Temelin (Czech Republic) in the mid-1990s (REC, 1994). A more shocking example is that of the closure of the town of Libkovice (Czech Republic) to enable the exploitation of coal deposits located beneath the town. The decision initially made by the former Communist government in 1987 was finally upheld by the new authorities despite protests by the local population and international and national NGOs (Pavlinek, 1996).

This problem is exacerbated by the inability of the courts of law to overcome the secondary role they used to play under the previous regimes. Judges have only on rare occasions in Hungary, Slovenia and the Czech Republic interpreted constitutional and legislative provisions to constrain conduct detrimental to the protection of the environment and, overall, the judiciary remains weak and disorganized (REC, 1998, p49). Moreover, restrictive standing rules, including the institution of class actions, and the costs related to legal proceedings act as potent barriers limiting citizens' access to justice (EBRD, 1999). Finally, the conviction of polluters is made difficult in the absence of clear guidelines and emission standards, while remedies, including injunctive relief, are not adequate (REC, 1998, pp48–49).

The unconditional freedom of the press contained in most CEE country constitutions does not prevent the enactment of state legislation regulating and sometimes significantly restricting broadcasting or journalistic activities (Henckaerts and Van der Jeught, 1998, pp487–89). NGOs find it difficult to marshal the necessary support to strengthen the environmental movement in the face of more pressing political and economic interests (Massam and Earl-Goulet, 1997, pp144–45). The environmental movement lacks both unity and appeal. Fundraising remains as important for NGOs as their work on the ground to protect the environment, while doubts are raised regarding the independence of NGOs financed by public funds (Jancar-Webster, 1993, p207). The recuperation of the environmental agenda by traditional political parties has also drained the fervour of environmentally sensitive individuals.

Economic considerations prevail over sound environmental practices

In reality, economic considerations have prevailed over sound environmental practices in CEE countries (OECD, 1999). Until recently both public opinion in CEE countries and international actors have overstated the importance of economic transition, failing to appreciate the urgency of social and environmental reforms. Election campaigns and opinion polls have showed that economic growth and the reduction of unemployment rank much higher on the list of priorities than the environment. The shift in citizens' preoccupation has legitimized a governmental political agenda embracing market economy at the expense of environmental and social progress (Massam and Earl-Goulet, 1997, p141). Membership in the World Trade Organization and in the Organisation for Economic Co-operation and Development (OECD) as well as the strong desire to join the EU at any cost have strongly influenced governmental policies. The eagerness in joining these organizations has created a distorted and asymmetrical relationship between CEE countries and their Western benefactors and has prioritized economic achievements.

The role of IFIs and foreign investors in the transition process has been essential in supporting the costly reforms. Their word has often sounded like gospel to the ruling political class and the new breed of entrepreneurs, often business cronies, who have reaped the rewards of transition while ordinary citizens have suffered from incomplete democratic and economic reforms (Meth-Cohn, 1999). IFIs have been criticized for turning a blind eye to corrupt or autocratic governmental practices of allegedly reformist governments while favouring market developments (Thomas, 1999, pp557–59). For instance, it took several years for the international community to realize that a sum of US$750 million of foreign aid allocated to the rehabilitation of the nuclear programme in Ukraine had not been accounted for (Manser, 1998).

Foreign investors have generally been welcomed to the region. Western investments have rapidly flown into CEE economies, fuelling the engine of economic growth and development. Multinational corporations (MNCs) have played an important role in the transition process. The expression MNC refers to 'cluster(s) of corporations or unincorporated bodies of diverse nationalities joined together by ties of common ownership and responsive to common management strategy' (Joseph, 1999, p172). The economic power of some MNCs enables them to dictate terms and conditions to national governments (Bowman and Hunter, 1992, pp977–78). MNC involvement in CEE economies has generated mixed results. In many cases, the activities of MNCs have led to job creation, the development of new industries and the rejuvenation of declining ones. Beneficial windfalls for the local economy and the local environment have been registered (EBRD, 1999). On the negative side, there is no doubt that MNCs may cause extensive environmental damage. Governments have often reconsidered their commitment to the protection of the environment when they felt that foreign investors might target countries with less burdensome regulatory and tax regimes. Governments have sometimes failed to obtain from potential investors an undertaking that they would assume liability for the decontamination of acquired sites.

Tragedies such as the recent cyanide spill from a Romanian mine into the Szamos, Tisza and Danube rivers illustrate the legal difficulties facing governments when trying to obtain compensation from investors who have only a limited presence in the country (Schwabach, 2000). The incident resulted from activities performed by a company jointly owned by an Australian firm and a state-owned Romanian enterprise. Not only did the Australian management of the company decline responsibility, but soon after the disaster, the Australian owners also filed for bankruptcy in Australia. It has since been difficult for the two most affected countries, Hungary and Yugoslavia, to take action against the company. The incident shows that in the absence of appropriate regulation by the company's home state, the state where the tragedy occurs might not be compensated and might have to cover on its own the enormous costs of the resulting ecological disaster.

The pressure on the natural environment has rarely been so strong (Beckmann, 2000). Expansion of urban areas, further development of the road system in the countryside, increase in mass consumption and waste production, extensive use of natural resources and the development of tourist attractions (Stursa, 1998, pp359–60) have contributed to the endangerment of unique ecosystems and natural habitats. If we accept that CEE countries are moving towards a model of liberal democracy, it is assumed that citizens will gradually exercise greater influence on projected governmental policies. It is also hoped that as the economy recovers from the harsh realities of the transition, citizens will demand more environmental protection. Unfortunately, there are few effective consultation mechanisms, institutionalized and efficient decision-making processes or appropriate accountability procedures relating to environmental issues at present (REC, 1998, pp11–13). It follows that without structural changes and complete reforms of a social, economic and political nature, it is pointless to discuss changes in production patterns, law enforcement, public participation, burden sharing and other means to tackle environmental problems. In the absence of comprehensive institutional and social transformations in CEE countries, no concept of justice – never mind environmental justice – will develop in the region.

TOWARDS A MODEL OF ENVIRONMENTAL JUSTICE FOR CEE

Surprisingly, the ailments afflicting the new democratic societies in CEE countries resemble those existing under the former Communist rule. Economic growth is pursued at the expense of social and environmental considerations, and the new political elite perceives any critique of its policies negatively. The management of public affairs is conducted along party lines rather than in the national interest. Interference with the media continues, even if to a lesser extent, and is now organized through more subtle means. At best, individuals nowadays play a limited role in the decision-making process.

Correcting this course of events necessitates time and fundamental adjustments. A successful transition requires political, economic, social and environmental (ie sustainability) concerns to be integrated in an open and transparent decision-making process that balances the costs and benefits of policies aimed at establishing sustainable development. Environmental considerations and economic development do not have to be incompatible. Accommodating them both is necessary in order to achieve sustainable development. Furthermore, the development of a culture of environmental justice would enable citizens' concerns regarding their environment and living conditions to be brought to the attention of decision-makers. A broad and genuine consultation process would highlight any contradictions in the interests at stake and would lead governments to devise innovative solutions that ensure the compatibility of economic growth, the protection of the environment and social equity. The conditions for environmental justice should be set not only within the region but also in conjunction with international actors influencing the reform process. These include foreign states, MNCs, IFIs and especially CEE's main partner, the EU.

How to achieve environmental justice in CEE countries?

For CEE countries the present analysis suggests the need for an integrated approach in protecting the environment and transforming the economy, the institutional and democratic structure of the state and society at large. All these aspects should be tackled in a parallel way. This does not mean, however, that the speed or the reform instruments used should be similar in all areas of public interest. Each country should find its own integrating formula. There is no doubt that the promotion of environmental justice relies on openness, accountability, civic participation and more specifically on good governance to integrate civil society. The implementation of good governance is fundamental to the transition towards a smoothly functioning market economy and a genuine democracy and, in any event, appears essential to eradicate the unofficial economy and corruption. Good governance aims at establishing the rule of law by limiting the discretion of civil servants, fostering transparency, stability, predictability and even-handedness in the manner of governing. It is based on the establishment of partnerships between the public and private sectors, the participation of all citizens and the development of interest groups through the emergence of an active civil society (EBRD, 1997). The establishment of good governance should prevent unnecessary clashes of interests or the emergence of inequity in the formulation and application of environmental policies. Good governance should, therefore, prevent major environmental crises and enhance the capacity to effectively manage them if and when they occur.

The main challenge in developing greater environmental justice, and a commitment to sustainable development therefore is to nurture good governance. No external pressure could ever provoke the required changes; these ought to emerge from within CEE societies. The reforms aim at fostering citizens' support for the common good and trust in the authorities through the equal application of laws and regulations, the effective enforcement of standards and the destruction of a model influenced by corruption and unnecessary state interference. It is, therefore, important to foster an environment where individuals are partners rather than simply recipients of reforms. For citizens to absorb the full meaning of the rights they enjoy, their duties and the consequences of their acts is a long process, in which public authorities and citizens must work hand in hand. One of the priorities, therefore, should be to organize environmental information and to make it available to citizens (Tickle and Clarke, 2000, p216). It is also important that interest groups find ways of working together and combine their energies and comparative strengths to achieve common goals of environmental and social protection. A strong environmental movement guarantees that environmental issues remain on the political agenda, while a strong social movement ensures more equitable and just policies.

A reform of the institutions concerned with the environment must take place by shifting responsibilities for environmental management from central governments to local authorities, improving the functional capacity of ministries dealing with the environment, and increasing coordination among all ministries in general. Ministries of the environment should concentrate on policy and supervision and leave the task of implementation to local authorities. In that respect, it is important that local authorities be given the necessary financial means and infrastructure to properly perform their activities. Furthermore, ministries of the environment and other ministries dealing with environmental issues (such as health, energy, transport) should work together to devise and implement coherent environmental and sustainability policies. This would not, in and of itself, be sufficient to ensure environmental justice. Indeed, ministries dealing with the environment and those concerned with social, economic, financial and industrial affairs should also build links to ensure coordination of their respective policies.

Strengthening state institutions will result in more efficient law enforcement not only through the improvement of the state's capacity to implement laws and environmental standards but also through the enhancement of its public image. A restored public image will enable the state to gain the citizens' trust and their participation in the decision-making process. This will finally ensure that, in addition to being fair, just and equitable, environmental legislation and policies are implemented more efficiently. State institutions must also be strengthened to combat the informal economy. The ultimate dissolution of the informal economy and a better functioning of market mechanisms will improve the efficiency of any economic instrument used to protect the environment.

Environmental justice: A pan-European perspective

The development of a concept of environmental justice in the CEE region must be assessed in the light of the more general context in which the transition is taking place. It is imperative to devise a pan-European approach to the protection of the environment managed as part of a common strategy. The pressures of a multifaceted transition have stretched financial resources and created priorities in the transformation process. Although gaps between CEE and EU legislation have been substantially reduced, even those countries that seem set to complete negotiations for EU accession by the end of the year 2002 still need to work hard towards the effective implementation of the environmental acquis (EC, 2001).

In a 1998 document the European Commission sketched out the environmental challenges lying ahead of the accession of CEE countries to the EU. The paper contained a series of considerations the EC believed the candidate countries needed to take into account in order to develop their national strategies for achieving full compliance with the environmental acquis (EC, 1998). The document highlighted the adoption of the acquis as the only way to protect the European environment effectively. The tensions caused by the pressures of shifting suddenly from one extreme (economic and democratic reforms) to another (environmental reforms) generated serious risks of creating further environmental injustices. With the current emphasis on the implementation of the acquis, it is imperative that the European Commission recognizes that the implementation of a large and complex body of EU directives and regulations in a region crippled by poor environmental conditions may only occur progressively.

Without bailing out CEE countries from their environmental obligations, the EU should lean towards a more understanding attitude to help CEE countries accommodate specific national circumstances. It should be borne in mind that the economic development of the EU has not been free from serious contradictions in its policies aiming towards sustainability. The standards expected of CEE countries should not be more stringent than those currently in force in Western Europe. The number of environmental disputes reaching the European Court of Justice bears witness to the difficulties that have so far marred the implementation and enforcement of EU standards by existing member states. In other words, it would be simply unjust to expect CEE countries to reach a level of environmental protection in hardly more than a decade that was achieved progressively by EU member states over a period of 40 years.

It is submitted that the current asymmetrical character of the relationship between East and Western Europe is not warranted by the rate of success in terms of environmental protection in the current EU member states. It should be remembered that despite the emergence of the first signs of a green movement in the 1960s, the environment only struck a chord with governments in the West when the link between continued economic growth and protection of the environment was clearly established (REC, 1994). Only then did comprehensive legislation to tackle pollution and resource related problems develop, mainly for economic purposes, although the role of anti-nuclear and social protests cannot be ignored. Despite a successful approach, the protection of the environment in Western Europe is also marked by tensions between economic growth and environmental protection (see Blowers' account of ecological modernization in Chapter 3). Limited success in formulating EU legislation on environmental liability, the negative environmental consequences of some current and past developments, including agricultural policies, and an imperfect understanding of the impact of their implementation in CEE countries, should make us wary of a blindfolded integration of EU standards into CEE law and policy.

However, the creation of an effective partnership between CEE countries and the EU is slowly replacing the asymmetrical character of their recent relationship. The participation of CEE countries as partners in the elaboration of a pan-European environmental strategy has been partly realized with the new partnership agreements implemented in 1999 (EC, 2000). These instruments allow CEE countries to voice their current concerns and implementation problems. This might help devise a policy that responds equally to the needs of the various regions of the continent. The progressive treatment of candidate countries on an equal footing to EU member states should promote mutual trust. Although any disengagement from further EU integration now appears impossible, careful planning of reasonable targets within a manageable time frame is necessary to maintain citizens' support for integration both in the EU and CEE.

Promotion of better institutional and corporate governance at the international level

In order to gain public support for environmental and economic policies, long term benefits must realistically and visibly outweigh the short term costs that sustainable development programmes impose upon citizens. More energy should be devoted to devise mechanisms through which economic reforms would lead to enhanced living conditions. Efforts should be put in providing a workable timetable for CEE countries to achieve their goals. International organizations must acknowledge that genuine environmental and social consciousness and progress cannot be imposed or imported from abroad. They should help CEE governments generate the necessary conditions for this consciousness to form within society by being more responsive to the needs of CEE.

IFIs and foreign investors should seriously reassess the principles that guide their actions in CEE countries in order to play a more constructive and responsible role in the transition. IFIs should incorporate good governance criteria in their lending activities and their periodic consultations with CEE countries. Support should be linked not only to the adjustment of economic and financial imbalances and to the implementation of structural economic reforms but also to the elimination of corruptive practices and mismanagement of public funds (Thomas, 1999, pp560–62).

The development of environmental standards applicable to MNCs would go a long way towards preventing them from moving their operations to countries where more lenient regulations are in force. As yet the regulation of MNCs under international law or simply under voluntary international codes of conduct has not been successful. The legal separation between the parent company and its subsidiaries prevents the country hosting a subsidiary from piercing the corporate veil and regulating the activities of the parent company. Moreover, in CEE countries, regulation of the subsidiary by the host country might be ineffective unless some of the problems analysed earlier in this chapter are overcome. In the absence of efficient regulations in the host country, it might be appropriate to prevent the parent company from exporting polluting technologies or to force subsidiaries of MNCs to comply with standards elaborated in the home country of the parent company. This option might entail an extraterritorial application of laws in the territory of another sovereign state and have political and legal implications (Fowler, 1995, pp27–28). It would, however, exercise sufficient pressure on MNCs not to engage in activities that might be harmful to local communities and their environment. It might be worth considering this option as a temporary measure until the emergence of strong political arguments forcing the industry worldwide to develop ethical business practices recognizing the importance of the environment in the governance of corporate affairs.

Fortunately, pressure from public opinion and NGOs is forcing MNCs to respond to calls for corporate responsibility by adopting voluntary corporate governance rules (Fowler, 1995, pp28–29). Internal environmental management codes demonstrate a company's commitment to society. The implementation of such codes aims at increasing public and state approval for the company's activities by showing that the company is interested in protecting the environment and in promoting ethical and transparent economic practices (Baram, 1994, pp60–61). Internal codes also serve as a tool for policy review and a measuring yardstick against which MNCs may compare themselves and keep employees abreast of a corporation's objectives. There is no reason why such internal codes should not be observed by subsidiaries of MNCs operating in CEE countries. After all, it is in the interest of ethically minded MNCs to convince CEE countries and their citizens that their motivation in operating in the region is not to take advantage of an improperly regulated environment.

As a short conclusion to this section, it appears obvious that in order to develop a model of environmental justice suited to the needs of CEE countries, it is necessary to balance the three key elements that have affected environmental policy since the end of the Communist era: political, economic and institutional transformation; international agreements; and assistance and the role of the public.

Conclusion: The Citizen as a Key Component of the Compatibility of Environmental Protection, Economic Growth and Environmental Justice in CEE

Are environmental protection, economic growth and environmental justice compatible in CEE countries, ie is sustainable development possible? This analysis suggests that at present, they are not, and it is not. The reasons are manifold and have been reviewed in the second section of this chapter. The third section however, leads us to conclude that there is no reason why these equally valuable objectives should not in principle be compatible. For them to be compatible, citizens should play a key role in the formulation of public policies.

Ultimately, citizens are the main beneficiaries of sustainable development: economic development and growth, environmental protection and social equity. It is therefore natural that citizens should be consulted, and play an active role, if not the principal role, in the decision-making process on matters regarding these three main components of their living environment. By acknowledging the central role of citizens as the ultimate providers and consumers of policies, and by ensuring that citizens be well informed and aware of their civic responsibilities, a balance may be struck between the costs and benefits of economic development and growth, environmental protection and social equity.

As illustrated in Figure 14.1, the concept of environmental justice incorporates all these elements and enables citizens to play this central role through the practices of good governance. The underlying assumption is that citizens are, after all, the only actors involved in the transition or development process who know what degree of economic growth, environmental protection and social equity is desired and the price they are willing to pay for achieving it. To give an informed judgement, citizens need education, access to information and a number of instruments and institutions to rely on. These include:

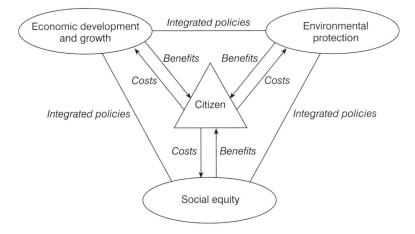

Figure 14.1 *A model of environmental justice for CEE*

- an established state apparatus with efficient legislative, executive and judiciary capabilities;
- effective democratic practices;
- a well-functioning market economy; and
- a strong and independent civil society.

Citizens, both as policy-makers and policy consumers, should take centre stage in the reform process. In addition, the decision-making process should also involve interest groups and NGOs. CEE countries should also muster the means to exercise increased influence on the activities of local subsidiaries of MNCs. Moreover, in CEE, the achievement of environmental justice should integrate a pan-European perspective and the involvement of CEE countries as equal partners in their relations with the EU and IFIs. A clear vision of the goals to be achieved and the alternative paths leading to them is also necessary.

This chapter only sketches a very general portrait of the current economic, social and environmental conditions in the region. It must be borne in mind that the level of protection of the environment varies greatly from one part of the region to another as CEE countries differ in legal traditions, stages of development and in the way they value the environment. Differences arise from cultural, social, political and economic factors and are partly determined by conditions prevailing before the transition (EBRD, 1999). Action at the regional level would foster greater uniformity of environmental legislation and eliminate discrepancies between national norms. The pollution problems are not local, they are regional; the causes of the deterioration of a state's environment might require action that must be undertaken beyond its boundaries. To develop a truly European environmental policy, cooperation among CEE countries and between EU and CEE countries is necessary as they share a common environmental heritage. First, however, it is imperative that each CEE country establishes its own equilibrium, its own version of sustainability. This will require the development of a balance between environmental protection, economic development and growth and social equity. Public acceptance can only be gained through the participation of citizens and the equitable sharing of the burden of reforms in CEE countries.

NOTES

1   The expression Central and Eastern Europe (CEE) in this chapter refers primarily to the ten countries that have concluded association agreements with the European Union: Bulgaria, Czech Republic, Estonia, Hungary, Latvia, Lithuania, Poland, Romania, Slovak Republic and Slovenia.
2   PHARE is the main channel for the European Union's financial and technical cooperation with CEE countries to support their economic and political transition (Casalino, 1995).
3   Accession shall occur when the applicant state will meet certain conditions laid down in June 1993 by the Copenhagen European Council:
   •   the applicant state must have a functioning market economy with the capacity to cope with competitive pressures and market forces within the EU;
   •   the applicant state must have achieved stability of institutions guaranteeing democracy, the rule of law, human rights and respect for and protection of minorities;
   •   the applicant state must be able to take on the obligations of membership, including adherence to the aims of economic and political union; and
   •   the EU must be able to absorb new members and maintain the momentum of integration.

REFERENCES

Baram, M S (1994) 'Multinational corporations, private codes, and technology transfer for sustainable development', *Environmental Law*, vol 24, pp33–65
Beckmann, A (2000) 'The upside of top-down', *Central Europe Review*, vol 2 (3), http: www.ce-review.org/00/3/beckmann3.html
Bowman, M and Hunter, D (1992) 'Environmental reforms in post communist Central Europe: From high hopes to hard reality', *Michigan Journal of International Law*, vol 13, pp921–80
Casalino, J (1995) 'Shaping environmental law and policy of Central and Eastern Europe: The European Union's critical role', *Temple Environmental Law and Technology Journal*, vol 13, pp227–56
Costi, A (1999) 'Reconciling environmental justice and development in transition economies: The Central and Eastern European reality' in K Bosselmann and B Richardson (eds) *Environmental Justice and Market Mechanisms: Key Challenges for Environmental Law and Policy*, Kluwer, London
European Bank for Reconstruction and Development (EBRD) (1997) *Transition Report 1997*, EBRD, London
EBRD (1999) *Transition Report 1999*, EBRD, London
European Commission (EC) (1998) *Communication from the Commission to the Council, the European Parliament, the Economic and Social Committee, the Committee of the Regions and the candidate countries in Central and Eastern Europe of 20 May 1998 on accession strategies for the environment: meeting the challenge of enlargement with the candidate countries in Central and Eastern Europe*, COM (98) 24 final

EC (2000) *Regular Reports from the Commission on progress towards Accession by each of the candidate countries*, http://www.europa.eu.int/comm/enlargement/report_11_00/index.htm#Progress

EC (2001) *Political Documents related to the Enlargement Process: Strategy Paper 2001*, http://www.europa.eu.int/comm/enlargement/report2001/index.htm#Strategy Paper 2001

Fowler, R J (1995) 'International environmental standards for transnational corporations', *Environmental Law*, vol 25, pp1–30

Grapska, G (1994) 'The legacy of anti-legalism', *Poznan Studies in the Philosophy of the Sciences and the Humanities*, vol 36, pp200–35

Grochowalska, J (1998) 'The implementation of Agenda 21 in Poland', *European Environment*, vol 8, pp79–85

Havlicek, P (1997) 'First steps towards a cleaner future', *Environment*, vol 39, pp16–42

Henckaerts, J-M and Van der Jeught, S (1998) 'Human rights protection under the new constitutions of Central Europe', *Loyola of Los Angeles International and Comparative Law Journal*, vol 20, pp475–506

Holmes, S (1997) 'Crime and corruption after communism', *East European Constitutional Review*, vol 6, pp69–70

Jancar-Webster, B (1993) 'The East European environmental movement and the transformation of East European society' in B Jancar-Webster (ed) *Environmental Action in Eastern Europe: Responses to Crisis*, M E Sharpe, Armonk, New York

Jendrocka, J (1996) 'Drafting new environmental law in Poland: Radical change or merely reform?' in G Winter (ed) *European Environmental Law: A Comparative Perspective*, Aldershot, Dartmouth

Joseph, S (1999) 'Taming the leviathans: Multinational enterprises and human rights', *Netherlands International Law Review*, vol 46, p172

Klarer, J, Francis, P and McNicholas, J (1999a) *Improving Environment and Economy: The Potential of Economic Incentives for Environmental Improvements and Sustainable Development in Countries with Economies in Transition*, Regional Environmental Center for Central and Eastern Europe, Szentendre, Hungary

Klarer, J, McNicholas, J and Knaus E-M (1999b) *Sofia Initiative on Economic Instruments: Sourcebook on Economic Instruments for Environmental Policy in Central and Eastern Europe: Abridged Version. A Regional Analysis*, Regional Environmental Center for Central and Eastern Europe, Szentendre, Hungary

Manser, R (1998) *The Squandered Dividend: The Free Market and the Environment in Eastern Europe*, Earthscan, London

Massam, B H and Earl-Goulet, R (1997) 'Environmental nongovernmental organizations in Central and Eastern Europe', *International Environmental Affairs*, vol 9, pp127–47

Meth-Cohn, D (1999) 'Was it worth it?', *Business Central Europe* , November, pp14–19

Organisation for Economic Co-operation and Development (OECD) (1999) *Environment in the Transition to a Market Economy: Progress in Central and Eastern Europe and the New Independent States*, OECD, Paris

Pavlinek, P (1996) 'Challenges of local environmental management in the Czech Republic', *The Bulletin*, vol 6 (2), http://www.rec.org/REC/Bulletin/Bull62/transition.html

Pickles, J and Smith, A (1998) *Theorizing Transition: The Political Economy of Post-Communist Transformations*, Routledge, London

Regional Environmental Center (REC) (1994) *Manual of Public Participation in Environmental Decisionmaking*, Regional Environmental Center for Central and Eastern Europe, Szentendre, Hungary

REC (1998) *Doors to Democracy: Pan-European Assessment of Current Trends and Practice in Public Participation in Environmental Matters*, Regional Environmental Center for Central and Eastern Europe, Szentendre, Hungary

Sajo, A (1996) 'Rights in post-Communism' in A Sajo (ed) *Western Rights? Post-Communist Application*, The Hague

Sands, P (1995) *Principles of International Environmental Law*, Manchester University Press, Manchester

Schwabach, A (2000) 'The Tisza cyanide disaster and international law', *Environmental Law Reporter*, vol 30, 10509

Stec, S (1998) 'Ecological rights advancing the rule of law in Eastern Europe', *Journal of Environmental Law and Litigation*, vol 13, pp275–358

Stursa, J (1998) 'Research and management of the Giant mountains' arctic-alpine tundra (Czech Republic)', *Ambio* vol 27, pp358–60

Szacki, J, Glowacka, I, Liro, A and Szulczewska, B (1993) 'Political and social changes in Poland: An environmental perspective' in B Jancar-Webster (ed) *Environmental Action in Eastern Europe: Responses to Crisis*, M E Sharpe, Armonk, New York

Szejnwald Brown, H, Angel, D and Derr, P (1998) 'Environmental reforms in Poland: A case for cautious optimism', *Environment*, vol 40, p10–38

Thomas, C (1999) 'Does the "good governance policy" of the international financial institutions privilege markets at the expense of democracy?', *Connecticut Journal of International Law*, vol 14, pp551–62

Tickle, A and Clarke, R (2000) 'Nature and landscape conservation in transition in Central and South-Eastern Europe', *European Environment*, vol 10, pp211–19

Weiner, D R (1988) *Models of Nature: Ecology, Conservation and Cultural Revolution in Soviet Russia*, Indiana University Press, Bloomington, IN

Weintraub, B A (1995) 'Environmental security, environmental management and environmental justice', *Pace Environmental Law Review*, vol 12, pp533–623

Chapter 15

The Campaign for Environmental Justice in Scotland as a Response to Poverty in a Northern Nation

Kevin Dunion and Eurig Scandrett

INTRODUCTION

Scotland is a country on the north-western periphery of Europe, with a population of about 5 million. Half of the population lives in a narrow band running from Glasgow to Edinburgh: the Central Lowlands, whereas the north and west of the country is made up of old glaciated mountains and islands with a degraded vegetation cover and a sparse, isolated population. The country is part of the so-called 'developed' world. But herein lies a paradox: resource consumption characteristics are like those of other countries in the North, yet poverty and inequality are significant – one child in three lives in poverty. Although part of the UK for nearly 300 years, Scotland has retained many institutional separations from the rest of the UK, such as the legal system, education, local government services and health; 1999 saw the election of a parliament with democratic control of these institutions, after years of struggle in civil society, in which Friends of the Earth participated.

An analysis of the new Scottish context exposes the maldistribution of environmental goods and costs at local, regional and global levels, as well as the opportunities which are emerging in these early stages of devolution. This context provides challenges for non-governmental organizations (NGOs) working for environmental justice, which in many cases are not addressed by the democratic participation models of Local Agenda 21, and may go beyond the comfort zone of much traditional 'environmental engagement'. Friends of the Earth Scotland, an independent NGO within the Friends of the Earth International umbrella, has recently adopted a clear position in favour of environmental justice. The strategy adopted may thus be of interest to NGOs working in other parts of the world.

Environmental justice incorporates both the struggle for an equitable distribution of the environmental benefits and costs of economic development, and a commitment to decreasing such costs, so that no one should suffer. Environmental injustices in Scotland are realized across a range of spatial and temporal levels.

First, as in the rest of the UK, poorer and less powerful social groups have an increased risk of living in a degraded environment than wealthier or more powerful groups (Walker, 1998, Boardman et al, 1999). There is a high incidence of sources of environmental damage close to communities with higher levels of poverty or social disadvantage, and the environment inhabited by poorer people tends to be of a lower standard than the environment of wealthier people (McBride, 1999).

Second, Scotland, being on the periphery of the European economic bloc, tends towards economic activities which are environmentally detrimental, in order to maintain a comparative advantage. In particular, Scotland acts as a source of natural resources for development which occurs elsewhere. Consequently it suffers disproportionately from associated environmental damage. In addition, there is evidence, like in the US, that the implementation of some environmental regulations is weaker than in England.

Third, Scotland shares responsibility for international environmental injustice. The elevated consumption of natural resources contributes to circumstances whereby globally, less powerful groups suffer disproportionate environmental damage and enjoy fewer benefits. Scotland's contribution to globally experienced damage is also disproportionate to needs. For example, despite being geographically well placed to benefit from renewable sources of energy (mountains, high rainfall, high wind levels, large coastline), Scotland emits nearly ten times its environmental space levels of carbon dioxide, largely from domestic and industrial fossil fuel generated energy use (Friends of the Earth Scotland, 1996).

Friends of the Earth Scotland has positioned itself clearly as an environmental justice organization. Acknowledging the debt to black communities worldwide for developing the environmental justice analysis, Friends of the Earth Scotland has attempted to recontextualize this analysis for the changing Scottish political situation. In our usage, environmental justice combines challenging the unequal social distribution of poor environments, with the requirement to meet resource reduction targets determined by the environmental space (Friends of the Earth Scotland, 1996, McLaren et al, 1998, McLaren, Chapter 1). This has been encapsulated in Friends of the Earth Scotland's vision statement and strapline 'Campaigning for environmental justice: no less than a decent environment for all; no more than our fair share of the Earth's resources'. Our conception of environmental justice therefore brings together the need for global and intergenerational equity in resource consumption and ecological health, with a priority to act with those who are the victims of that inequality in the present.

This chapter will provide case studies of community action, initial outcomes of research, and an account of the influence of an environmental justice analysis on Friends of the Earth's approach to the developing policy context in Scotland.

SCOTLAND'S ECONOMIC CONTEXT

The global accumulation of capital in the late 20th century has shifted the dynamics of environmental conflict. It has weakened both the power of the state to regulate, and of the labour movement to resist economic, social and environmental fragmentation (Stephens, 1998). It has also created a series of regional economic blocs with marginalization occurring outside central areas of growth. In Europe, the central area of economic activity increasingly occurs within a triangle between London, Berlin and Turin (European Commission, 1995). This is surrounded by a corona of supportive industrial development, beyond which is a peripheral economic area which includes Scotland. The periphery is characterized by poor economic development and capital flight, overdependence on inward investment and public grants, and lower regulation standards and practices. It is also used as a source of primary natural resources for the benefit of the economic development within the centre.

With its peripheral economic and geographical location within Europe, Scotland now has some similarities with post-colonial countries of the South (Dunion, 1996, Gray, 1996). There are high levels of poverty: one quarter of Scottish households live below half-average income. Much of the poverty relates to post-industrialism, in both urban and rural areas, but also much rural poverty is associated with a pre-industrial history of concentration of power and the geographical and political marginalization of the population. Large parts of Scotland rely on diminishing redistributional aid from Europe. The distribution of land ownership is the most unequal in Europe, with approximately one fifth of land owned by 66 landowners in estates of 30,700 acres or more (Wightman, 1999).

It may be difficult for some Southern groups to accept that there is structural poverty in the North, which needs to be addressed. Equally, what is difficult for many groups in the North to grasp is that we cannot address such poverty by requisitioning even more of the planet's finite resources. Traditionally the political Right has presumed unfettered access to unlimited resources resulting in a trickle down of growth which would benefit the poor. However, there is also a specific challenge to the Left in Scotland, which has accepted a perspective that environmentalism is a middle-class preoccupation and which regards environmental action as, at best a desirable objective to be pursued once socio-economic priorities have been addressed, and at worst as an obstacle to industrial development and conventional growth. What needs to be recognized therefore is first, as Scotland must in total reduce its share of the world's resource consumption to contribute to global sustainable development, it must therefore redistribute access to a share of such resources nationally and internationally. Second, it is the poor in our society who suffer from the most degraded environments, conditioning their quality of life. Far from accepting that environmental deterioration is the price to be paid for economic development, they resist its effects, and often the presence of the economic activities which compounds this degradation.

POLITICAL DEVELOPMENTS

Scotland's devolved parliament was instituted with a declared commitment to openness and accessibility in policy-making, particularly through greater power of the parliament and its committees relative to the executive. There was an expectation that the parliament would lead to an era of progressive policies given that the Scottish electorate had traditionally supported social democratic parties (Dunion, 1999). The first executive was formed by an alliance of centre-left parties, Labour and Liberal-Democrat, with the main opposition from the traditionally left-leaning Scottish Nationalists (SNP). Proportional representation has also allowed for the election of Britain's first Green parliamentarian, and of a Scottish Socialist Party member. A broad liberal consensus in mainstream civil society has demonstrated a capacity to organize through the Scottish Civic Forum (Scandrett, 2000a).

There is some evidence of the executive's commitment to progressive change such as legislation in its first year on land reform, National Parks and freedom of information. It has, however, followed an ideological approach close to Tony Blair's New Labour government in the UK, characterized by economic non-intervention and constitutional modernization, although with greater sympathy with the rhetoric of social justice. However, the parliament is vulnerable to populism. The SNP's tactics of opportunism cause it to alternate between espousing green principles (ie anti-nuclear) while at times aligning itself with reactionary groups in civil society. Along with the smaller, right-wing Conservative Party, it has led a backlash against specific attempts at progressive environmental and social change. For example, legitimate concerns about mobility in rural and remote parts of Scotland (fuel costs are higher than in urban areas and public transport is poor or absent) have been exploited to resist progressive environmental measures to tackle climate change. This is consistent with their opposition to road tolling in the metropolitan Central Lowlands, and the climate change levy in industry, so that, while claiming to accept the environmental challenge of tackling climate change, they oppose the fiscal and regulatory measures to cut emissions and offer no alternatives in return (Dunion, 2000). Moreover, the development of a conservative alliance in civil society between the road lobby, industry, rural interests and sections of the press serve to provide resistance to progressive developments in the future.

Among the first pieces of legislation enacted by the Scottish parliament were laws aimed at land reform. Land reform struggles have been a prominent manifestation of class conflict, (land grabbing in Honduras, the Landless Peoples' Movement in Brazil (Kane, 2000)), and of post-colonial settlements (ie recent developments in Zimbabwe), and are primarily aimed at redistribution of this resource. They have been little associated with Northern nations in recent history. Land reform legislation in Scotland has abolished the 500 year old feudal land tenure system and permitted limited community ownership, but the detail of the legislation does not allow for redistribution of land and even jeopardizes redistribution by conferring absolute ownership on existing landlords. As Wightman (1999) puts it 'Scotland, in one of the first acts of our new Parliament, is to be given away to those, who in many cases have inherited their interest from the ancient theft of the land many centuries ago'.

Although the Scottish Executive has a commitment to place sustainable development at the heart of policy, it has failed to articulate a sustainable development strategy. It has set no targets or objectives, for example on resource reduction, climate change emissions, waste minimization and recycling or road traffic reduction. Until recently it had resisted demands for a set of sustainability indicators (Dunion, 2000). However, it has clearly set itself a trajectory which would allow it to focus on those areas of environmental action which most impinge upon issues of environmental justice by deciding to establish a framework of executive action around waste, energy and transport. As Pearson (2000), on behalf of the Scottish Executive, notes, 'we felt a need to present sustainable development in a way which made it more of a day to day concern, a matter for mainstream business. Waste, energy and travel represent major human impacts on the environment and on sustainable development. We can readily relate them to effects on biodiversity, climate change, health and, with thought, to the social changes we wish to see'.

In the context of a dominant culture of liberal progressivism, and powerful conservative populism, the Scottish environmental movement needs to position itself carefully. Some British environmentalists have argued that a discourse of environmental modernization is an effective means of promoting the environmental agenda within existing Labour ideology (Jacobs, 1999). This approach emphasizes the common interests between environmental concerns and economic development (in particular, the interests of business) and advocates an expansion of the use of market mechanisms of environmental protection and a reduction in regulation. While this may be convincing to UK policy-makers, the discourse of environmental justice may be more conducive to the ideology of the Scottish Executive, and to popular concerns of fairness (see Blowers critique of ecological modernization in Chapter 3). Markets are poor distributors of costs and benefits, and environmental justice requires a greater range of instruments, including those inimical to the interests of business.

CASE STUDIES

The impacts of the maldistribution of environmental costs and benefits are primarily experienced at a local level. The causes and impacts of environmental injustices are frequently geographically separated, so it is by locating environmental action in the communities who are the victims that a coalition for environmental justice can be built. The following case studies give examples of how NGO action at this level can serve to expose wider processes and make alliances with communities.

Greengairs landfill

Greengairs is a working-class community in North Lanarkshire, surrounded by opencast coal mines and landfill sites, which take waste from the Glasgow conurbation (population 1.8 million) and beyond. Planning preferences for siting waste operations on devastated areas, combined with a history of unemployment and exclusion from decision-making, led to a concentration of environmental damage around the community. In 1998 evidence was discovered in the Environmental News Digest Service (ENDS, 1998a), that soil contaminated with polychlorinated biphenyls (PCBs) was being imported from Hertfordshire in southern England to one of the landfill sites operated by Shanks Waste Solutions. Negligence on the part of the Scottish Environment Protection Agency led to a failure to upgrade Shanks' licence at Greengairs, and the company exploited this loophole quickly.

Friends of the Earth Scotland released the information to the community and provided technical and organizational support as the community mobilized to blockade the site. The company was pressed into accepting a number of concessions, including an end to the toxic dumping, an independent inspection of the landfill site and improvements to its safety. The community's now ongoing environmental forum monitors environmental conditions in the area and secures funds for environmental improvements.

Harris Superquarry

In March 1991 Redlands Aggregates Limited (now owned by multinational Lafarge), applied for permission to extract 10 million tons of rock per annum over 60 years by quarrying Roineabhal, a mountain on the south tip of the Isle of Harris. Roineabhal is one of 20 sites in Scotland identified for the potential development of these self-styled 'superquarries'. The aggregate was to be exported by bulk carriers from an expanded harbour at Lingerbay, for construction projects in the south-east of England and Europe. The application led to the longest running public inquiry into any development in Scotland which, unusually, was held some 60 miles from the site, in Stornoway, the main town of the Western Isles (Mackenzie, 1998).

Traditionally, aggregates have been a low-value product which have been quarried relatively close to their end-use point. Projected high demand in the 1990s for quarried materials has since been found to be erroneous (Cowell et al, 1998). This led to the industry abandoning any pretence to the proximity principle and favouring instead large scale operations in so-called 'remote areas' to avoid known local hostility to quarries in the south of England. Instead of exploring how to manage demand, and to look to alternative and recycled sources, aggregate companies simply looked for the lowest common denominator: more isolated areas with poorer populations. During the inquiry, local opinion in the Western Isles changed from 62 per cent in favour, to 68 per cent against, as knowledge of the scale of the operation, the nuisance and health impact on local communities and the negative economic effect on tourism and fisheries, became more widely known. On 3 November 2000 the Environment Minister rejected the application. However, at the time of writing, this decision was subject to legal appeal by Lafarge Redlands.

Douglas Incinerator

In 1996 the waste incinerator at Baldovie Industrial Estate in Dundee closed, following Europe-wide improvements in emission standards. There had been ongoing protests and complaints from residents of the neighbouring public sector housing estate about emissions of dioxin and other pollutants. Plans for an improved 'waste to energy' incinerator were opposed both by the community and Friends of the Earth Scotland, but permission was granted and the plant was built in 1999 under a public private partnership company Dundee Energy Recycling Limited (DERL).

During construction there were local concerns, not just over emissions, but also truck movements through the estate and near schools, and a minor explosion provoked anxieties about the implications of an accident. Despite opposing incineration as a means of waste management, Friends of the Earth worked with the community to explore how the company could be made more accountable. Meetings with residents' associations, local environmental groups, trade union council representatives and community workers raised the possibility of a 'Good Neighbour Agreement', a model originating in the US to provide a degree of accountability from large industries to their neighbouring communities (Lewis and Henkels, 1998).

Contact with made with the plant manager and the local elected councillor, and support was obtained from the local authority and the DERL board. In May 2000 a 'Good Neighbour Charter' was signed by residents' representatives and the chair of DERL, committing the company to negotiate a Good Neighbour Agreement including access to information, preparation for accidents, improving environmental performance and safer transportation of waste. Within weeks of the signing of the charter, newspaper coverage of research findings about the health risks from dioxin levels below UK standards precipitated a meeting of the liaison group, which led to a doubling of the frequency of monitoring for dioxins to ensure that emission levels stayed low.

CLASS, RACE AND ENVIRONMENTALISM

The focus of environmental action on the economically marginalized victims of environmental injustice highlights questions of the class orientation of environmental NGOs. 'Mainstream' environmental NGOs have often been criticized, including by the environmental justice movement, for white middle-class domination (Bullard, 1993, O'Leary, 1996, Lean, 1998). Lowe and Goyder (1983) in their study of British environmentalism highlight the domination of middle-class membership, and Yearley (1994) focuses this environmental class allegiance more specifically to the professional (*sensu* Williams, 1989) class, as distinct from the bourgeoisie. Guha and Martinez-Alier (1997) contrast post-materialist, Northern middle-class environmentalism with a 'materialist environmentalism of the poor'. This latter environmental movement is primarily in the South, but includes Northern groups such as the US Environmental Justice movement. Their caricature of mainstream environmentalism, while helpful in highlighting subaltern environmental groups, inadequately explains phenomena such as environmental movements in Central and Eastern Europe (see Costi, Chapter 14), and the wide range of political allegiances of environmentalists in the West, from European green-socialist parties (Wainwright, 1994) to solidarity action by environmental NGOs. As the case studies above show, materialist environmentalists in the North can make common cause with NGOs such as Friends of the Earth. While Friends of the Earth Scotland's membership is predominantly professional (50 per cent of membership has a household income of over £20,000 (US$30,000)), the organization's work is frequently in working-class, isolated and poor areas, which reflects the communities defending their own environment against polluting, toxic and dangerous facilities.

Williams' (1973) analysis of the relationship between class (structure) and culture (superstructure) is useful in understanding such changing relationships. The range of values, meanings and practices which make up environmentalism is shaped, but not determined by, the class position of the membership of NGOs and their intellectuals. They will include ideologies which reinforce the interests of the dominant class, but also which support other, subaltern groups. A tension therefore arises in the environmental movement between cultural incorporation by dominant ideologies and solidarity with the interests of working-class and other oppressed groups (Sklair, 1994, Scandrett, 1999). This has implications for the political praxis of NGOs.

A similar question may be directed in relation to race. The US environmental movement has been clearly divided along race lines, between the white traditional NGOs and the people of colour environmental justice movement. The Scottish minority ethnic population is relatively small (1.3 per cent of population at 1991 Census), predominately Asian, and not geographically segregated. These communities experience the 'working class plus' (Agyeman and Evans, 1996) effects of racial discrimination, institutionalised in a range of forms including environmental. In Scotland, religion has traditionally divided communities, with roots in anti-Irish racism, and the perception of some communities is that sectarianism has given rise to environmental injustices. Agyeman (1990, 2000) has analysed institutional racism in the UK environmental movement, and argues for a broader understanding of environmental racism, as 'a perspective which both creates and legitimizes environmental degradation and poor living and working environments globally for those "others" who are viewed as economically, politically and socially subordinate' (Agyeman and Evans, 1996).

OPPORTUNITIES FOR ENVIRONMENTAL JUSTICE IN SCOTLAND

Improving access to environmental rights and procedural justice is a tool which is frequently used by NGOs, and constitutional environmental rights have been advocated as a means of delivering environmental justice (Hayward, 2000). The European Convention on Human Rights applies to all Scottish legislation, and whilst this does not institute constitutional environmental rights, Article 8 of the Convention (which protects the individual's rights to respect for his/her home, private and family life) has been used successfully in a few European cases to protect environmental rights (Thornton and Tromans, 1999).

More usefully, the Convention provides an opportunity for procedural rights which are currently denied in Scotland, such as the right of appeal in planning decisions, which is currently available only to developers. Friends of the Earth Scotland campaigns for widespread and proactive rights to information such as a Toxic Release Inventory (available in the US and in England and Wales) as well as to strengthen the Scottish Executive's Freedom of Information legislation.

However, justice is not simply the existence of rights, but also their delivery. Many studies demonstrate a culture in which access to existing rights is routinely denied, and offenders inadequately policed (Friends of the Earth Scotland, 1992, Matthews, 1994, ENDS, 1998b, 1999, McBride, 1999). The failings in the current system may be partially addressed by the Convention, by the implementation of the Aarhus Convention on 'Access to Information, Public Participation in Environmental Decision-Making and Access to Justice in Environmental Matters' (see Introduction) and the establishment of a freedom of information Commissioner. However, a paradigm shift is needed, whereby simply asking for information from a government agency does not mark the citizen out as a troublemaker and prompt the official, when in doubt, to say 'no'.

Access to environmental rights is necessary but inadequate for delivering environmental justice. Despite its failings, Scotland enjoys a democratic and relatively open planning system, which nevertheless regularly delivers the worst environments to the poorest communities. Equal procedural, legal and constitutional rights for all individuals still results in unfair outcomes. On the other hand, a social understanding of environmental justice acknowledges existing inequality and asserts the possibility of collective means of redress.

POPULAR EDUCATION

In addition to the campaigning work for legislative change and policy development focused on the structures which can contribute to greater environmental justice, Friends of the Earth Scotland has adopted a strategy of direct work with communities experiencing the effects of environmental injustice, employing the methodologies of popular education. This approach goes beyond the provision of information about legal rights, to a curriculum and practice which incorporates a political option in favour of environmental victims, joining them in their struggle for justice (Crowther et al, 1999, Scandrett, 2000b). Drawing on the liberation philosophy of Freire and the practice of popular movements in Latin America, popular education involves work with communities which promotes participative and collective learning relevant to their engagement in conflicts with powerful agents (Kane, 2001). Through a series of community-based projects, Friends of the Earth Scotland has aimed for a relationship in which both NGO and community are challenged to develop a critical understanding of each others' interests, and a political practice which can move towards environmental justice.

This is not a simplistic belief that the community is always right, nor a case of retaining an environmental purism while cherry picking communities which support a fixed agenda. The communities of South Harris initially tended to support, and remain divided over the superquarry, however, our commitment was to dialogue with them to seek a shared aim. The community of Douglas, and Friends of the Earth, lost the battle to prevent the Dundee incinerator: the Good Neighbour Charter remains a second best option which retains an ongoing commitment to the concerns of the community and exacts a little more accountability from the industry. Greengairs continues to live with the worst collection of environmental devastation in Scotland, and continues to struggle for its improvement. Moreover, the victims of environmental injustices in Scotland are invited to join a global struggle for sustainability, recognizing that the solutions involve an international redistribution of resource exploitation and environmental damage.

This is not an easy option for an environmental NGO. There is a risk to Friends of the Earth Scotland of alienating our traditional supporters in the professional classes for whom environmental injustice, is, quite literally, 'someone else's problem (Agyeman, 2000, p1). Many environmental victims, even if they are successful in their campaigns, do not move beyond their local struggles. Solidarity between environmentalists and community activists is a journey, not a prerequisite for participation.

CONCLUSIONS: ENVIRONMENTAL JUSTICE AND SUSTAINABLE DEVELOPMENT

Sustainable development has provided a useful nexus for the convergence of discourse(s) allowing parties with conflicting interests to address shared issues. The aim has been to develop win-win solutions, particularly embracing the interests of businesses. The consensual model implicit in sustainable development has led to real environmental improvements, but does not challenge existing power relations (see Blowers, Chapter 3). In practice, sustainable development is often weak on social justice and resource limits, and comprises business as usual with some environmental protection. By contrast, the environmental justice discourse originates in the struggles of oppressed social groups and is developing through dialogue with the environmental movement.

This is crucial. Sustainable development is a vertical dialogue that seeks to incorporate environmental interests into the practice of powerful groups. It is an essential activity but will inevitably be limited by the marginalization or destruction of oppositional elements (Sklair, 1994). The environmental justice discourse however, may be understood as a horizontal dialogue between social movements representing oppressed interests (poor/black) and environmentalism. Dialogue here requires cultural 'negotiation', so that environmentalists genuinely take on the interests of the powerless, and these groups genuinely take on the interests of environmentalists in their own political struggles. The aim of environmental justice therefore becomes a synthesis of political practice of diverse oppressed groups.

For environmental NGOs to adopt this strategy is a risk. There are very real differences between environmental demands and the interests of other groups, and it is never easy to distinguish between a bottom line and sacred cow (see Dobson, Chapter 4). However, the environmental justice discourse does present a challenge, especially to environmentalists in the North, not to accept a version of sustainable development which lets down the poor at home, globally and in the future.

REFERENCES

Agyeman, J (1990) 'Black people in a white landscape: social and environmental justice', *Built Environment*, vol 16 (3) pp232–36

Agyeman, J (2000) 'Environmental justice: From the margins to the mainstream', *TCPA Tomorrow Series*, TCPA London

Agyeman, J and Evans, B (1996) 'Black on green: Race, ethnicity and the environment' in S Buckingham-Hatfield and B Evans (eds) *Environmental Planning and Sustainability*, Wiley, London

Boardman, B, Bullock, S and Mclaren, D (1999) 'Equity and the environment: Guidelines for a green and socially just government', *Friends of the Earth/Catalyst Pamphlet 5*, Aldgate, London

Bullard, R (1993) 'Anatomy of environmental racism and the environmental justice movement' in R D Bullard (ed) *Confronting Environmental Racism: Voices from the Grassroots*, South End Press, Boston

Crowther, J, Shaw, M and Martin, I (1999) (eds) *Popular Education and Social Movements in Scotland Today*, NIACE, London

Cowell, R, Jehlicka, P, Marlow, P and Owens S (1998) 'Aggregates, trade and the environment: European perspectives', a report for the IUCN UK Committee

Dunion K (1996) *Living in the Real World*, Scottish Education and Action for Development, Edinburgh

Dunion K (1999) 'Sustainable development in a small country: The global and European agenda' in E McDowell and J McCormick (eds) *Environment Scotland: Prospects for Sustainability*, Ashgate, Aldershot

Dunion, K (2000) 'On the Scottish road to sustainability?' in A Wright (ed) *The Scottish Parliament: The Challenge of Devolution*, Ashgate, Aldershot

Environmental Data Services (ENDS) (1998a) 'Inconsistent regulation pushes polluted soil into Scotland', *The ENDS Report*, vol 277, pp14–15, Environmental Data Services, London

ENDS (1998 b) 'Scotland's feeble fines for industrial pollution', *The ENDS Report*, vol 285, p51, Environmental Data Services, London

ENDS (1999) 'Prosecution rate, fines remain at low level in Scotland', *The ENDS Report*, vol 295, p51, Environmental Data Services, London

European Commission (1995) 'The Prospective development of the northern seaboard', *Regional Development Studies 18*, Commission of the European Communities, Directorate-General for Regional Policy, OOPEC, Luxembourg

Friends of the Earth Scotland (1992) *Come Clean*, FoES/Consumer Council Edinburgh

Friends of the Earth Scotland (1996) *Towards a Sustainable Scotland*, FoES, Edinburgh

Gray, L (1996) 'Development education in Scotland: a model from SEAD', *Development Education Journal*, vol 4, pp16–18

Guha, R and Martinez-Alier, J (1997) *Varieties of Environmentalism: Essays North and South*, Earthscan, London

Hayward, T (2000) 'Constitutional environmental rights: A case for political analysis', *Political Studies*, vol 48, pp558–72

Jacobs, M (1999) *Environmental Modernisation*, Fabian Society, London

Kane, L (2000) 'Popular education and the landless peoples' movement in Brazil (MST)', *Studies in the Education of Adults*, vol 32(1), pp36–50

Kane, L (2001) *Popular Education and Social Change in Latin America*, Latin America Bureau, London

Lean. G (1998) 'It's the poor that do the suffering', *New Statesman*, London, 16 October, pp10–11

Lewis, S and Henkels, D (1998) 'Good neighbor agreements: A tool for environmental and social justice' in C Williams (ed), *Environmental Victims*, Earthscan, London

Lowe, P and Goyder, J (1983) *Environmental Groups in Politics*, Allen and Unwin, London

McBride, G (1999) *Scottish Applications of Environmental Justice*, unpublished thesis, University of Edinburgh

Mackenzie, A (1998) ' "The cheviot, the stag ... and the white, white rock?": Community identity and environmental threat on the Isle of Harris', *Environment and Planning D: Society and Space*, vol 16, pp509–32

McLaren, D, Bullock S and Yousuf, N (1998) *Tomorrow's World: Britain's Share in a Sustainable Future*, Earthscan, London

Matthews, P (1994) *Watered Down: Why the Law is Failing to Protect Scotland's Water*, Friends of the Earth Scotland, Edinburgh

O'Leary, T (1996) ' "Nae fur the likes of us": Poverty, Agenda 21 and Scotland's environmental non-governmental organizations', *Scottish Affairs*, vol 16, pp62–80

Pearson, G (2000) 'Sustainable Scotland: Sustainable development in the Scottish Executive priorities and progress', public draft document, 20 July 2000, Scottish Executive, Edinburgh

Scandrett E (1999) 'Cultivating knowledge, environment, education and conflict' in J Crowther I Martin and M Shaw (eds) *Popular Education and Social Movements in Scotland Today*, NIACE, London

Scandrett, E (2000a) 'Scotland's parliament, civil society and popular education' in J Thompson, M Shaw and L Bane (eds) *Reclaiming Common Purpose*, Aontas/Concept/NIACE, Dublin, Edinburgh, London

Scandrett, E (2000b) 'Community work, sustainable development and environmental justice', *Scottish Journal of Community Work and Development*, vol 6, pp7–13

Sklair, L (1994) 'Global sociology and global environmental change' in M Redclift and T Benton (eds) *Social Theory and the Global Environment*, Routledge, London

Stephens, S (1998) 'Reflections on environmental justice: Children as victims and actors' in C Williams (ed) *Environmental Victims*, Earthscan, London

Thornton, J and Tromans, S (1999) 'Human rights and environmental wrongs – Incorporating the European Convention on Human Rights: Some thoughts on the consequences for UK environmental law', *Journal of Environmental Law*, vol 11 (1), pp35–57

Wainwright, H (1994) *Arguments for a New Left*, Blackwell, London

Walker, G (1998) 'Environmental justice and the politics of risk', *Town and Country Planning*, vol 67 (11), pp358–59

Wightman, A (1999) *Scotland Land and Power: The Agenda for Land Reform*, Democratic Left Scotland/Luath Press, Edinburgh

Williams, R (1973) 'Base and superstructure in Marxist cultural theory', *New Left Review*, vol 82, pp3–16

Williams, R (1989) *Resources of Hope*, Verso, London

Yearley, S (1994) 'Social movements and environmental change' in M Redclift and T Benton (eds) *Social Theory and the Global Environment*, Routledge, London

Towards Just Sustainabilities: Perspectives and Possibilities

Julian Agyeman, Robert D Bullard and Bob Evans

INTRODUCTION

We believe that the foregoing chapters in *Just Sustainabilities: Development in an Unequal World* contribute to an important and emerging realization: that a sustainable society must also be an equitable society, locally, nationally and internationally, both within and between generations and between species. As McLaren emphasized in Chapter 1 'equity considerations are embedded in all conceptualizations of sustainable development but rarely unpacked'. In unpacking sustainable development, and undertaking a project such as this, which extends and builds on previous work in both environmental justice and sustainability, our original aims were:

- to map some of the key conceptual and practical challenges confronting both the ideas of *sustainability* and *environmental justice* to understand if and how we might see greater linkages between these ideas and their practical actions in the future;
- to address different aspects of the three dimensions of the multiscalar links between environmental quality and human equality.

In doing this, each chapter has extended and developed issues which address these aims which are at the core of sustainability and environmental justice discourses. So the question we must now ask ourselves is 'have we achieved our aims?' This is clearly a subjective question, and one which has many potential frames. In this concluding chapter, we seek to address this question by drawing together some of the central themes and debates that have been addressed by our authors. We will do this first, by reflecting on the linkages between environmental justice and sustainability; second, by making some remarks on the compatibility of these two; and third, by examining the potential for transforming environmental justice activists towards sustainability and that for embedding the principles of environmental justice into the sustainability policy process. Finally, we offer some perspectives and possibilities for environmental justice and sustainability.

Our authors have provided a range of insights into the linkages between environmental justice and sustainability. In particular, Dobson (Chapter 4) has offered a critical perspective on the compatibility of the objectives of each of these. We particularly welcome this critical reflection as part of the debate, which has helped us to focus our thoughts more clearly. In the remainder of this conclusion we take the opportunity to respond to Dobson's critique as an integral part of our overall discussion.

ENVIRONMENTAL JUSTICE AND SUSTAINABILITY

As we stated in our introduction, it is not our task to mount a discussion of the concept of justice. However, we do need to make clear our usage of the term 'justice' within the context of 'environmental justice' and our own book title of '*Just Sustainabilities*'.

We are clear that more sustainable societies will only emerge if those societies begin to demonstrate greater levels of material, social, economic and political equality. Moreover, the same principle will apply between as well as within nations. Our reasoning is that inequality within societies effectively excludes large proportions of citizens from a sense of citizenship and collective responsibility. The sharing of common futures and fates and a concern for 'unseen others' or future generations is unlikely to be engendered in a situation where the gulf between the 'haves' and 'have nots' is substantial. As we reiterate below, there is a growing body of research which supports this thesis.

We recognize that a 'just' society, for example in Rawlsian terms, might produce a materially unequal society, and for this reason, our emphasis in this discussion is not upon 'social justice' as conceived by Dobson. Our usage of the term 'justice' is a direct consequence of our concern to understand 'environmental justice', and we need to emphasize that our principal focus here is upon questions of equity and equality, as opposed to social justice. This is not to underplay the importance of debates around justice, but simply to recognize the inextricable equity–environment link.

The questioning of the proposition that 'social justice is a precondition for environmental sustainability' is both linguistically at odds with and is opposed to our, and most of our contributors' theses. Indeed, Middleton and O'Keefe (2001, p100) argue 'that sustainability can mean nothing unless development is socially just'. While we do not use the word '*precondition*', but the phrase 'inextricably linked', we, like Middleton and O'Keefe, find it hard to see how better quality environments for all, a foundation of a sustainable community, will come about in the midst of growing ecological segregation ('eco-apartheid'): the dire and increasing poverty illustrated so graphically by Rees and Westra (Chapter 5). In supporting our position, we cited three pieces of empirical research in our Introduction, which addressed the multiscalar links between *environmental quality* and *human equality*. Torras and Boyce (1998) looked at the quality–equality link on an international basis; Boyce et al (1999) looked at it between the 50 US states and Morello-Frosch (1997) looked at it in counties of of the State of California. We feel that these pieces of research, which produced the same conclusions at different scales, and our following arguments, allow us to conclude, *in Dobson's terminology*, that social justice and environmental sustainability are inextricably linked, and that the achievement of the latter without greater commitment to the former will be exceptionally difficult.

In support of this, Rixecker and Tipene-Matua (Chapter 12) argue that 'indigenous peoples' resistance to bio-piracy and their desire to control their own futures is *simultaneously* about social, cultural and ecological justice; they are inextricably linked. Using the modern, environmental lexicon, this struggle is fundamentally about sustainability – the ability of peoples to sustain themselves and future generations by ensuring that the diversity, integrity and life force between all organic life forms is nurtured and protected'. Wickramasinghe (Chapter 11) is clearly saying the same from a woman of colour eco-feminist perspective. She argues that 'environmental justice for women whose livelihood systems are local resource-based, will not be achieved as long as stratifications in the conventional social, economic and political systems alienate them'. This linkage between community, the treatment of women and the management of natural resources, is a point well made, and one which has not received nearly enough attention. She argues that it is only through *justice*: through *legal* rights, not *customary* rights, that women's environmental citizenship can be assured, and that they can then play a full role in the management of local environments, resources and spaces.

Another way in which the two concepts are closely interlinked is suggested by McLaren (Chapter 1), who uses the concepts of *environmental space* and *ecological debt* to look both forward and backward respectively, at the concepts of sustainability and justice. He asks, 'does redistribution – or perhaps compensation or reparation – provide a guide for us here too?' Whatever the answer, this past–future reflexivity is essential because it provides a robust analytical framework through which to study the essential *reactivity* of the environmental justice project, and the *proactivity* of the sustainable development project. He notes that: 'environmental space (sustainability with equity) and ecological debt (environmental space with history and justice) offer valuable tools, not only to the campaigner and activist, but also to the academic and policy-maker. Fundamentally they offer a joined-up framework for understanding and promoting both sustainable development and environmental justice'. This a similar point to that of Roberts (Chapter 9) who argues that 'a key difference between the two approaches is the weight assigned to issues of *inter*-generational versus *intra*-generational equity. Sustainable development requires that we give consideration to our own developmental needs, as well as those of generations still to come, while environmental justice prioritizes accountability to those currently alive. This is an important distinction as it influences the manner in which we will respond to the challenges that face our species in the years ahead'.

Using a different tool Rees and Westra (Chapter 5) argue that 'given the size of their citizens' per capita eco-footprints, it should come as no surprise that many high-income countries exceed their domestic bio-productivity by 100 per cent or more. Indeed, many industrial nations impose ecological footprints on the Earth several times larger than their political territories. In effect, the enormous purchasing power of the world's richest countries enables them to finance massive "ecological deficits" by appropriating through commercial trade or natural flows the unused productive capacity of other nations and the global commons'. In this way, today's unjust *ecological deficits* are the contemporary equivalent of the massive *ecological debts* that McLaren argues the developed world has amassed historically. This is clearly an unsustainable and unjust situation, which, as we argued in the Introduction, is not helped by the so-called 'flexible mechanisms' proposed to let countries in the North off the hook regarding their greenhouse gas emissions.

Another facet of ecological debt and deficit is *bio-prospecting* and *bio-piracy*, which the Maori of Aotearoa New Zealand are contesting, using the 1975 Waitangi Tribunal (the WAI262 claim), the Mataatua Declaration on the Cultural and Intellectual Property Rights of Indigenous Peoples and the Resource Management Act (1991). Rixecker and Tipene-Matua (Chapter 12) situate their analysis of injustice within the context of state power, but this time deriving from British colonialism, through the Treaty of Waitangi/Te Tiriti O Waitangi. They note that 'Maori have been proactive about and attentive to such nefarious neo-colonial bio-politics in order to retain their sovereignty and resist (bio)cultural assimilation. In addition to the long standing attempts to regain their lands and resurrect their cultural belief systems, various members in the Maori community have worked tirelessly campaigning for their right to decide their futures within the realm of bio-politics. Alongside their ongoing defence of tino rangatiratanga, or self-determination, Maori have contributed to the national dialogue and politics surrounding the ethics, policies, and implementation of biotechnology'.

ENVIRONMENTAL JUSTICE AND SUSTAINABILITY: COMPATIBLE OBJECTIVES?

Dobson's (Chapter 4) fundamental position is 'that social justice and environmental sustainability are not always compatible objectives'. He states that 'the working hypothesis – to be confirmed or denied – might be that the environmental justice movement is not a movement for environmental sustainability. Let us assume, here, that "environmental sustainability" refers to a reduction in the aggregate tonnage of waste consigned to landfill sites. On the face of it the environmental justice movement's objective is not to reduce aggregate tonnage but to have existing tonnage more fairly distributed across wealth and racial cleavages. This suggests that the sustainability and justice movements have rather different objectives'.

There are two parts to this hypothesis which need to be addressed separately.

The first part of the hypothesis is that 'the environmental justice movement is not a movement for environmental sustainability'. The ideological bedrock upon which environmental justice and the 'environmental justice paradigm'

(Taylor 2000) rests, are the 'Principles of Environmental Justice' (Principles of Environmental Justice, 1991, see Appendix 1). Taylor (2000, p538) states that they 'show a well developed environmental ideological framework that *explicitly* links ecological concerns with labor and social justice' (our emphasis). Indeed, Principle 1 states that: 'environmental justice affirms the sacredness of Mother Earth, ecological unity and the interdependence of all species, and the right to be free from ecological destruction'. This indicates that 'environmental sustainability' is a clear focus for the environmental justice movement, and that the movement's *intent* is non-anthropocentric.

However, we concede that much environmental justice *practice* does differ from this non-anthropocentric stance, and that this is related to two considerable issues. First, in part it is related to what the early environmental justice activists saw as a mainstream environmental movement which cared more about wilderness and wildlife than it did about (especially African-American) urban residents. Roberts (Chapter 9) rightly reminds us of the dangers inherent in many current sustainability and equity formulations, which are selective and anthropocentric. She warns that 'it is this obsession with our own self-importance and the relative unimportance of the non-human world that lay the foundations for the social and economic world order that has produced global environmental degradation and injustice as its by-products. The key question posed by the deep ecologists is: how can we hope to deliver justice or sustainability to members of our own species if we are capable of denying it to the millions of others with whom we share the Earth?' Citing the social ecologist's and eco-feminist critiques, she opts 'for an approach that advocates the need for equity and sustainability that addresses and serves the human and non-human world equally'. We fully endorse this position. Second, most 'environmental justice communities' are in effect 'communities of resistance' (see Peña Chapter 7) who, because of issues of social location, lack of advocacy and resource mobilization, in most cases *react* to issues, rather than anticipating them and developing *proactive* strategies. Most of our authors addressed this issue in some way. Peña does it through his case study of the Sangre de Cristo land grant in the Rio Culebra watershed in Colorado's San Luis Valley, and through 'place-based identity politics'. He cites Castells' (1997) three identity formations: legitimizing, resistance, and project identities, and examines how these different identities are used by farmers to deal with 'lifelong experiences with racial discrimination and racist ideology'. Eady (Chapter 8) looks at 'communities of resistance' through a discussion of the demarcation and other issues inherent in the proposed designation by the state of '*environmental justice communities*' in Massachusetts. Indeed, the proponents behind the Environmental Justice Designation Bill (Senate 1060), which would create 'Areas of Critical Environmental Justice Concern', model this on the fact that the state already has an 'Areas of Critical Environmental Concern' law. Sometimes anthropocentrism is justified as a remedial action.

Agbola and Alabi (Chapter 13) bring another 'communities of resistance'-based concept to the table, that of 'selective victimization'. They argue that 'realizing their rights, the people of the Niger Delta have created a series of resistance movements in order to prevent their environment from further degradation. There is a link between human rights abuse and environmental degradation, which can be explained through the process of selective victimization. With selective victimization, the Niger Delta region is losing critical resources as well as a healthy environment, thereby exposing residents to hazardous environmental conditions while the non-oil producing regions which received the lion's share of the oil revenue are free to live, recreate, procreate and die in a healthy setting'. Taylor (2000, p534) adds that 'their (people of colour) activism revolves around the struggle for civil and human rights-the desire to be treated fairly and with human dignity. It is not surprising therefore, that the environmental discourses of people of colour are framed around concepts like autonomy, self-determination, access to resources, fairness and justice, and civil and human rights'. If this is so, and we have no reason to doubt it, then the *intent* of Principle 1 will only be *practised* when greater social justice is visited upon communities of colour.

Blowers (Chapter 3), arguing from a European perspective notes that 'inequality, and especially the problem of localized disadvantage, is exposed as a stumbling block to the cooperative relationships promoted by ecological modernization or the need for collective reflexivity espoused in the risk society analysis. Environmental justice, in its broader definition, provides a compelling empirical and theoretical counter perspective to modernization theories'. This, he argues, is given visibility by people from affected local communities and the environmental movement using the mechanisms of civil society to shift power relations away from acquiescence and defence towards demands for greater environmental and economic justice.

The second part of Dobson's hypothesis is that 'the environmental justice movement's objective is not to reduce aggregate tonnage but to have existing tonnage more fairly distributed across wealth and racial cleavages'. This is the 'equity vs justice' argument which was debated in the early 1990s, and resulted in the name change from the 'environmental equity movement' to the 'environmental justice movement'. Cutter (1995) notes that 'for many, the phrase "environmental equity" implies an equal sharing of risk burdens, not an overall reduction in the burdens themselves (Lavelle 1994). Environmental justice is a more politically charged term, one that connotes remedial action to correct an injustice imposed on a specific group of people, mostly people of colour in the USA (Bullard 1994)'. There are two aspects to this. One is, as Bullard (1994) and others have argued, the *corrective* need for *both* procedural and substantive justice in order to improve living conditions among those affected. The other is Faber's (1998, p14) point that 'the struggle for environmental justice is not just about distributing risks equally but about preventing them from being produced in the first place' (see also Faber and McCarthy, Chapter 2). In this he is criticising those in the movement who are content to focus solely on *distributional* equity, ie 'NIMBY', rather than on the procedures which produce the problems in the first place ie on *procedural inequity* and *productive justice*. Similarly, Heiman (1996, p120) has observed, 'if we settle for liberal procedural and distributional equity, relying upon negotiation, mitigation, and fair-share allocation to address some sort of disproportional impact, we merely perpetuate the current production system that by its very structure is discriminatory and non-sustainable'. Both are acknowledging that *sustainability* issues such as clean production, toxics use reduction and pollution prevention are legitimate concerns for *environmental justice* activists who are no longer content to push them into someone else's backyard, but are moving towards co-activism around source reduction (see below).

Indeed, as we argued in our Introduction, environmental justice and sustainability are linked and interdependent, certainly at the *problem* level, and increasingly at the *solution* level *through* issues like toxics use reduction, waste reduction and re-use and recycling. Moreover, in December 2001, one of the US's premier environmental justice centres, The Deep South Center for Environmental Justice at Xavier University, and one of its premier centres for sustainability research, the Lowell Center for Sustainable Production at the University of Massachusetts, together with the newly formed Clean Production Network, held a two-day training workshop on clean production for environmental justice advocates. This workshop was organized specifically in recognition of the fact that it is an area of justice and sustainability where co-activist solutions are both possible and they are happening. Its conception echoes Principle 6 of the Principles of Environmental Justice: 'environmental justice demands the cessation of the production of all toxins, hazardous wastes, and radioactive materials, and that all past and current producers be held strictly accountable to the people for detoxification and the containment at the point of production' (see below).

JUSTICE AND SUSTAINABILITY: TRANSFORMING THE PARTICIPANTS?

Schlosberg (1999) argues that environmental justice activists will become politicized towards sustainability – in Dobson's words this means 'that their involvement has the transformative effect of shifting the objective from sharing out landfill sites more fairly to campaigning for a reduction in the aggregate production of waste'. Dobson argues that he has not found any empirical studies to refute this counter hypothesis. Does the lack of empirical study mean that this transformative effect isn't happening? Forester (1999, p116), in his study of deliberation and participation in the planning process, argues that 'the transformations at stake are those not only of knowledge or of class structure, but of people more or less able to act practically together to better their lives, people we might call citizens'. Similarly, Cole and Foster (2001, p15), focusing specifically on transformation and environmental justice activists note that 'the transformation of environmental justice participants and their local communities, ultimately lies in the forging of coalitions and the networking of grass-roots organizations across substantive areas. Environmental justice groups are networking with other groups to provide information and technical expertise to grass-roots constituencies on various issues of interest to disenfranchised communities, beyond environmental justice'. Clearly, we would like to see empirical studies of transformation, but until then we have any number of practical examples.

One leading environmental justice organization in the US, the Concerned Citizens of South Central Los Angeles (CCSCLA), has experienced this transformation. CCSCLA is a public benefit, community-based organization whose mission is to work for social justice and economic and environmental change within the South Central community. CCSCLA was formed in 1985 to help organize against the development of a mass waste incinerator (LANCER) which was planned for construction in the neighbourhood backed by a US$535million bond issue sponsored by the city of Los Angeles. After defeating the LANCER project, CCSCLA stayed together to 'go beyond environmental justice' (Cole and Foster 2001, p15). They used their transferable skills to work on other issues impacting their community such as affordable housing, banking, planning and land use, and recycling. Furthermore, the organization has remained representative of its community: 70 per cent of its board members live in the organization's target area bounded by the Santa Monica Freeway to the north, the Harbor Freeway to the west, Slauson Avenue to the south and Alameda to the east. This area includes both the Vernon-Central neighbourhood and the Central Avenue Corridor.

Over the course of the 1980s as many as 10,000 local groups, including many in poor and minority communities, had contacted or affiliated with one of the then two major networks in the anti-toxics movement: the Citizens' Clearinghouse on Hazardous Waste (CCHW), founded by Love Canal housewife Lois Gibbs in 1981, and the former National Toxics Campaign (NTC), founded by community organizer and asbestos survivor, the late John O'Connor in 1984. These organizations used what Di Chiro (1995, p300) calls their 'elevated environmental consciousness' and participated fully in the 1980s public policy shift towards toxics use reduction and pollution prevention.

The Center for Health, Environment and Justice (CHEJ) (formerly CCHW) believes the most effective way to win environmental justice is from the bottom up through community organizing and empowerment. CHEJ aims to help local citizens and organizations come together and take an organized, unified stand in order to hold industry and government accountable and work toward a healthy, environmentally sustainable future. CHEJ is also part of the Global Anti-Incinerator Alliance (GAIA), a new international initiative to end the incineration of all forms of waste and to promote sustainable waste prevention and management practices. GAIA is an expanding alliance of individuals, voluntary associations, non-governmental organizations, community-based organizations and others who are working to promote economic and environmental sustainability and justice throughout the world. GAIA developed the Lowell Statement, which was drafted by participants in the International Clean Production Training Event at the Toxics Use Reduction Institute (TURI) at the University of Massachusetts, in Lowell, Massachusetts, in June 2000. One of the Statement's calls is that:

> *We, the undersigned individuals and organizations, call upon our governments to halt proposals for new incinerators and phase out existing incinerators. Instead we call for the implementation of production and waste management systems which are based on the principles of clean production and*

environmental justice. These include toxics use reduction; waste reduction, re-use and recycling; guaranteed public access to information and public involvement; and sustainable, equitable and just consumption patterns.
(Center for Health, Environment and Justice, 2001)

In the Introduction we mentioned that, in relation to *problem-focused*, as opposed to *solution-focused campaigning*, there is 'a far greater ease in identifying and (co-) organizing around problems (ie being reactive), than around solutions (ie being *proactive*). Proactivity in this case requires the sharing of values, visions and vocabularies among groups of people who are for the most part from different social locations: it is difficult to do, but not impossible. The structures required to build bridges, which represent a direct challenge to the dominant social paradigm (Milbrath 1989), are only now being identified by organizations and individuals around the world and put into place. This is the single greatest challenge to developing greater theoretical linkages and practical co-activism between the two areas. Perhaps the greatest hope for what Cole and Foster (2001, p164) call 'movement fusion' lies in initiatives like the two-day training workshop on clean production co-organized by The Deep South Center for Environmental Justice at Xavier University, the Lowell Center for Sustainable Production at the University of Massachusetts, and the newly formed Clean Production Network. Here, theoretical linkages, and practical co-activism were explored, led by trainers from both the environmental justice and sustainability/clean production fields. Sessions included: tools for clean production; life cycle assessment; design for environment; sustainable product design; policies and resources for clean production; extended producer responsibility; ecological taxes; product lifecycle labelling; applying clean production in campaigns; brownfields redevelopment; and developing a vision for clean production. In her opening, Beverley Wright, Director of The Deep South Center for Environmental Justice at Xavier University posed the question: 'why is clean production so important to the environmental justice community?' Very few people didn't know the answer.

The discussion so far has focused upon the processes of transformation of participants in the environmental justice project. We also need to recognize, however, that environmental justice thinking needs to permeate the discourse and practice of sustainability. As some of our contributors have shown (eg Eady, Chapter 8), there is evidence that this thinking has already begun to inform policy and practice, particularly in the US. Cole and Foster (2001, p160) show how Greenpeace USA has 'been transformed by their interaction with the movement' and that 'Greenpeace restructured not only its national policy but its personnel as a result of involvement with the Indigenous Environment Network and the environmental justice movement'. Funding problems in the 1990s however, effectively ended this with the closure of all but the Washington, DC office. The clearest recent expression, as McClaren (Chapter 1) and Dunion and Scandrett (Chapter 15) have shown, is in the growing transformation of sustainability campaigning agendas towards environmental justice in the UK (see also Agyeman, 2000, ESRC, 2001).

ENVIRONMENTAL JUSTICE AND SUSTAINABILITY: PERSPECTIVES AND POSSIBILITIES

In closing we wish to emphasize three interrelated points that we see as fundamental to the further development of theoretical linkages and practical co-activism between the environmental justice and sustainability movements:

- The cradle of the environmental justice movement is the US, and it will undoubtedly continue to provide much of the impetus and inspiration for the rest of the world. However, it must be recognized, as many of our chapter authors demonstrate, that there are now both parallel and different developments around the world. Different contexts and worldviews will generate different interpretations of, and approaches to, environmental justice, but it is increasingly apparent that pressure groups, governments, organized interests and increasingly businesses (for example through Good Neighbour Agreements and Corporate Social Responsibility) are beginning to recognize the importance of an environmental justice perspective to the development of sustainable communities and futures.

- The issue of environmental and human rights has been addressed by several of our authors (eg McLaren, Chapter 1; Rees and Westra, Chapter 5; Wickramasinghe, Chapter 11; Rixecker and Tipene-Matua, Chapter 12, Agbola and Alabi, Chapter 13). It is encouraging to see that these 'human environmental rights' are beginning to be embedded in international law, specifically and firstly through the 1999 Aarhus Convention on Access to Information, Public Participation in Environmental Decision Making and Access to Justice in Environmental Matters. While the law has frequently been used in environmental injustices in the US, it now looks likely that its use will increase at the international level too.

- The challenge for governments – local, regional, national and supra-national – is to recognize the pivotal importance of environmental justice to sustainability policy-making. This is by far the most under-theorized and researched aspect of sustainability, as well as being the area with the least practical action, when compared with environmental and economic sustainability initiatives worldwide. Consistently, Local Agenda 21 surveys (see, for example ICLEI, 2001) show that, for any region in the world, issues of equity and justice are the least likely focus of community-based Local Agenda 21s. Yet, as has been consistently argued throughout this book, equity and justice are integrally connected to the achievement of more sustainable communities. At the time of writing the 2002 World Summit on Sustainable Development was about to be held in Johannesburg, South Africa, with an agenda that prioritized questions of poverty alleviation, governance and environmental protection. If the ten years since the Rio Earth Summit have delivered so little in terms of developing greater commitments at all levels to equity and justice within sustainability, then that challenge most certainly lies ahead.

Appendix I

The following list was adopted as the 'Principles of Environmental Justice' at the People of Colour Environmental Leadership Summit. This list was adopted on 27 October 1991, in Washington, DC.

1 Environmental justice affirms the sacredness of Mother Earth, ecological unity and the interdependence of all species, and the right to be free from ecological destruction.
2 Environmental justice demands that public policy be based on mutual respect and justice for all peoples, free from any form of discrimination or bias.
3 Environmental justice mandates the right to ethical, balanced and responsible uses of land and renewable resources in the interest of a sustainable planet for humans and other living things.
4 Environmental justice calls for universal protection from nuclear testing, extraction, production and disposal of toxic/hazardous wastes and poisons and nuclear testing that threaten the fundamental right to clean air, land, water, and food.
5 Environmental justice affirms the fundamental right to political, economic, cultural and environmental self-determination of all peoples.
6 Environmental justice demands the cessation of the production of all toxins, hazardous wastes, and radioactive materials, and that all past and current producers be held strictly accountable to the people for detoxification and the containment at the point of production.
7 Environmental justice demands the right to participate as equal partners at every level of decision-making including needs assessment, planning, implementation, enforcement and evaluation.
8 Environmental justice affirms the right of all workers to a safe and healthy work environment, without being forced to choose between an unsafe livelihood and unemployment. It also affirms the right of those who work at home to be free from environmental hazards.
9 Environmental justice protects the right of victims of environmental injustice to receive full compensation and reparations for damages as well as quality health care.
10 Environmental justice considers governmental acts of environmental injustice a violation of international law, the Universal Declaration on Human Rights, and the United Nations Convention on Genocide.
11 Environmental justice must recognize a special legal and natural relationship of Native Peoples to the US government through treaties, agreements, compacts, and covenants affirming sovereignty and self-determination.
12 Environmental justice affirms the need for urban and rural ecological policies to clean up and rebuild our cities and rural areas in balance with nature, honouring the cultural integrity of all our communities, and providing fair access for all to the full range of resources.
13 Environmental justice calls for the strict enforcement of principles of informed consent, and a halt to the testing of experimental reproductive and medical procedures and vaccinations on people of colour.

14 Environmental justice opposes the destructive operations of multinational corporations.

15 Environmental justice opposes military occupation, repression and exploitation of lands, peoples and cultures, and other life forms.

16 Environmental justice calls for the education of present and future generations which emphasizes social and environmental issues, based on our experience and an appreciation of our diverse cultural perspectives.

17 Environmental justice requires that we, as individuals, make personal and consumer choices to consume as little of Mother Earth's resources and to produce as little waste as possible; and make the conscious decision to challenge and reprioritize our lifestyles to insure the health of the natural world for present and future generations.

REFERENCES

Agyeman, J (2000) *Environmental Justice: From the Margins to the Mainstream?*, Town and Country Planning Association, London

Boyce, J K, Klemer, A R, Templet, P H and Willis, C E (1999) 'Power distribution, the environment, and public health: A state level analysis', *Ecological Economics*, vol 29, pp127–40

Castells, M (1997) *The Power of Identity*, Blackwell, London

Center for Health, Environment and Justice (2001) *'Global Anti-Incinerator Alliance'* http://www.chej.org/globalalliance.html

Cole, L and Foster, S (2001) *From the Ground Up: Environmental Racism and the Rise of the Environmental Justice Movement*, New York University Press, New York and London

Cutter, S (1995) 'Race, class and environmental justice', *Progress in Geography*, vol 19 (1), pp111–22

Di Chiro, G (1995) 'Nature as community: The convergence of environmental and social justice' in W Cronon (ed) *Uncommon Ground: Rethinking the Human Place in Nature*, WW Norton, New York, pp298–320

Dowie, M (1995). *Losing Ground: American Environmentalism at the Close of the Twentieth Century*, MIT Press, Cambridge, MA

ESRC Global Environmental Change Programme (2001) *Environmental Justice: Rights and Means to a Healthy Environment for All*, Special Briefing Number 7, University of Sussex

Faber, D (ed) (1998) *The Struggle for Ecological Democracy: Environmental Justice Movements in the United States*, The Guilford Press, New York and London

Faber, D and O'Connor, J (1993) 'Capitalism and the Crisis of Environmentalism' in R Hofrichter (ed), *Toxic Struggles: The Theory and Practice of Environmental Justice*, New Society Publishers, Philadelphia

Forester, J (1999) *The Deliberative Practitioner, Encouraging Participatory Planning Processes*, MIT Press, Cambridge MA

Heiman, M (1996) 'Race, waste, and class: New perspectives on environmental justice', *Antipode*, vol 28, pp111–21

International Council for Local Environmental Initiatives (ICLEI) (2001) *Local Agenda 21 Survey Report*, ICLEI, Toronto

Lavalle, M (1994) 'Environmental Justice' in World Resources Institute (ed) *The 1994 Information Please Almanac*, Houghton-Mifflin, Boston, MA

Middleton, N and O'Keefe, P (2001) *Redefining Sustainable Development*, Pluto Press, London

Milbrath, L (1989) *Envisioning a Sustainable Society*, SUNY Press, Albany

Morello-Frosch, R (1997) *Environmental justice and California's 'Riskscape', The distribution of air toxics and associated cancer and non cancer risks among diverse communities*, unpublished dissertation, Department of Health Sciences, University of California, Berkeley

Principles of Environmental Justice (1991) Ratified at the First People of Color Environmental Leadership Summit, Washington, DC, October

Schlosberg, D (1999) *Environmental Justice and the New Pluralism*, Oxford University Press, Oxford

Shiva, V (1989) *Staying Alive*, ZED Books, London

Taylor, D E (2000) 'The rise of the environmental justice paradigm: Injustice framing and the social construction of environmental discourses', *American Behavioral Scientist*, vol 43 (4), pp508–80

Torras, M and Boyce, J K (1998) 'Income, inequality and pollution: A reassessment of the environmental Kuznets curve', *Ecological Economics*, 25, pp147–60

Index

Page references in *italics* refer to tables, figures and boxes